D0936650

DISCARD

SNACK FOOD TECHNOLOGY

some other AVI books

Food Science and Technology

BAKERY TECHNOLOGY AND ENGINEERING, 2ND EDITION   *Matz*
BASIC FOOD CHEMISTRY   Cloth and Soft Cover   *Lee*
COOKIE AND CRACKER TECHNOLOGY   *Matz*
ENCYCLOPEDIA OF FOOD ENGINEERING   *Hall, Farrall and Rippen*
ENCYCLOPEDIA OF FOOD TECHNOLOGY   *Johnson and Peterson*
FABRICATED FOODS   *Inglett*
FOOD CHEMISTRY   *Aurand and Woods*
FOOD CHEMISTRY   Soft Cover   *Meyer*
FOOD COLORIMETRY: THEORY AND APPLICATIONS   *Francis and Clydesdale*
FOOD DEHYDRATION, 2ND EDITION, VOLS. 1 AND 2   *Van Arsdel, Copley and Morgan*
FOOD PACKAGING   *Sacharow and Griffin*
FOOD PROCESS ENGINEERING   Cloth and Soft Cover   *Heldman*
FOOD PROCESSING OPERATIONS, VOLS. 1, 2, AND 3   *Joslyn and Heid*
FOOD PRODUCTS FORMULARY, VOL. 1   *Komarik, Tressler and Long,*VOL. 2   *Tressler and Sultan*
FOOD TEXTURE   *Matz*
FUNDAMENTALS OF FOOD PROCESSING OPERATIONS   *Heid and Joslyn*
FREEZING PRESERVATION OF FOODS, 4TH EDITION, VOLS. 1, 2, 3, AND 4   *Tressler, Van Arsdel and Copley*
NUTRITIONAL EVALUATION OF FOOD PROCESSING, 2ND EDITION   Cloth and Soft Cover   *Harris and Karmas*
PACKAGING: A SCIENTIFIC MARKETING TOOL   *Raphael*
POTATO PROCESSING, 3RD EDITION   *Talburt and Smith*
PRINCIPLES OF PACKAGE DEVELOPMENT   *Griffin and Sacharow*
SYMPOSIUM: SWEETENERS   *Inglett*
TECHNOLOGY OF FOOD PRESERVATION, 3RD EDITION   *Desrosier*
THERMAL PROCESSING OF FOODS   *Goldblith, Joslyn and Nickerson*
WATER IN FOODS   *Matz*

Dairy Science and Technology

BY-PRODUCTS FROM MILK, 2ND EDITION   *Webb and Whittier*
DRYING OF MILK AND MILK PRODUCTS, 2ND EDITION   *Hall and Hedrick*
FLUID MILK INDUSTRY, 3RD EDITION   *Henderson*
FUNDAMENTALS OF DAIRY CHEMISTRY, 2ND EDITION   *Webb, Johnson and Alford*
JUDGING DAIRY PRODUCTS, 4TH EDITION   *Nelson and Trout*

Agriculture

AN INTRODUCTION TO AGRICULTURAL ENGINEERING   Cloth and Soft Cover   *Roth, Crow and Mahoney*
CEREAL SCIENCE   *Matz*
CEREAL TECHNOLOGY   *Matz*
COMMERCIAL VEGETABLE PROCESSING   *Luh and Woodroof*
CORN: CULTURE, PROCESSING, PRODUCTS   *Inglett*
DRYING CEREAL GRAINS   Cloth and Soft Cover   *Brooker, Bakker-Arkema and Hall*
PEANUTS: PRODUCTION, PROCESSING, PRODUCTS, 2ND EDITION   *Woodroof*
POTATOES: PRODUCTION, STORING, PRODUCTS, 2ND EDITION   *Smith*
PRINCIPLES OF FARM MACHINERY, 2ND EDITION   *Kepner, Bainer and Barger*
WHEAT: PRODUCTION AND UTILIZATION   *Inglett*

SNACK FOOD TECHNOLOGY

by SAMUEL A. MATZ, Ph.D.

Vice President
Research, Development, and Regulatory Affairs
Ovaltine Products Division
Sandoz, Inc.
Villa Park, Illinois

WESTPORT, CONNECTICUT
THE AVI PUBLISHING COMPANY, INC.
1976

664.6
M446 a

© *Copyright 1976 by*

THE AVI PUBLISHING COMPANY, INC.

Westport, Connecticut

Library of Congress Catalog Card Number: 75-43297
ISBN-0-87055-193-0

Printed in the United States of America
BY MACK PRINTING COMPANY, EASTON, PENNSYLVANIA

Preface

This book is an attempt to fill a need for a technological guide in a field that has experienced an almost explosive increase in the last decade. No other book available to food scientists provides detailed coverage of the ingredients, processes, products, and equipment of nearly every type of snack food made today.

One of the main purposes of this volume is to provide a source for answers to problems which the technologist encounters in the course of his daily work. Extensive bibliographies, in which the emphasis is on recent publications (extending into 1975), should permit the reader to resolve more complex or new questions. With these bibliographies as guides, the food technologist can delve as deeply as he wishes into specialized aspects of the subject while at the same time the reader who is interested in the broad overall picture will not be distracted by excess detail.

This book is not directed solely to food technologists working directly with snack foods. Scientists, technologists, and engineers connected with companies which supply ingredients, packaging materials, and equipment used in the industry should find the information helpful in guiding new product development and product improvement studies. Marketing and technical sales personnel should find the book of value in educating themselves with respect to factors affecting product quality and customer requirements. Administrators and general management personnel who are not technically trained will find the volume can make the reports of their research, development, and quality assurance departments more meaningful.

The plan of the book is to discuss ingredients, products and processes, equipment, and technical services in separate groups of chapters. In each of the chapters dealing with ingredients, the characteristics of the various raw materials will be covered as well as the quality control tests applied to them, the effects they have on the finished product, bulk handling details (if applicable), and the effects on quality of storage conditions. Products and the processes used to make them will be discussed in the second set of chapters. In the next group, the specialized equipment of the snack food industry will be described, their sources identified, and their effects on the product analyzed. Packaging material is also covered in the third section. Finally, a group of three chapters is devoted to technical functions and nutritional supplementation.

The following persons reviewed one or more chapters and, in some cases provided valuable supplementary material.

David M. Strietelmeier
  Morton Salt Company
Charles E. Holaday
  National Peanut Research
  Laboratory, U.S.D.A.
Charles Staff
  Fairmont Foods
Ray Mykleby
  Land O'Lakes, Inc.
James Moncrieff
  Glidden-Durkee
James E. Mack
  Peanut Butter Manufacturers
  and Salters Association
Richard J. Alexander
  Krause Milling Co.
Roy Shaw
  Red River Valley Potato
  Research Center

Ernest L. Semersky
  Clinton Corn Processing Co.
Edward E. Nuebel
  Associated Milk Producers, Inc.
Paul Perry
  Warner-Jenkinson
Lee Brazier
  Milprint, Inc.
Dr. Ben Borenstein
  Roche Chemical Division
John Butler
  American Can Co.
A. V. Petersen
  Wright Machinery Co., Inc.
Dr. Rulon Chappell
  Land-O-Lakes, Inc.

Illustrations, previously unpublished data, and other information for incorporation into the text of this book were supplied by the following persons.

Thomas E. Walmsley
  E. I. du Pont de Nemours & Co.
David Strietelmeier
  Morton Salt Co.
Dr. Ora Smith
  Potato Chip Institute, International
S. T. Jacobson
  Krispy Kist Korn Machine Co.

Harry M. Pancoast
  Consultant
James H. Colabine
  Sortex Co.
Marshall F. McBeth
  MacBeth Engineering Corp.

A special acknowledgment is due Mrs. Ludmila Ulanowski, who provided encouragement as well as important editorial assistance.

SAMUEL A. MATZ

September 1975

Contents

Snack Food Ingredients

CHAPTER 1

Popcorn and Other Cereal Products

INTRODUCTION

Although it could probably be shown that nearly every available kind of cereal ingredient is or has been used in some form of snack product, popcorn and corn meal are undoubtedly utilized in the largest quantities by the industry. Substantial amounts of whole corn are also made into tortilla-style snacks by way of masa, the alkalized dough which is a traditional processing intermediate.

POPCORN

The use of corn as the expanded or puffed kernel is very ancient, examples of grain suitable for this purpose having been found not only in the Toltec pyramids of Central America, but also in 4,000 yr old deposits in the Bat Cave of New Mexico. Popcorn has always been predominantly an American food, although fairly large quantities are sold in England and other European countries, principally as caramel-coated confections.

About 55% of the popcorn consumed in the United States is popped at home. Of the remaining 45%, approximately ⅔ is sold through retail outlets, theaters, and the like, while the rest is used as an ingredient in the confection industry or is factory popped and packaged for distribution in the grocery trade.

In the United States, popcorn growing and processing are big businesses. Commercial production ranges from 175,000 to 200,000 acres. The leading states in popcorn production are Iowa and Indiana. In addition, Ohio, Illinois, Nebraska, and Kentucky each normally plant more than 10,000 acres annually.

Cultivation procedures for popcorn are not much different from those used for other corn crops. Time of seeding, optimum soil and climate conditions, diseases, and harvesting procedures are much the same as those for dent corn. Popcorn growers usually have more trouble with weeds because their crop grows slower than dent corn and the plants are

1

lower in stature at maturity. The new herbicides alleviate this problem considerably. The conventional 40-in. distance between hills used for dent and flint corn is also followed in planting popcorn, but the popcorn grain is usually seeded at a higher rate per hill, 5 kernels per hill being a common rate for check-planted popcorn.

Popcorn is a favorite crop with home gardeners; probably as much is produced by them as by commercial growers, although figures for the small plots do not show up in the yield records. Commercial producers of popcorn almost always operate under contract with a large merchandising concern which provides the seed and offers an assured outlet for the product. Because of this close control by relatively few persons, annual production figures are subject to great fluctuations, whereas field corn production is more inflexible because of the greater inertia of a larger body of independent growers. Popcorn sells for considerably more per bushel (70 lb of ear corn) than field corn, but the yield per acre is less and cultivation procedures are more time consuming.

Some of the old favorite varieties of popcorn were Japanese hulless, South American, White Rice, Tom Thumb, and Yellow Pearl. Hybrid popcorns are now sown to the almost complete exclusion of the pure varieties. Most are single-cross hybrids because the plants have greater hybrid vigor and are much more uniform than the double-cross hybrids that have been so successful in field corn. Purdue and Iowa State College carry on extensive breeding programs for the development of superior hybrids, each organization growing more than 2,000 crosses annually. Hybrids are carefully graded for yield per acre, lodging tendencies, disease resistance, and amount of expansion of the kernels.

Concessionaires and other vendors of popcorn to the consuming public are primarily concerned with "eatability" and "expansion." These two factors are related to some extent. The eatability or taste satisfaction is dependent upon such factors as flavor, tenderness, and uniformity of the popped kernels as well as the proportion of oil, salt, and other adjuncts applied to the kernels.

Although commercial distributors of popcorn claim that they consider the kernel texture, flavor, color, and other subjective characteristics in selecting seed varieties, the chief basis for judging popcorn desirability has always been the relative amount of kernel expansion that is obtained when the corn is popped. This is a natural tendency, since expansion is closely related to the price received by the processor. Popped corn is usually sold and bought by the consumer (in amusement parks, theaters, etc.) on the basis of volume rather than weight. The concessionaire must fill a box or bag of given volume with popped corn, regardless of whether he has a raw popcorn of high or low expansion. Naturally, he prefers a corn of high expansion since this means he will have to purchase fewer pounds of raw material to fill the required number of containers. Furthermore, corn of higher expansion potential generally has a more tender texture when popped than does corn of lower expansion.

The trade measures expansion by popping a given weight or volume of corn under controlled conditions, dumping the popped kernels into a graduated plastic cylinder of standard diameter and reading the height of the column. Formerly, a standard volume of kernels was used for test popping, but measurement in this manner did not always accurately express the contribution of kernel weight to the results, so a change was made in the procedure, and, currently, a standard weight of grain is used in the test. Results are now expressed as cubic inches of popped corn obtained per pound of raw corn. In the volume-versus-volume test the volume of unpopped corn used fills the graduated tube to a height of 1 in., and the height of the column of popped corn in inches is the expansion ratio of the corn. Most commercial corn gives an expansion between 30- and 35-fold, although a few recently developed hybrids give expansion ratios as high as 40-fold (Nelson 1955).

It has been suggested (Richardson 1957) that pericarp thickness should be used as another indicator of popcorn quality. In this test, the pericarp sections are mounted on edge in modeling clay and observed under a compound microscope equipped with an ocular micrometer. Hull sections are more easily obtained from popped samples than from unpopped kernels. The pericarp thickness is expressed in microns, and in a large series of hybrid corn samples varied between 41 and 65. The measurement is thought to give an indication of quality because it is related to the amount of tough, horny material remaining on the popped corn and thus is related to the acceptability of the product. Pericarp thickness is largely influenced by genetic factors, although changes in the environment have some effect (Richardson 1958).

The optimum moisture percentage for popping has been recommended as 13.5%. In working with four hybrids, Huelsen and Bemis (1954) found that increasing the moisture, at least to about 14.0%, increased expansion. However, the response to increasing moisture differed with respect to maturity at harvest and popping temperatures. Corn harvested at greater maturity (e.g., 15.65% moisture) gave higher popping expansion than a lot harvested at 31% moisture. The temperature of the popper giving optimum expansion varied with the different hybrids. Reducing the moisture content at popping likewise reduced the required popping time. There was an inverse relationship between popping expansion and time required to pop. The temperature at which popping began decreased with increasing moisture content.

Until hybrid seed corn started to become available in the early 1940's, virtually all commercial popcorn was of the open pollinated type. By 1947, hybrid seed developed at various agricultural research institutes had displaced nearly all of the older type of seed. Many new strains of popcorn are developed and tested each year by midwestern university scientists. Most of these are found wanting in one or more critical characteristics and achieve no commercial acceptance. Others, being more prolific or more resistant to disease, yielding a more desirable popped kernel, or having other significant advantages without being inferior in any of the important qualities to existing varieties, will displace some of the acreage formerly planted to less desirable corn. This continuing flow of new seed should lead to a gradual increase in overall quality and yield.

Some specialized terms used in the popcorn trade should be defined in order to avoid misunderstandings. Popped corn with a highly irregular pronged appearance is known as "butterfly" while kernels that are predominantly spheroidal with few projections are called "mushroom" or "ball." Sometimes the popped kernels are called "flakes."

Buyers recognize four major types of popcorn: (1) white hull-less, primarily used for home popping; (2) yellow hull-less, which is also sold in kernel form to consumers for home popping, but is used in large quantity by some concessionaires, especially in hot and humid regions, because it retains good texture at higher moisture contents than other hybrids; (3) large kernel yellow, popular for factory popping and the theater trade because it pops into large flakes with good eye appeal and is resistant to rough handling even though it requires more oil than the other types and gets tough in the presence of moisture; and, (4) medium yellow, a compromise having good appearance when popped and reasonably good texture. Further details on some of these follow, but it should be understood that there is considerable overlapping of characteristics between the types.

Large kernel yellow hybrids pop out large in size—they have a creamy yellow appearance. Under the usual popping conditions a ratio of about 25% mushroom or ball kernels to 75% butterfly kernels will be observed. Large kernel hybrids are popular with the confection industry and for factory popping because they resist breaking better than any of the other kinds of hybrids. They do, however, require more oil and humidity-resistant packaging because they get tough and elastic when they absorb even a small amount of moisture.

Small yellow hull-less popcorn is preferred for consumer preparation because it pops at a relatively low temperature and has superior eating quality. It has also been widely used in low volume on premises popping locations in the South and Southeast where the consumer requires a corn that is tender and palatable under a wide variety of atmospheric conditions. This type of corn will remain acceptable at a higher moisture content than any of the other hybrids. Its main disadvantages are the small kernel size and the high degree of fragility.

Medium kernel yellow was developed as a compromise between the yellow hull-less and the large yellow hybrids. In appearance and response to moisture uptake it occupies an intermediate level between the last two types. It will form a reasonably high percentage of mushrooms, not as many as a good large kernel hybrid, but enough to make good caramel corn and cheese corn with only moderate breakage.

Tables 1.1 and 1.2 give examples of specifications for yellow shelled unprocessed popcorn and nonprocessed white shelled popcorn.

TABLE 1.1

YELLOW, SHELLED, UNPROCESSED POPCORN

Classes: Large Kernel—52 to 67 kernels per 10 gm; Medium Kernel—68 to 75 kernels per 10 gm; Small Kernel—76 to 105 kernels per 10 gm.

	Minimum Limits		Maximum Limits		
Grade	Test Weight per Bushel	WVT Volume	Moisture (%)	Foreign Material and Cracked Kernels (%)	Damaged Kernels (%)
#1	66	1100	14.0	1.0	1.0
#2	65	—	14.5	2.0	1.5
#3	63	—	15.5	3.0	2.0
#4	61	—	16.5	4.0	2.5

Sample Grade
Sample grade shall be corn which does not meet the requirements for No. 1 to No. 4 inclusive; or which contains stones or rodent excreta; or which is musty, or sour, or heating, or which has any commercially objectionable foreign odor; any stained kernels from crushed weed seeds in excess of 1%, moisture content of less than 12.5%, insect damaged kernels in excess of 1% or infested with live weevils or other insects injurious to stored popcorn, or which is otherwise of distinctly low quality.

Definitions
Foreign material shall be popcorn kernels and pieces of popcorn kernels and all other matter than popcorn which readily pass through a 12/64 round hole screen from large kernel and medium kernel popcorn. A 10/64 round hole screen to be used on small yellow. Also all foreign material such as stones, dirt, etc. that remains in the popcorn sample.

Cracked Kernels shall be pieces of kernels, broken, externally cracked, or chipped kernels (kernels having a portion of the endosperm area missing from either mechanical or insect causes) remaining on the 10/64 sieve.

Damaged kernels shall be kernels of popcorn that are heat damaged, sprouted, frosted, badly ground damaged, badly weather damaged, moldy, diseased, or otherwise materially damaged. Kernels with tip cap missing will be considered damaged kernels. This test to be made basis of the sample remaining after removal of the cracked kernels and foreign material.

TABLE 1.2

NONPROCESSED, WHITE, SHELLED POPCORN

Classes: Large Kernel—60 to 95 kernels per 10 gm;
Small Kernel—96 to 115 kernels per 10 gm.

| | Minimum Limits | | | Maximum Limits | |
Grade	Test Weight per Bushel	WVT Volume	Moisture (%)	Cracked Kernels and Foreign Material (%)	Damaged Kernels (%)
#1	64	950	14.0	1.0	1.0
#2	63	—	14.5	2.0	1.5
#3	62	—	15.5	3.0	2.0
#4	61	—	16.5	4.0	2.5

Sample Grade
Sample grade shall be corn which does not meet the requirements for No. 1 to No. 4 inclusive; or which contains stones or rodent excreta; or which is musty, or sour, or heating, or which has any commercially objectionable foreign odor; any stained kernels from crushed weed seeds in excess of 1%, moisture content of less than 12.5%, insect damaged kernels in excess of 1% or infested with live weevils or other insects injurious to stored popcorn, or which is otherwise of distinctly low quality.
Definitions
Foreign Material shall be popcorn kernels and pieces of popcorn kernels and all other matter than popcorn which readily pass through a 10/64 round hole sieve, and all matter other than popcorn which remains in the sieved sample. Cracked Kernels shall be pieces of kernels, broken, externally cracked, or chipped kernels (kernels having a portion of the endosperm area missing from either mechanical or insect causes) remaining on the 10/64 sieve.
Damaged Kernels shall be kernels of popcorn that are heat damaged, sprouted, frosted, badly ground damaged, badly weather damaged, moldy, diseased, or otherwise materially damaged. Kernels with tip cap missing will be considered damaged kernels. This test to be made basis of the sample remaining after removal of the cracked kernels and foreign material.

DRY-MILLED CORN PRODUCTS

Large quantities of corn meal are used in puffed extruded snack products. The principles governing quality of corn as grain and as meal will be discussed in the following paragraphs.

Standards

The market quality of field corn is established by referring to standards published by the U.S. Dept. of Agr. (1957). According to these standards, corn is any grain consisting of 50% or more of shelled corn of the dent or flint varieties containing not more than 10% of other grains for which standards have been established under the provisions of the United States Grain Standards Act.

Color of the grain furnishes the basis for three fundamental market classes for corn. Yellow corn includes all varieties of yellow and may not include more than 5% by weight of kernels of other colors. A slight tinge of red on kernels which are predominantly yellow shall not affect their classification as yellow corn. White corn must contain at least 98% by weight of white kernels. Grains showing slight tinges of light straw color or of pink shall be graded white. Mixed corn includes lots of corn not falling into the classes of white or yellow.

These three basic classes are further qualified as Flint, if 95% or more of the kernels (by weight) are of the flint varieties. If it meets this requirement, the grain would be classed as Yellow Flint Corn, White Flint Corn, or Mixed Flint Corn, as appropriate. "Flint and Dent Corn" is a mixture of the flint and dent varieties which includes from 5 to 95% flint. Dent corn requires no modifying adjective signifying the variety.

Each named category must also be classified into one of five numerical grades or into sample grade.

Inspection

It is clear that many of the tests required in the Standards (e.g., those for color) can only be applied by visual inspection. Their performance requires the services of qualified inspectors, and, in case of litigation based on a difference of opinion (such litigation is rather rare), the final decision must of necessity be based upon a weighing of the evidence of two or more experts. Tests of this sort are clearly less satisfactory from a theoretical standpoint than objective criteria, but they seem to function well in practice, perhaps because the producer-vendor is generally disinclined to employ an expert of his own.

When grading a lot of shelled corn, a trier or probe is used to draw a fair average sample. If it is noticed that the interior of the grain mass is appreciably warmer than the surroundings, the grain is said to be "heating" and the whole batch is graded "Sample Grade." Grain removed by the probe is examined carefully for insects. If insects (or their larvae) destructive to grain are found, the word "Weevilly" is added to the grain grade. Accidental contamination with insects of other types (e.g., ants) may be ignored. The odor of the sample is noted. If it smells musty or sour, indicating extensive microbiological activity, it is graded "Sample Grade." The same rating may be assessed if it smells noticeably of insecticides, hydrocarbons, or other foreign odorants.

Weight per bushel is determined by weighing a known volume (rarely a bushel) of grain and calculating the weight of 2,150.42 cu in.

A sieve with round holes $^{12}\!/_{64}$ in. in diameter is used to remove cracked corn and foreign material. Any material other than corn remaining on top of the sieve is removed and added to the screenings. The material removed and that passing through the screen is weighed and the percentage by weight is calculated to determine the cracked corn and foreign material.

The kernels which have been damaged mechanically or by heating are picked from a 250-gm portion of the cleaned corn and are weighed and the percentage calculated. Heat-damaged kernels are segregated from the total damaged portion and weighed separately. Heat damage is manifested by a darkening of the germ and other changes. It destroys the viability of the grain and reduces its milling quality.

Only the tests required for fixing the lowest possible grade need be made. Thus, corn which is sour or musty immediately falls into Sample Grade and the tests for damaged kernels, bushel weight, foreign material, etc., need not be considered.

A water oven method of determining moisture percentage is the legally prescribed technique, but in practice, moisture may be determined in several ways. The AOAC Methods (Anon. 1970) requires heating a sample at 266 ± 5°F and atmospheric pressure for 1 hr, or at 208 to 212°F and 25 mm pressure until constant weight is attained (about 5 hr).

Other "Official" tests not required by the Standards, but which are occasionally run for special purposes, are ash, extract soluble in cold water, crude fiber, crude fat, fat acidity, and total protein. Details of these tests may be obtained by referring to the Official and Tentative Methods of the Assoc. of Offic. Agr. Chemists (1970).

The Dry Milling of Corn[1]

In contrast to the milling of rye, durum, and flour, there are many mills, particularly in southern United States, where corn milling is still done on mill stones. Such stone-ground corn meal is made from the total corn seed and is an unbolted whole corn product. These mills (frequently referred to as grist mills) are small and they are local in distribution. Rancidity and infestation are problems which limit the distribution of stone-ground whole corn meal.

There is a rather logical explanation for the continued existence of these small milling units in the corn milling industry. Until 1906 there was no suitable method for the removal of corn germ from the corn berry. Since roller mills have a tendency to break the corn germ into very small pieces, rendering an unsatisfactory product, a new system of degermination was needed. In 1906, the Beall corn degerminator which answered this need was introduced to the dry corn milling industry.

The Beall corn degerminator is still used in most of the dry corn mills producing degerminated corn meal or corn flour products. This device is

---

[1] Some of this material is attributable to Larsen (1959).

a cone-shaped shell rotating around a stationary inner cone. Corn is fed in the smaller end of the cone and works its way down to the larger end. During the passage, the corn is rubbed between the stationary cone and the rotating cone, both cones having special knob-like and auger-like surfaces. During the rubbing, the corn is dehulled and the germ loosened and knocked out. In other words, the Beall degerminator is an attrition device.

Entoleters can also be used for dehulling and degerming. The Entoleter differs from the Beall in that it is an impact rather than an attrition machine. Corn is fed through a center opening in the Entoleter and falls on a rapidly rotating disk containing pins on the surface. The pins of the rotor throw the corn against a stationary wall thereby dehulling and degerming it.

In either case, that is, after the Beall or the Entoleter, the mixture of corn endosperm, germ, and bran resulting is sent through rolls to flatten the germ so that it, along with the bran, can be removed from the endosperm by sifting.

Besides the problem of degerming, another problem in corn milling is the extreme irregularity in the size and shape of the corn kernels from the same ear of corn. This irregularity is even greater among different types of corn. Moreover, elevators commonly mix varieties before they reach the miller so that the different problems of corn particle size become compounded. This problem has been solved in part by separating the corn into large-kernel and small-kernel fractions which are then degerminated in separate degerminators adjusted for the specific size of grain.

As with other types of dry cereal milling, the object of dry corn milling is to make the cleanest separation possible of endosperm, bran, and germ. During this operation, one wishes to recover the greatest amount of endosperm.

Tip caps are found at the end of the corn germ and are a problem peculiar to corn. Since they are an intense black color, their presence in the corn flour in small amounts badly discolors the flour. Tip caps are to be avoided in dry corn milling and are removed with the corn bran by aspiration.

With these exceptions which have been mentioned, dry corn milling is in general like wheat flour milling. Upon receipt, the corn first goes through a cleaning process. Both dry cleaning and wet cleaning processes are used. The use of an electrostatic separator deserves special treatment.

This instrument finds particular use in the dry corn milling industry for the removal of rat pellets. Because of the similarity of the size and weight of corn kernels to the size and weight of rat pellets, separation by sifting is difficult. Electrostatic separation fortunately works very well. Usually the electrostatic separator is added at the tail end of reels and disk separators and just before the wet stoner and washer.

After cleaning comes conditioning. Normally, the moisture content of the corn is raised not to 17% as with wheat but to about 21%. This is done because the germ of corn tends to be more friable than the germ of wheat, and if corn germ is too dry it will break into small flour-sized pieces during degerming. If enough water is added, not only the bran is toughened, but so is the germ. Conveyors and whizzers both add moisture to the corn.

Degerming follows the conditioning according to procedures already discussed. Following the degerming (and dehulling), the corn must be dried so it can be handled on roller mills and in sieves. The moisture is brought down to about 15%. Inclined barrel-shaped rotary driers are used and the product is air cooled after drying in louvered coolers. At this point, the product is aspirated to remove the bran and then is ready for the main part of the milling system.

The milling system consists of grinding, sifting and classifying, purifying, aspirating, and possibly final drying. The normal flow is through break rolls and then to plansifters. The fines go to the next break roll and the coarse goes to purifiers and then to the germ rolls. The germ rolls flatten the germ so that it is easily removed by sifting. The break rolls are followed by reduction rolls which produce the final fine flour.

The break system is longer in a dry corn mill than the reduction roll system. This is quite understandable since corn flour is not as important a product as corn grits or corn meal. Hence, a long reduction system is not needed to reduce the size of the endosperm particles (see corn products specifications).

Careful treatment of the germ in the dry corn milling industry is accomplished by a special system of rolls which is set aside for the cleaning of the corn germ. In wheat, the germ makes up about 2.5% of the wheat berry. In corn, the germ comprises 12% of the berry. Moreover, the fat content of wheat germ is about 8% while the fat content of corn germ is higher—nearly 34.5%. Thus, it is economical and also desirable to extract the oil from corn germ and sell it as corn oil. This requires clean germ and hence the emphasis of the corn germ milling system in contrast to wheat milling.

Every dry corn milling plant has an oil extraction plant. Normally, they use an oil press to extract the oil. The extraction and refining of oils are discussed in a subsequent chapter and will not be considered further here. Several excellent references are available (Stimmel 1941; Stiver 1955; Neenan 1951; Gehle 1937).

Specifications for Corn Products

The corn kernel is somewhat unique in that the endosperm opposite the tip of the kernel is quite soft while the other parts of the corn seed endosperm are hard. When ground the soft or floury endosperm portion easily breaks into flour (called soft flour or break flour) while the horny endosperm fraction of the corn stays as larger pieces. These larger pieces can be classified according to size or ground into flour. Such is called sharp flour or reduction flour.

The yield of endosperm products in corn milling varies between 65 and 70%. The residual is bran, germ, and high fat endosperm fines. These are combined to produce a product for animal feeding called hominy feed.

Before the germ is added to the hominy feed, however, the valuable corn oil in it is removed by expression and the use of filter presses (Stiver 1955). Thus, the germ in corn millfeed is a defatted product.

Classified by size, the products of corn milling are these:

Grits	6 to 16 USBS Sieve
Meals	
Coarse	16 to 24 USBS Sieve
Medium	24 to 40 USBS Sieve
Cones	40 to 70 USBS Sieve
Flours	
Sharp	100% through 70 mesh
Soft	and 70% through 100 mesh

Soft (Break) Corn Flour.—This material comes mostly from the endosperm opposite the tip of the kernel, which lies between the flinty (Horny) endosperm and the hull. It is separated from the meals and grits by sieving the early break products.

Sharp (Reduction) Corn Flour.—This flour is produced by grinding to the proper size the horny endosperm particles that come from early break rolls and bolting the product to take out the flour-like material. The specifications are similar to those of the soft corn flours, but the feel of the product is harsh rather than soft. This is caused by the sharp edges of these particles in contrast to the rounded edges of the soft flour particles. Hence, the name sharp flour.

Cones.—This product is similar to the sharp flour except that it is coarser. It has the same feel to the touch as does the sharp flour.

Cones are often used by the brewing, breakfast cereal, and snack food industries and are used where the dust from flours must be avoided in other food industries.

Corn Meals.—Corn meal is a product somewhat smaller than corn grits but still much coarser than a corn flour. Its chief use is for table consumption. In these cases, white corn products are used more often.

TABLE 1.3

SPECIFICATION FOR "K" GRIND GRITS

	Moisture (%)	Fat (%)	Protein (%)	Ash (%)	Fiber (%)
"K" Grind Grits	13.5–14.5	0.60–1.00	6.5–8.5	0.25–0.50	0.35–0.75

Granulation
  Over U.S. 16 Wire: 0–7%
  Over U.S. 20 Wire: 75.0% min.
  Over U.S. 25 Wire: 5–15%
  Over U.S. 30 Wire: 0–1.5%
  Over U.S. 40 Wire: 0–1.5%
  Thru U.S. 40 Wire: 0–1.5%

Corn meals are used to make gelatinized corn flour. Such flours find their way into industrial as well as food uses.

Corn Grits.—Grits of screen sizes such as numbers 8, 10, and 12 are produced. These products are used in making breakfast cereals and some snack foods. Particle sizes ranging from 14 grits all the way to corn meals are used in breweries to make alcoholic products by fermentation. Each brewery has its own partiality for particle size; to change this would change the brewing conditions based upon definite cooking times and temperatures using predetermined formulas. Thus, close attention is paid to particle size.

Visual inspection is important in these grit products since, with such coarse materials, it is hard to remove adventitious substances by sifting techniques.

While formerly only white corn was used for the production of corn grits, the corn milling industry is turning to cheaper yellow corn. The brewers' antagonism to this type of grits largely disappeared during World War II.

Pearl Hominy.—This product is sometimes called number four grits because it will pass through a number four screen but will remain on top of any finer screen. It is also called "flaking grits" since much of the production is used for making corn flakes, which require yellow grits.

Visual inspection is of great importance. Freedom from foreign particles, such as yellow corn, soybeans, and insect refuse, is a necessity before further processing to breakfast cereals is possible. Many of these foreign materials are of the proper size to concentrate with the number four grits during sizing. Thus, avoidance depends upon good milling practice and quality control by visual appearance.

Hominy Feed.—Corn millfeed is sold as one product, hominy feed. Hominy feed is a combination of corn bran, corn shorts, and defatted corn germ. It is used as an ingredient in animal feeds.

TABLE 1.4

SPECIFICATION FOR 104 MEAL

	Moisture (%)	Protein (%)	Fat (%)	Ash (%)	Fiber (%)
104 Meal	11.0–13.0	7.0–8.0	0.75–1.5	0.25–0.75	0.25–0.75

Granulation
Over 20 Wire: 0.0–0.5%
Over 25 Wire: 0.0–1.0%
Over 30 Wire: 25.0–36.0%
Over 35 Wire: 45.0–60.0%
Over 60 Wire: 4.0–14.0%
Thru 60 Wire: 0.0–2.5%

According to Hamdy (1972), a "K" grind meal is the most used by snack manufacturers because its granulation range allows it to perform satisfactorily with most extruders. See Table 1.3. Another type of corn product identified by Hamdy (1972) as "104 Meal" is also used for extruded or cooked snacks. The specification for this milled corn product is given in Table 1.4.

BIBLIOGRAPHY

ASSOC. OF OFFIC. AGR. CHEMISTS. 1975. Official and Tentative Methods, 12th Edition. Assoc. of Offic. Agr. Chemists, Washington, D.C.

BRUNSON, A. M., and RICHARDSON, D. C. 1958. Popcorn. U.S. Dept. Agr. Farmers Bull. 1679.

ELDREDGE, J. C. 1954. Factors affecting popping volume. Popcorn Merchandiser 9, No. 4, 20–32.

ELDREDGE, J. C. and THOMAS, W. I. 1959A. Popcorn in the home. Iowa State University. Coop. Extension Serv. Pamphlet 260.

ELDREDGE, J. C., and THOMAS, W. I. 1959B. Popcorn—its production, processing, and utilization. Iowa State Univ. Bull. P127.

GEHLE, H. 1937. Conditioning and grinding of corn. Muhle 74, 361–362.

HAMDY, M. 1972. Personal communication. Sept. 21.

HUELSEN, W. A., and BEMIS, W. P. 1954. Temperature of popper in relation to volumetric expansion of popcorn. Food Technol. 8, 394–399.

LARSEN, R. A. 1959. Milling. In Chemistry and Technology of Cereals as Food and Feed, S. A. Matz, (Editor). Avi Publishing Co., Westport, Conn.

NEENAN, J. L. 1951. The degerminating corn mill. Am. Miller 79, 44–45.

NELSON, O. E., Jr. 1955. Purdue hybrid performance tests for 1955. Popcorn Merchandiser 10, No. 3, 3–9.

RICHARDSON, D. L. 1957. Purdue hybrid performance trials encouraging. Popcorn Concessions Merchandiser 12, No. 4, 10–17.

RICHARDSON, D. L. 1958. Two factors of early harvesting contribute to popcorn quality. Concessionaire Merchandiser 13, No. 4, 5, 12-B.

STIMMEL, E. P. 1941. Dry corn milling. Am. Miller 69, No. 10, 30–33.

STIVER, T. E. 1955. American corn milling systems for de-germed products. Bull. Assoc. Operative Millers (June).

THOMAS, W. I. 1958. Relation of sample size and temperature to volumetric expansion of popcorn. Food Technol. 12, 514–517.

U.S. DEPT. OF AGR. 1957. Official Grain Standards of the United States. U.S. Dept. Agr. Regulatory Announcements No. AMS-177.

CHAPTER 2

Fats, Oils, Emulsifiers, and Antioxidants

INTRODUCTION

The unifying characteristic of the ingredients described in this chapter is that the materials are either glycerides of fatty acids or are compounds added to modify the function or storage stability of such substances. Manufacturing processes will be discussed briefly, because an understanding of the elements of these techniques is important in developing meaningful specifications and quality control tests.

Frying fat is both a processing agent (heat transfer medium) and an ingredient in many important snack products such as potato chips, etc. In other snacks, it functions as an ingredient only. In any case, it has significant effects on the appearance, flavor, and texture of the product and is often the ingredient which establishes the limit on shelf-life. It may be the most expensive ingredient in the product.

Fats and oils are triglycerides of fatty acids, that is, 3 fatty acid molecules are combined chemically by ester bonds with 1 glycerol molecule to yield a molecule of fat or oil. The fatty acids are straight chain compounds bearing a single carboxyl group. Naturally occurring fatty acids almost always have an even number of carbon atoms between 4 and 26. Melting point increases with chain length, and fatty acids with 12 or more carbon atoms are solid at room temperature.

If the fatty acid contains double-bond linkages, it is classified as unsaturated. Such compounds tend to be considerably more reactive than saturated fatty acids and they melt at lower temperatures than their saturated counterparts. Polyunsaturated fatty acids are thought to be beneficial in the diet and contain more than one unsaturated linkage.

So far as melting points and reactivity to a number of reagents are concerned, fats reflect in a general and approximate way the properties of the uncombined fatty acids which have gone into their makeup. Table 2.1 gives the fatty acid composition of a number of food fats.

NATURAL FATS AND OILS

Butter

Butter contains over 80% butterfat, about 16% water, 0.5% lactose, and 0.1 to 3.0% ash (mostly from added salt). The structure of butter consists of a continuous phase of solid butterfat enclosing globules of liquid fatty material and drops of aqueous solution. Crystals or agglomerates of solid fatty substances may also be present in forms sufficiently large to be detected microscopically. Some authorities claim that a second continuous phase composed of aqueous solution exists in butter.

Butterfat is a mixture of many different glycerides. The relative proportions of these compounds are controlled by the breed of cattle, the season of the year, and particularly, the type of feed. The aqueous phase contains milk proteins, native minerals plus any added salt, lactose, and skim milk and wash water components. The aqueous phase is the principal source of flavor in butter.

Since butter is quite expensive relative to most other fats, its use is restricted to those products in which its flavor makes a significant contribution to the acceptability or in which its use permits advertising claims of marketing value. Low-score butter is often preferred to the blander high-score products. Specially "ripened" butter of extremely high flavor can be used to reduce cost. Butter is classified in the following grades according to its flavor, color, etc.

(1) U.S. Grade AA or U.S. 93 Score
(2) U.S. Grade A or U.S. 92 Score
(3) U.S. Grade B or U.S. 90 Score
(4) U.S. Grade C or U.S. 89 Score

The relatively low-melting point of butterfat leads to the occurrence of greasiness in products containing moderate to high levels of this ingredient. Although this greasiness causes annoyance when the product is handled and tends to smear packaging material and to be related to early development of rancidity, there are indications that it has certain organoleptic attractions for the consumer.

Lard

Lard has a distinctive natural flavor that is thought to be desirable in some foods, although it is not a common ingredient in snacks because of its limited stability.

Lard is classified on the basis of the rendering method used as either prime steam, dry rendered, open-kettle rendered, or continuous process lard. Lard sold commercially to bakers and other food processors is called refined pure lard and may be made by any of the preceding methods. The characteristics of lard are governed by the composition of the hog fat from which it is made, by the method of rendering, and by the refining processes applied to the extracted fat.

Variations in hardness of the lard depend upon the body location of the fats rendered. For example, internal fats, such as leaf fats, are always higher in melting point than fat from the external portion of the carcass. The refiner controls and standardizes his product by selecting

TABLE 2.1

COMPOSITION AND PROPERTIES OF FATS AND OILS[1]

Acid (Commonly Refd. to: Predominant Specie)	G.C. Commonly Designation	Babassu	Butter Fat	Cocoa Butter	Coconut	Corn	Cottonseed	Lard	Olive	Palm
Caprylic	C8:0	7	1.5	—	8	—	—	—	—	—
Capric	C10:0	5	3	—	7	—	—	—	—	—
Lauric	C12:0	45	4	—	48	—	—	—	—	—
Myristic	C14:0	15	12	0.5	18	0.2	0.9	3	—	1
Palmitic	C16:0	9	25	25	8.5	12	23.5	23	14	46
Stearic	C18:0	3	9	35	2.3	2.2	2.5	13	2.5	4
Oleic	C18:1	13	32	37.5	6	27	18	46	68	38
Linoleic	C18:2	2	—	2	2	57	54	14	13	10
Arachidic	C20:0	0.1	1	—	—	0.3	0.3	0.2	0.4	0.4
Linolenic	C18:3	—	—	—	—	1 ⎞	0.3 ⎞	1 ⎞	0.7 ⎞	0.3 ⎞
Gadoleic	C20:1	—	—	—	—	— ⎠	— ⎠	— ⎠	— ⎠	— ⎠
Behenic	C22:0	—	—	—	—	—	—	—	0.2	—
Lignoceric	C24:0	—	—	—	—	—	—	—	—	—
Others:	C4:0	—	3	—	—	—	—	—	—	—
	C6:0	—	1	—	—	—	—	—	—	—
	C12:1	—	0.4	—	—	—	—	—	—	—
	C14:1	—	1.5	—	—	—	—	—	—	—
	C16:1	—	4	—	—	—	—	—	2	—
	C22:1	—	—	—	—	—	—	—	—	—
Iodine no. (WIJS) typical		16	30	40	9	125	110	73	85	50
Iodine no. (WIJS) range		15–19	25–35	35–43	8–12	120–128	105–116	65–80	80–88	45–55
Saponification value range		247–250	216–240	190–200	254–262	189–193	189–198	190–198	188–196	196–200
Wiley M.P.°F		79	82–95	79–99	76	—	—	88–110	—	104–110

Source: Courtesy of PVO International, Inc.
[1]The Values are typical gas chromatographic fatty acid analyses, normalized and rounded off. Individual samples of some fats (particularly lard) can differ considerably from those listed.

and blending different fats. Special grades are produced by segregating certain fats and by special rendering methods. Pure leaf lard, which has the highest melting point, is made only from leaf fat. If a still firmer shortening is required, hydrogenated lard flakes can be added to the natural fats. By varying the amount of hydrogenated material, any reasonable melting point can be obtained.

Beef Fats

Beef tallow is made from edible fatty tissues of cattle. As with lard, its physical properties vary considerably depending upon the history of the

Palm Kernel	Peanut	Rape-Seed	Rice Bran	Saf-flower	High Oleic Saf-flower	Sesame (USA)	Sor-ghum	Soy-bean	Sun-flower	Tal-low-beef	Tal-low-mutton
4	—	—	—	—	—	—	—	—	—	—	—
4	—	—	—	—	—	—	—	—	—	—	—
50	—	—	—	—	—	—	—	—	—	—	—
16	0.1	—	0.5	—	—	—	—	—	—	2	1
8	11	3	17	8	5	9	12	11	8	35	21
2.5	3	1.5	2.5	3	1.2	5	1	4	3	16	30
12	46	32	46	13	84	42	31	25	20	44	43
3	31	19	32	75	10	43	53	50	67.8	2	5
0.1	1.5	—	0.5	trace	trace	trace	0.1	0.4	0.5	—	—
0.1⎱	1.5⎱	10	1⎱	1⎱	trace⎱	0.5⎱	2⎱	8⎱	0.5⎱	0.4⎱	—
—⎰	—⎰	10	⎰	⎰	⎰	⎰	⎰	⎰	⎰	⎰	—
—	3.5	0.5	—	—	—	—	—	0.3	0.2	—	—
—	1.5	—	—	—	—	—	—	—	—	—	—
—	—	—	—	—	—	—	—	—	—	—	—
—	—	—	—	—	—	—	—	—	—	—	—
—	—	—	—	—	—	—	trace	—	—	—	—
—	—	23.5	—	—	—	—	—	—	—	—	—
17	98	101	110	132	93	110	115	130	130	40	40
16–20	90–110	95–108	100–120	127–140	90–100	100–120	105–120	125–140	120–140	35–50	35–46
244–255	180–195	183–194	188–192	190–194	185–195	188–195	188–195	188–194	188–195	193–195	192–197
80	—	—	—	—	—	—	—	—	—	100–110	101–115

cattle from which the fat is taken. It is normally a hard plastic fat having a melting point of about 110 to 120°F. Because of its hardness, it may be subjected to further processing rather than used in its native form.

Beef fats rendered by special methods are separated by fractional crystallization into oleo oil (low melting fraction) and oleostearin (high melting fraction). The short plastic range (from about 70 to 80°F) and relatively low melting point make oleo oil a fair substitute for coconut oil in some applications.

Vegetable Oils

Soybean and cottonseed oil are common frying fats and are the principal raw materials for hydrogenated vegetable oil shortenings. Where bland flavors (or vegetable origins) are important, hydrogenated fats prepared from soybean or cottonseed oil are the shortenings of choice. Coconut oil is widely used as a spray fat, and in fillings and coatings. Cocoa butter is an essential part of pure chocolate coatings and is used in a few other special formulas. Peanut, corn, and palm oils are less frequently utilized.

Vegetable oils are pressed or extracted from the seed and contain many nonglyceride components (impurities). Among these may be found free fatty acids, amino acid polymers, phospholipids, resins, pigments, mucilaginous substances, and carbohydrates. Crude oils are not usable directly in snack products because of these impurities and must be processed to remove the more objectionable compounds. The refining process involves mixing the oil with alkali and then washing it. Deodorizing by steam or vacuum stripping may follow this treatment.

All vegetable oils offered to snack manufacturers have been refined and deodorized. They are usually bleached. Cottonseed oil and partially hydrogenated soybean oil may also be treated to remove the higher melting fraction to give a "winterized" or salad oil. Although for many purposes soybean oil and cottonseed oil may be used interchangeably, the latter has the reputation of being more resistant to oxidation or flavor reversion.

Refined soy and cottonseed oils are usually further treated by blending, hydrogenation, addition of emulsifiers and antioxidants, and other processes.

Peanut Oil

Peanuts yield a high-quality cooking and frying oil which is, however, rarely used in the snack food industry because of its high price. The crude peanut oil output in the United States for the marketing year ended September 30, 1974, has been estimated at 150 million pounds.

Peanut oil has some definite advantages for deep fat frying. It has a high smoke point and it is the lightest of the common unsaturated vegetable oils. Because it contains natural antioxidants, it has a fair resistance to the development of rancidity. The flavor is pleasant and mild but distinctive.

Coconut Oil

Coconut oil is one of the so-called lauric acid fats. These fats, although relatively highly saturated, have melting points not far above room temperature because of their high content of short chain fatty acids. Coconut oil is available mostly as 76° and 92° oils, the latter being partially hydrogenated. A fully hardened or 110° oil can also be obtained. The unhydrogenated or 76° oil is much used for the frying of nuts and snacks since its low level of unsaturated fatty acids gives it a high degree of stability against oxidative rancidity, and its low melting point tends to prevent waxy mouth feel.

Coconut oil seems to be more susceptible than most fats to hydrolytic rancidity. If hydrolysis of the oil occurs (usually due to enzymatic action) the short chain fatty acids that are released give rise to an unpleasant soapy flavor in the product. Some unblanched nuts contain active lipases responsible for catalyzing these deteriorative reactions.

Popcorn popping oil for small operations or retail sale is generally 76° coconut oil to which a high level of heat-stable carotenoid pigments (and sometimes butter flavors) has been added. Coconut oil is also used for spraying finished snack pieces of many kinds, including crackers.

MODIFIED SHORTENINGS

Hydrogenation

Hydrogenation is the process by which hydrogen is added directly to points of unsaturation in the fatty acids. Hydrogenation of fats has developed as a result of the need to convert oils to solid fats and to increase the stability of the fat or oil to oxidative rancidity.

Interesterification

Molecular rearrangement or interesterification is another method for changing the chemical and physical structure of fats. This process causes the reshuffling of the fatty acid moieties between the glycerol residues so that a more random distribution exists. Interesterification is accomplished by catalytic techniques at relatively low temperatures and the net effect is usually to increase the plastic range. Lard so treated has many of the desirable textural characteristics of vegetable oil shortenings and contributes less of a greasy sensation when used at high levels. Molecular rearrangement is not commercially applied to fats other than lard.

ANALYTICAL TESTS APPLIED TO SHORTENINGS

As with most ingredients, the most nearly definitive test for fats and oils is a performance test under actual conditions of use and consumption. Since tests of this sort are often impractical and are too expensive and time consuming for routine use, other procedures have been developed to evaluate certain characteristics of shortening likely to affect the quality of the finished product. Some of the most useful of the present-day analyses are described briefly in the following section. A detailed description of all common tests applied to fats and oils can be found in

Official and Tentative Methods published by the American Oil Chemists' Society (Link 1973).

Melting Point

Fats, being mixtures, do not have the sharp melting points of pure compounds. Any natural fat will contain some liquid and some solid material over a wide range of temperatures. Increases in the proportion of liquid component as the temperature is raised causes a gradual softening and liquefaction. Conversely, on cooling the fluid, clouding is first noticed followed by a pasty condition and, finally, by development of a hard solid. "Plastic range" is a term used to describe the range of temperature in which the fat appears to be solid but can still be readily deformed. Because of these considerations, establishment of the melting point is rather inexact in most cases. There are several methods in use to establish this datum and they differ primarily in definition of the end point.

The *capillary melting point* is the temperature at which a sample of the fat contained in a small glass tube becomes completely clear. The *open tube melting point slip point,* or softening point, is the temperature at which a solidified sample in an open tube immersed in water will soften sufficiently to rise under the buoyant effect of the water.

In the *Wiley melting point test,* a molded tablet of the fat is allowed to float at the interface of a water and alcohol bath. The temperature of the batch is gradually raised and the end point is taken to be the temperature at which the tablet assumes a spherical form.

Ordinarily each of these "melting points" will be different for a given fat. The closed tube melting point will be the highest of the three temperatures while the open tube will be the lowest, in most cases. The congealing point, determined by cooling liquid fat until it becomes cloudy, then transferring it to a 68°F air batch and observing the highest temperature reached as the fat congeals, will be different from any of the melting points.

Stability Tests

The degree of resistance to rancidity development is an extremely important characteristic of fats and oils. The fresh product can be tested for free fatty acids or peroxides to get a general, rough idea of the extent to which deterioration has already occurred. Tests based on bringing the sample to a stage of detectable rancidity by controlled heating are perhaps more meaningful from a practical point of view.

The *active oxygen method* or AOM test is widely used for determining stability to oxidative deterioration. In this procedure, air is bubbled through fat held at 208°F. The end point is the time at which a peroxide value of 100 meq per kilogram is reached. Other peroxide value end points may be specified for certain fats to give a better correlation with sensory perception of rancidity. A modification of the AOM test using a temperature of 230°F is widely used because deterioration occurs more than twice as fast at the higher temperature, so that results are available much sooner than they would be using the lower temperature.

The *Schaal test* consists of holding a sample of the fat, or a product containing it, in an oven maintained at 145°F. Other temperatures can also be used. The sample is examined daily, or more frequently, and the time required to reach a condition of detectable rancidity is recorded. Ground or crumbled snacks made with the fat in question can be tested in this manner and the results can frequently be related to problems associated with distribution and shelf-life. The interaction of product and packaging can also be estimated by placing strips of the packaging material in contact with crumbs of the product.

Peroxide value is determined by reacting a sample of the fat with potassium iodide and titrating the excess iodide with potassium thiosulfate. It is expressed as milliequivalents of oxygen per kilogram of fat. Peroxide value is an indication of the extent to which the fat has already reacted with oxygen and thus indicates approximately how much storage life remains. The deodorization process applied to shortenings reduces the peroxide value to zero.

The free fatty acid content of a fat or oil is essentially a measure of the amount of hydrolysis which has occurred. Since hydrolysis and oxidation are the reactions leading to rancidity, the peroxide value and free fatty acid content together give a reasonably good picture of current status and future prospects of a shortening. Free fatty acid is determined by titration with a standard solution of alkali.

Solids Fraction Index (SFI)

The proportion of solid to liquid fat in a shortening at a given temperature has an important relationship to the performance of shortening at that temperature. This proportion cannot be deduced from the melting point of the fat or from the consistency of the shortening at the given temperature. However, it can be accurately determined by a technique called dilatometry. The principle of the test is measurement of the change in volume of samples of constant mass held at a series of temperatures. Density of a liquid glyceride will differ from that of the solid form, so, as the temperature of a fat is progressively raised, additional molecular species will liquefy and lead to changes in volume of the total sample. The situation is complicated somewhat by the mutual solubility of the various glycerides which also varies with temperature.

An apparatus of the type described by Pontius (1965) is used in conjunction with a series of water baths to determine the solids fraction index (SFI; also called the solids factor index). The sample of melted fat is placed in a chamber attached to a calibrated and graduated capillary tube, then solidified. The instrument containing the solidified fat is allowed to stand (temper) at certain standard temperatures, then heated gradually to melt the sample. The increase in volume is read by observing the movement of an indicator fluid in the capillary tube. Examples of the percentage of solids found in different shortenings by this method are shown in Stingley et al. (1961). These figures can be converted into chart form. Fats with a wide plastic range will produce a flatter curve with small differences in the SFI at each temperature. Fats having a shorter plastic range will have a steep curve, generally being high at the low temperatures and low at the high temperatures.

Flavor and Odor

Except for butter and lard, it is generally desired that shortenings have a bland flavor, although, as Thomas (1968) points out, all shortenings do have a characteristic flavor even though it is usually faint. Taste panels consisting of 2 to 10 experienced members can be used to verify the absence of off-flavors in samples of shortenings received for use as ingredients. Evans (1955) describes procedures for setting up, conducting, and evaluating panel tests. Experience has shown that good agreement and reproducibility can be obtained in these tests.

Other Tests

Iodine value is a measure of the number of double bonds present in the fatty acids. It indicates the degree of unsaturation of the fat or oil, and is a rough measure of the storage stability. If the material is unhydrogenated, the iodine value is evidence of the type of oil. Hydrogenation decreases the iodine value, as does oxidation of the double bonds.

Smoke point, flash point, and *fire point* are valuable tests for frying fats. They involve gradual heating of the fat in standard equipment under rigidly specified conditions and noting the temperature at which the material emits smoke or catches fire.

FRYING FATS

Fats used for frying snack products must have different properties from the fats intended for use as shortenings or as coatings. A high smoke point and excellent stability to hydrolysis and oxidation at elevated temperatures are essential characteristics. In most markets, the frying fat is expected to contribute little flavor of its own to the finished product, while other regions are accustomed to the flavor of lard. Hydrogenated cottonseed oil is the most common frying fat and it is bland.

The fresh fat, as it is received, may not be completely satisfactory for frying because the heat transfer characteristics—partly a function of the viscosity—are not optimal. If the kettle is replenished from time to time with small additions of fresh fat, an equilibrium state can be maintained in which the finished product always has satisfactory properties. When the frying equipment contains all fresh fat, it must be "broken in" by heating the fat at frying temperatures in contact with product until a free fatty acid content of about 0.4% is reached, or until the product is being cooked as desired.

Fat is broken down to free fatty acids, glycerine, etc., by steam from the product, by oxygen from the air, and by heat. Excessive deterioration is evidenced by smoking, darkening, and foaming. Methyl silicone is often effective in reducing foaming. Filtration is effective in removing particles which can char and discolor product, but it does not retard oxidation and it does not effect the free fatty acid composition or the viscosity of shortening (Moyer 1965).

Copper ions greatly accelerate the development of oxidative rancidity, so fats should be prevented from contacting copper, brass, or bronze utensils at any time.

EMULSIFIERS

Emulsifiers are natural or synthesized substances which promote the formation and improve the stability of emulsions, e.g., dispersions of fat droplets in water or vice versa. In some cases, they can be used to improve the wetting properties of water or aqueous solutions. Snack food manufacturers sometimes find it advantageous to employ emulsifiers in their processes.

The unifying characteristic of emulsifiers is presence of a hydrophilic group (dissolving in aqueous solutions) and a lipophilic group (dissolving in lipids) on the same molecule, but not all such molecules have practical effects as surface-active agents. It is not necessary that the hydrophilic and lipophilic groups have equal effectiveness, and, as a matter of fact, it is usual to find that one or the other dominates the actions of the emulsifier. The variability in performance of different emulsifiers is due to the relative potency of the two kinds of regions, their spatial relationship, the size of the entire molecule, and certain other factors. There are some very potent emulsifiers which cannot be used in foodstuffs because of legal restrictions. A few are acceptable in most standardized foods, however, and several more are permitted for nonstandardized foods.

Without some kind of screening system, the time required to select an emulsifier of optimum function for a given application might prove to be prohibitive. A prescreening system or emulsifier rating scale which is rather widely used is the HLB system (Anon. 1963). In brief, this method provides rules for assigning an HLB (hydrophile-lipophile balance) number to the combination of ingredients which is to be emulsified, and then directs the selection of an emulsifier or blend of emulsifiers having the same number. The HLB numbers of all common emulsifiers can be found in the literature. Unfortunately, it is not as easy to secure the HLB numbers of many of the ingredients in doughs and batters. Apparently, the HLB numbers are of little or no value for predicting the starch-complexing ability of surfactants.

Some emulsifiers form complexes with starch, and particularly amylose. It appears that the starch must be gelatinized before the full effect of the complexing action is observed. The exact form of the reaction is a subject of continuing debate, but the most recent opinions seem to favor the view that a kind of clathrate is formed, with the long amylose molecule wrapping around an extended molecule of, e.g., monoglyceride. It is possible that some interesting results in texture modification could be observed if such starch-reacting emulsifiers were added to the meal used in extruded snacks.

Lecithin

Lecithin, a mixture of phosholipids, is found widely distributed in nature, but is commercially prepared almost exclusively from soybean oil at the present time. It exists preformed as a contaminant in crude soybean oil, and the commercial method of preparation involves precipitation from the oil and subsequent purification. It may be further processed by bleaching, etc. Lecithin is the least expensive of the emulsifiers and is relatively potent. It does contribute a flavor which may be objectionable in some products.

Commercial grades of lecithin are classed according to total phosphatides, color, and fluidity. The amount of phosphatides in commercial lecithin ranges between 54 and 72%. The product specifications usually report the concentration of phosphatides as "percent acetone insolubles." Color is stated as unbleached, single bleached, and double bleached. The bleaching process tends to reduce the effectiveness of lecithin as a surface-active material. The consistency will be "plastic" or "fluid." Lecithin can be dispersed in water to form hydrates; similar responses occur with propylene glycol, glycerine, etc.

Flavor is an important consideration since the off-flavors due to the lecithin itself or to the carriers mixed with it often are apparent in the finished product.

Monoglycerides and Diglycerides

These emulsifiers consist of fatty acids chemically combined either 1 or 2 on a glycerol residue. The uncombined-OH groups on the glycerol moiety provide the hydrophilic portion of the molecules. It has been shown that monoglycerides are far more effective than diglycerides in reducing surface tension, but the nature of the manufacturing process is such that some diglycerides (and, for that matter, some triglycerides, or fats, plus free glycerol) are inevitably included in the reaction mixture. Molecular distillation can be resorted to as a means of separating the monoglycerides from other components of the mixture. Diglycerides are said to be helpful in dispersing the monoglycerides.

The common basic raw materials for monoglyceride manufacture are lard and vegetable oils such as cottonseed oil (usually hydrogenated). This results in a mixture of fatty acid moieties combined in random fashion with glycerol. Special mixtures and more or less purified fatty acids can be used to give monoglycerides of specific composition.

Monoglycerides, but not diglycerides, exhibit a starch-complexing reaction. In bread and rolls, an important manifestation of this reaction is a retardation of the rate at which the interior of the loaf, or crumb, becomes firm. Since snack products do not undergo the texture staling typical of bread, because of their low moisture content, monoglycerides are not commonly added to these products.

Other Surfactants

Among the surfactants which have been applied to nonstandardized foods are sorbitan monostearate, polyoxyethylene sorbitan monostearate, ethoxylated mono- and diglycerides, propylene glycol monostearate, glycerolactopalmitate, sodium stearoyl lactylate, calcium stearoyl lactylate, sodium stearoyl fumarate, succinylated monoglycerides, and polyglycerol esters of fatty acids.

ANTIOXIDANTS

Although there is a great variety in the composition of snack foods, they all contain some fat. All of these fats are subject to oxidative and hydrolytic rancidity which can cause the development of objectionable odor and flavor.

Antioxidants are materials which can retard the development of rancidity during storage of foods containing fat. Natural antioxidants are found in many "nonpurified" fats, such as cocoa butter, and certain chemical compounds can be added to fats to accomplish this purpose.

From a chemical standpoint, rancidity is of two types: (1) hydrolytic rancidity, which can lead to the occurrence of soapy flavors, and (2) oxidative rancidity which causes the pungent or acrid odor characteristic of badly deteriorated fat. When hydrolytic rancidity occurs, oxidative deterioration is facilitated. Oxidative rancidity is unquestionably the most important of these two mechanisms so far as effects on food acceptability are concerned. The susceptibility of a fat to oxidation depends to a considerable extent upon the number of unsaturated bonds in the fatty acid moiety. Polyunsaturated fats are very prone to the development of oxidation while fully saturated fats and oils are much more resistant.

In hydrolytic rancidity, moisture and enzymes cause splitting of the triglyceride molecule into glycerol and free fatty acids. The rate of lipolysis of the fat is strongly influenced by temperature. Optimum temperature for this reaction is near 100°F. Hydrolytic rancidity can be controlled by inactivation of the enzymes through sterilization, low moisture content, and low storage temperatures. The high processing temperatures to which most snacks are subjected effectively prevent hydrolytic rancidity in these items.

The predominant fatty acids in cereals are palmitic, stearic, oleic, linoleic, and linolenic, with the unsaturated acids, oleic and linoleic, accounting for perhaps ⅔ of the oil content. The reactions leading to oxidative rancidity attack the unsaturated portion of the fatty acid. The steps in this process are not fully understood, but probably there is autoxidation by a free radical mechanism which is catalyzed by heat, light, and trace quantities of metal ions, especially copper and iron. When the fatty ester free radical captures a hydrogen atom, the action of the free radical is terminated and autoxidation is inhibited. The chain reaction can be broken by adding phenolic-type antioxidants, which readily give up a hydrogen atom to the free radicals.

A great deal of research activity has been directed toward finding substances which will retard the development of oxidative rancidity in foods and at the same time be acceptable to federal regulatory agencies as food additives. At the present time, only four chemical compounds are commercially important as antioxidants for foods. They are butylated hydroxyanisole (BHA), butylated hydroxytoluene (BHT), and tertiary butylhydroquinone (TBHQ), and propyl gallate. Citric acid or phosphoric acid may be added to improve the effectiveness of the antioxidants by chelating ions of copper or iron but they do not themselves function directly to prevent fat oxidation.

Tertiary butylhydroquinone (TBHQ) is a relatively new antioxidant which has been approved by the FDA for use in foods. The permitted concentration must be such that the combined total of BHA, BHT, propyl gallate, and TBHQ does not exceed 0.02% of the weight of the fat and oil, including essential oil. The regulations concerning antioxidants in animal fats, as promulgated by the USDA, are slightly different. According to data released by the manufacturer (see Table 2.2), TBHQ seems to be somewhat more effective than BHA in retarding the devel-

TABLE 2.2

EFFECTIVENESS OF TBHQ IN DRY CEREALS

		Days to Develop Rancid Odor at 145°F			
Antioxidant	% Added	Mixed Cereal[1]	Corn Cereal	Oat Cereal	Wheat Cereal
None	—	12	12	10	10
TBHQ	0.001	72	17	13	18
TBHQ	0.005	275	56	38	54
TBHQ	0.010	[2]	113	50	88
BHA	0.001	62	18	17	16
BHA	0.005	84	36	34	24
BHA	0.010	106	76	45	32

Source: Anon (1973).
[1]This cereal contained a mixture of corn, rice, and wheat.
[2]Sample not rancid at the end of 375 days.

opment of rancidity in dry cereals (which reasonably can be assumed to react similarly to many snack products).

Companies such as Eastman Chemical Products offer mixtures of antioxidants in convenient diluted forms. They can also provide advice as to the legality of proposed uses of the materials.

The FDA limits the addition of butylated hydroxyanisole (BHA) butylated hydroxytoluene (BHT), and propyl gallate (PG) in food products to 0.02% based on the weight of fat and oil (including essential oils) as 100%. The antioxidant concentration applies to the total amount present, whether the total includes one or more compounds.

Formulations containing propyl gallate are not recommended if there is substantial contact of the treated material with iron equipment, since PG becomes discolored as it picks up iron.

In the antioxidant treatment of nuts or any other foodstuff subject to oxidative rancidity, it is important to add the preservative before oxidation begins in the oil phase of the kernel. When added to fresh nuts, the antioxidant is available to terminate the free radicals as they form. The antioxidant cannot reverse the fat oxidation that has already occurred.

Antioxidant can be applied by spraying a more or less dilute solution on the nuts, popcorn, chips, etc., or, more commonly, is added to the frying or roasting oil. As described further in Chapter 5, Salt, application of salts containing antioxidants is a means of treatment. Packaging materials are frequently treated with antioxidants so that oils which contact and penetrate into the boards or films will not exhibit the rapid rancidification which would otherwise occur. There is also reason for believing that the antioxidant-treated packaging material may give off vapors which can affect noncontacting product.

BIBLIOGRAPHY

ANON. 1963. The Atlas HLB Systems, 3rd Edition. Atlas Chemical Industries, Wilmington, Del.

ANON. 1973. Tenox TBHQ Antioxidant for fats, oils, and fat containing foods. Eastman Chemical Products Publ. ZG-201.

CHANG, S. S. 1967. Chemistry and technology of deep fat frying: an introduction. Food Technol. 21, 33–34.

COCKS, L. V., and VAN REDE, C. 1966. Laboratory Handbook for Oil and Fat Analysts. Academic Press, New York.

DEUEL, H. J., JR. 1951. The Lipids—Their Chemistry and Biochemistry. Interscience Publishers, New York.

ECKEY, E. W. 1954. Vegetable Fats and Oils. Reinhold Publishing Corp., New York.

EVANS, C. D. 1955. Flavor evaluation of fats and oils. J. Am. Oil Chemists Soc. 32, 596–604.

HAWLEY, H. K., and HOLMAN, G. W. 1956. Directed interesterification as a new processing tool for lard. J. Am. Oil Chemists' Soc. 33, 29–35.

JACOBSON, G. A. 1967. Quality control of commercial deep fat frying. Food Technol. 21, 147–152.

LINK, W. E. 1973. Official and Tentative Methods of Analysis, 3rd Edition. American Oil Chemists' Soc., Champaign, Ill.

MEHLENBACHER, V. C. 1960. The Analysis of Fats and Oils. Garrard Press, Champaign, Ill.

MOYER, J. 1965. Selection, maintenance, and protection of frying fats. Proc. Am. Soc. Bakery Engrs. 1965, 273–278.

POHLE, W. D. 1964. A study of methods for evaluation of the stability of fats and shortenings. J. Am. Oil Chemists' Soc. 30, 186–190.

PONTIUS, W. I. 1965. The Meaning of Solids Factor Index. Armour and Co., Chicago.

ROBERTSON, C. J. 1967. The practice of deep fat frying. Food Technol. 21, 34–36.

SHERWIN, E. R. 1968. Methods for stability and antioxidant measurement. J. Am. Oil Chemists' Soc. 45, 632A, 634A, 646A, 648A.

STINGLEY, D. V., VANDER WAL, R. J., and WHEELER, F. E. 1961. The solids content of shortening and its significance to the baker. Bakers Dig. 35, No. 8, 16–19.

THOMAS, B. L..1968. Specifications: What do they really mean? Snack Food 57, No. 12, 30–32.

WEISS, T. J. 1963. Fats and oils. In Food Processing Operations, Vol. 2, M. A. Joslyn, and J. L. Heid, (Editors). Avi Publishing Co., Westport, Conn.

WEISS, T. J. 1970. Food Oils and Their Uses. Avi Publishing Co., Westport, Conn.

WOERFEL, J. B. 1960. Shortenings. In Bakery Technology and Engineering, S. A. Matz, (Editor). Avi Publishing Co., Westport, Conn.

CHAPTER 3

Sweeteners

INTRODUCTION

Sweeteners are used in large quantities for glazed popcorn and extruded collets. Although they are not "sweeteners" in the strict sense of the word, corn syrup solids of low D.E. (dextrins or maltodextrins) are useful as carriers for flavors which are to be applied to salty snacks by dusting or in an oil spray. Various kinds of formulated snacks depend upon sugars and corn syrups as both texturizing and flavoring ingredients. This chapter contains a brief survey of the sweeteners and similar products which may be encountered by the snack food technologist.

Commercial sucrose—refined cane or beet sugar—is one of the purest ingredients available to the food manufacturer. The composition of the fractional percentage of nonsucrose material in beet sugar differs slightly from that in cane sugar, but in practice the two sugars can be used interchangeably.

Sugars from various manufacturers do not differ significantly in composition, although syrups are not always as consistent. Nearly all commercially available sugar contains in excess of 99.8% sucrose, with less than 0.05% moisture, about 0.05% invert sugar and other carbohydrates besides sucrose, and a trace of ash. Different granulations of sucrose are offered for sale, and products having similar names may vary in particle size distribution between manufacturers.

Table 3.1 gives a typical sieve analysis of the granulations of sugar commercially available (Bohn and Junk 1960). The common varieties of powdered sugars are ultrafine (Confectioners' 10X), very fine (Confectioners' 6X), fine (Confectioners' 4X), medium, and coarse.

Sugars of the fine particle sizes tend to cake badly if stored for long periods of time. This problem can be alleviated by adding about 3% of cornstarch or, less often, 1% tricalcium phosphate during grinding.

Some food processors make their own powdered sugar by pulverizing granulated sucrose in a Schutz-O'Neill or Mikro Pulverizer mill. Automated grinding systems can be set up to provide a constant supply of powdered sugar as it is required with minimal attention from plant personnel. Some of the newer designs of pulverizing systems deliver cooled sugar and operate substantially dust-free. Cornstarch can be automatically added at about the 3% level if the sugar is customarily held for more than 24 hr before use.

TABLE 3.1

TYPICAL SCREEN ANALYSIS OF GRANULATED SUCROSE

Sieve Analysis		Confectioners AA or Medium Granulated	Sanding	Bottlers, Manufacturers or Fine Granulated	Standard Granulated or Extra-fine Granulated	Bakers' Special or Fruit Granulated	Standard Powdered or 6X	Extra-fine Powdered or 10X	Fondant and Icing	Dehydrated Fondant, Drivert or Drifond
Tyler Screen, % on	U.S. Screen % on									
8 mesh	8 mesh	—	—	—	—	—	—	—	—	—
10 mesh	10 mesh	5.6	—	—	—	—	—	—	—	—
14 mesh	16 mesh	59.0	9.3	—	—	—	—	—	—	—
20 mesh	20 mesh	27.4	49.2	4.3	—	—	—	—	—	—
28 mesh	30 mesh	7.4	37.6	74.5	0.1	—	—	—	—	—
35 mesh	40 mesh	0.4	3.3	18.6	13.8	0.4	—	—	—	—
48 mesh	50 mesh		0.3	2.3	40.2	1.7	—	—	—	—
80 mesh	80 mesh				40.6	24.8	—	—	—	—
100 mesh	100 mesh					32.3	0.3	0.1	—	—
150 mesh	140 mesh					31.6	1.8	1.4	—	—
200 mesh	200 mesh						6.6	2.4	—	—
270 mesh	270 mesh						8.2	3.0	—	—
325 mesh	325 mesh						10.8	7.0	—	—
Through last sieve		0.2	0.3	0.3	5.0	9.2	72.3	86.1	99.0[1]	99.0[1]

[1] The extremely fine grain size of fondant and icing sugar makes the regular screen analysis impractical. However, practically all particles will pass through a 325 mesh standard sieve having 0.0017 in. openings. The average particle size of fondant and icing sugar is about 20 μ, i.e., 0.0008 in.

Sugar should be stored under dry conditions (less than 60% RH) and variations in humidity should be avoided. When the sugar is exposed to atmospheres of high water activity, moisture is absorbed on the surface of the particles and a thin layer of concentrated syrup forms bridges between the granules. When the humidity decreases, the water may evaporate leaving the crystals cemented together in large lumps. Since relative humidity or water activity is directly related to air temperature, sugar should not be stored where a wide variation in temperature is apt to occur.

The obvious advantages, in many applications, of handling sugar in dissolved form have led to the extensive distribution of sucrose as syrups. Sucrose syrup is available at 66.5 to 68% solids content, or at the limits of solubility of sucrose at ordinary temperatures. Refinery syrups or liquid brown sugars are also available in some localities. Two or three grades, varying in color and composition, are usually offered. For example, water white grade is a sparkling clear syrup with very few impurities while light straw can be used whenever a small amount of color and a slightly higher percentage of nonsugars will not adversely affect the final product. Table 3.2 gives typical composition of refinery syrups.

TABLE 3.2

APPROXIMATE COMPOSITION OF REFINERY SYRUPS

	Moisture (%)	Sucrose (%)	Invert (%)	Ash (%)
West Coast				
Light	23	25	47	3
Medium	23	40	27	6
Dark	25	46	17	7
East Coast				
Light	23	24	47	3
Medium	23	43	22	5
Dark	22	42	21	6

Source: Junk and Pancoast (1973).

Refiners' syrups are obtained by special processing of material drawn off from the intermediate steps in the manufacture of cane sugar. They have a relatively strong flavor and a dark color which can be advantageous in certain food uses and tolerated in some others. Since they are usually priced lower than pure granulated sugar, substitution is desirable wherever it is feasible. These syrups contain a total of 50 to 75% sucrose and invert solids combined, and other substances making up 70 to 80% solids.

Liquid sugars are usually cheaper, on a solids basis, than bagged

TABLE 3.3

MAXIMUM SOLUBILITIES OF SUCROSE, INVERT SUGAR, AND THEIR MIXTURES IN WATER AT VARIOUS TEMPERATURES

Temp °C	Temp °F	Solutions of Sucrose (% by Wt)	Solutions of Invert Sugar (% by Wt)	Mixed Solutions of Sucrose and Invert Sugar		
				Total Sugars (% by Wt)	Sucrose (% by Wt)	Invert Sugar (% by Wt)
0	32.0	64.18	50.8	70.9	43.7	27.2
10	50.0	65.58	56.6	72.7	40.9	31.8
15	59.0	66.33	59.8	73.9	39.1	34.8
20	68.0	67.09	62.6			
23.15	73.7			76.2	36.3	39.9
25	77.0	67.89	66.2			
30	86.0	68.70	69.7	79.0	33.6	45.4
35	95.0	69.55	72.2			
40	104.0	70.42	74.8	81.8	31.1	50.7
45	113.0	71.32	78.0			
50	122.0	72.25	81.9	85.7	27.7	58.0

Source: Junk and Pancoast (1973).

sugar, if they can be received in bulk. In comparison with bulk dry sugar, the situation is not as clear, and may vary with the date and the location of the consumer. If they must be received and handled in drums, the syrups are more expensive. Suppliers of liquid sugar will finance receiving and bulk storage installations and amortize them with a surcharge on each pound of syrup purchased.

INVERT SYRUPS

When a sucrose solution is heated in the presence of an acid or certain enzymes, each carbohydrate molecule reacts with water to give a molecule of D-glucose and a molecule of D-fructose, often called dextrose and levulose, respectively. This mixture of dextrose and levulose in equal weights is known as invert sugar. Because of the chemical combination with a water molecule, there is a net gain of over 5 lb of sugar for each 100 lb of the original sucrose. In addition, the sweetness of the solution increases considerably, and more of the solids will dissolve in any given amount of water. These obvious advantages have resulted in a considerable market developing in invert syrups and in syrups containing invert combined with sugar.

Although sucrose syrup has an upper limit of 66.5 to 68% solids content at ordinary temperatures, if part of the sugar is inverted, the resulting syrup will retain higher concentrations of solids in solution. Common types of these mixed syrups contain 73 to 76% solids with 30 to 60% invert. "Totally" inverted syrups contain 72 to 73% solids with perhaps 5% of this being sucrose. Table 3.3 shows the solubility relationships of sucrose and invert mixtures.

All of these products are reasonably resistant to microbiological spoilage during storage, but the invert syrups are somewhat superior in this regard due to their higher osmotic pressure. There has been some controversy about the relative sweetness of sucrose and invert sugar in foods and beverages, but, when solutions are compared directly, it is generally agreed that invert sugar is definitely sweeter on a pound-for-pound basis. For each individual application, a separate evaluation should be made.

MOLASSES

The juices extracted from sugar-bearing plants and concentrated by boiling are called molasses. These thick syrups contain other substances besides sugar that are present in the plant juice, and the ash content is often quite high. Sulfur dioxide is sometimes added as a preservative. Open-kettle molasses is produced in the West Indies by boiling cane juice until most of the water has been evaporated, and is an edible grade as opposed to some industrial grades used mainly for the production of

TABLE 3.4

CHARACTERISTICS OF DIFFERENT TYPES OF MOLASSES

Type	Total sugars (%)	Sucrose (%)	Invert Sugar (%)	Ash (%)	Color
Open-kettle	68–70	43–45	24–26	1–2	Reddish yellow
First centrifugal	60–66	48–50	13–15	13–15	Light yellow
Second centrifugal	56–60	36–38	20–22	20–22	Reddish
Blackstrap	52–55	25–27	27–29	27–29	Dark brown

alcohol, as feed ingredients, etc. Characteristics of some of the molasses types which may be available commercially are shown in Table 3.4.

BROWN SUGAR

Brown sugar is made in two different ways. Brown sugar from cane syrups results from stopping the purification process short of the final steps so that some of the molasses flavoring ingredients are retained in the finished product. Such a procedure does not yield an acceptably flavored product when applied to beet sugars, so in this case brown sugar is made by adding cane molasses to fully refined sugar crystals. Total sugar content ranges from 90 to 95%, and moisture is about 2 to 4%. Brown sugars are composed of very small crystals covered with a film of highly-refined, dark-colored, cane flavored syrup. On a solids basis, these sugars are always slightly more expensive than ordinary granulated sugar. The syrup contributes a characteristic color and flavor which is considered desirable in certain kinds of confections, e.g., caramel corn and baked goods.

At one time, refiners produced 15 grades of brown or "soft" sugars, ranging from No. 1, with a slight creamy tint, to the very dark brown No. 15. Because of the limited demand for the finer gradations of color, most refiners now produce only 4 grades, No. 6, 8, 10, and 13. The retail varieties are light brown (No. 8), dark brown (No. 13), and medium brown (No. 10).

Moisture content ranges from about 2% in the lightest type to about 4% in the darkest type. If most of this moisture is lost through exposure to atmospheres of low relative humidity, the sugar crystals will be cemented into a solid block. It is this familiar change which creates numerous difficulties in commercial use of brown sugar and which has led many food processors to turn to various syrups, or even to flavor concentrates to be used in combination with granulated sugar. At relative humidities of 65 to 70% brown sugars will retain their moisture and remain soft.

TABLE 3.5

APPROXIMATE COMPOSITION OF BROWN OR SOFT SUGAR

		Sucrose (%)	Invert (%)	Ash (%)	ONS[1] (%)	Moisture (%)
West Coast	Med Brown[2]	87.5	4.4	2.1	2.3	3.6
West Coast	Dark Brown	86.2	5.0	2.3	2.8	3.7
East Coast	Med Brown[2]	91.0	3.0	1.3	2.2	2.5
East Coast	Dark Brown	89.0	3.2	1.7	2.3	3.8
Southern	Med Brown[2]	90.0	3.5	1.3	1.1	4.1
Southern	Dark Brown	91.0	3.0	1.4	0.8	3.8
Beet	Med Brown	88.5	8.0	0.7	0.8	2.0

Source: Junk and Pancoast (1973).
[1] ONS—Organic nonsugars which are usually determined by difference.
[2] Medium Brown is frequently referred to as Golden Brown.

The total sugar content varies from about 95% in No. 6 to 91% in No. 13. From 1 to 5% of the sugar is present as invert. Table 3.5 lists the approximate composition of some commercially available varieties of brown sugar.

SWEETENERS DERIVED FROM CORNSTARCH

Vast quantities of liquid and dried corn syrups are consumed by the food industry. Significant amounts are used by manufacturers of sweet glazed snacks. Corn syrups, which consist of cornstarch hydrolyzed to varying degrees, have the outstanding advantage of being cheaper, on a solids basis, than refined cane or beet sugar. These sugars are not interchangeable on an *ad lib* basis, but substitution of the cheaper material is often successful if minor adjustments are made in formulas and slight differences accepted in the finished product. Too, there are instances in which a corn syrup has functional advantages over sugar.

The two specifications of prime importance in corn syrup procurement are the dextrose equivalent (DE) and the degrees Baumé. The former figure indicates the extent of conversion, and thus is related to the sweetness, viscosity, and preservative quality of the material, while the latter datum is related to the solids content. For some special syrups, additional information is required, since the distribution of molecular species is not always the same for syrups having similar DE.

Cornstarch may be hydrolyzed by acids, by enzymes, or by a sequence of acid hydrolysis followed by enzyme conversion. The normal end product of acid hydrolysis is dextrose while the acid-enzyme process leads to a syrup of relatively high maltose content. The dextrose-maltose ratio can be varied as desired, within limits. The ratio is influenced both by the type of enzyme employed and by the extent of the preliminary acid conversion (see Table 3.6).

TABLE 3.6

CARBOHYDRATE COMPOSITION OF ACID-ENZYME
DUAL-CONVERTED CORN SYRUPS

DE	Mono	Percentage Saccharides[1] Di	Tri	Tetra	Higher
26–30	8.0–9.0	10.0–13.0	10.0–12.0	6.0–9.0	60.0–63.0
35–36 HM[2]	6.3	23.2	—	—	70.5
42–43 HM	5.0–10.0	39.0–48.0	13.0–18.0	3.0–6.0	24.0–33.0
48–50 HM	9.0–9.4	51.5–52.0	13.5–15.0	1.7–2.0	21.0–24.3
62–63	35.0–40.0	25.0–30.0	13.0–14.0	4.0–8.0	8.0–15.5
64–66	37.0–39.0	31.5–33.5	10.0–12.0	5.0–5.5	13.0–16.0
68–71	40.0–43.0	32.0–40.0	2.0–10.0	2.0–5.5	8.5–15.5

Source: Junk and Pancoast (1973).
[1] Approximate values.
[2] High maltose.

Dextrose equivalent (DE) is based on an analysis for total reducing sugars, the results of which are expressed as glucose calculated as a percentage of the total dry substance; i.e., the DE is the percentage of pure glucose which would give the same analytical effect as is given by the total of all the different reducing sugars actually present. Commercially available corn syrups and sugars fall into the following classes (Hoover 1964; also see Table 3.7).

(1) Low conversion corn syrup: at least 28 but less than 38 DE.
(2) Regular conversion corn syrup: at least 38 but less than 48 DE.
(3) Intermediate conversion corn syrup: at least 48 but less than 58 DE.
(4) High conversion corn syrup: at least 58 but less than 68 DE.
(5) "Extra" high conversion corn syrup: DE of 68 or higher.
(6) Crude corn sugar "70": DE of 80 to 89.
(7) Crude corn sugar "80": DE of 90 to 94.
(8) Dextrose: DE of 100

TABLE 3.7

NOMENCLATURE OF CORN SYRUPS BASED ON TYPE OF CONVERSION
AND DEXTROSE EQUIVALENT

Acid-Converted	Acid-Enzyme Converted
28–30 DE	26–30 DE
35–38 DE	35–36 DE high maltose
42–43 DE	42–43 DE high maltose
52–55 DE	45 DE high maltose
Enzyme-Enzyme Converted	48–50 DE high maltose
68 DE	62–63 DE
75 DE levulose containing corn syrup	64–66 DE
92–95 DE levulose containing corn syrup	69–71 DE

Source: Junk and Pancoast (1973).

DE can vary over a wide range. The "regular" corn syrup of most manufacturers is a 42-DE product. Lower conversion syrups are available for users desiring less sweetness, greater viscosity, or other special qualities. Many food processors prefer the high conversion syrups, available up to 71 DE or even higher, because of the added sweetness, greater fermentability, and enhanced browning effects. The lower viscosities facilitate transfer (flow and pumping) of the material. The hydrolysis reaction can be continued to give almost 100% conversion to dextrose, but crystallization of glucose from the reaction mixture makes impractical the commercial handling of syrups higher than 71 DE. If the material can be stored and dispensed in excess of 130°F, practically pure dextrose can be handled in syrup form.

Corn syrups may be specially treated. For example, ion exchange techniques are sometimes applied to reduce the mineral content of the material. The low ash content contributes to the clarity of hard candy since salts do not precipitate out as the candy batch moisture is reduced. Removal of the ash (mostly sodium chloride) also improves the flavor of the syrup.

Special high maltose syrups are available. To make these products, the hydrolysis is "guided" by special techniques so as to yield a larger percentage of disaccharides (maltose) and less monosaccharides (glucose) than normal.

As rapid control tests, determination of either refractive index or density will serve as indications of solids content. Manufacturers' specifications usually quote degrees Baumé. The latter figure is an indication of the specific gravity and can be determined by special hydrometers. It is impractical to take hydrometer readings at room temperature because of the very high viscosity of corn syrup under these conditions. It is customary, therefore, to make the determination at 140°F and add an arbitrary correction of 1.00° Baumé to the reading. The resultant figure is called "commercial Baumé" and is approximately equal to the value that would have been obtained at 100°F, the Baumé basis on which the syrup is sold.

When the degrees Baumé and the dextrose equivalent of a syrup are known, the dry-substance concentration can be determined by reference to published tables. A 43° Baumé high-conversion corn syrup has about 2% more solids per unit volume than does the 43° Baumé regular conversion corn syrup.

Degrees Baumé of commercially available syrups vary from 41 to 46. In a regular 42 DE syrup these figures are roughly equivalent to a range of 76 to 84% solids. Most corn syrups are now being distributed at 43° Baumé, equivalent to a specific gravity of 1.42 or a weight per gallon of 11.82 lb. Higher concentrations lead to exceedingly viscous syrups which are difficult or impossible to handle with ordinary equipment, especially if the syrup has a low DE.

All corn syrup manufacturers offer low conversion (in the range of 28 to 38 DE), regular conversion (38 to 48 DE), intermediate conversion (48 to 58 DE), and high conversion (58 to 68 DE) products. Additional varieties are frequently available, but not from all suppliers.

Corn syrup solids and dextrose are available for those users who desire a higher concentration of solids or who do not want to handle the liquid form. As the name suggests, corn syrup solids (as well as dextrose) are prepared by drying corn syrups. The normal moisture content of these products lies between 3 and 3.5%. Corn syrup solids are offered at several dextrose equivalents, but the common type is 42 DE. The usual dextrose is a monohydrate of glucose, but one manufacturer offers anhydrous glucose.

Complete hydrolysis of the starch is required to produce dextrose. The conditions for hydrolysis, therefore, are adjusted so as to give a maximum dextrose content in the hydrolyzate. After purification and concentration, as in the manufacture of corn syrup, the highly refined dextrose liquor is seeded with dextrose crystals from a previous batch. Crystallization is then carried under carefully controlled rates of cooling in large crystallizers equipped with spiral agitators.

After approximately 100 hr in the crystallizers the crystalline dextrose monohydrate (one molecule of water held in the crystal with each molecule of dextrose) is separated from the "mother liquor" in centrifuges, washed, dried, and packed in 100-lb paper bags. The mother liquor separated from the crystals by the centrifuges is reconverted, clarified, concentrated, and then passed through another crystallization cycle. As before, the dextrose crystals which form during this second cycle are removed by centrifuges. The dextrose so obtained is redissolved in water, added to new starch hydrolyzate and recrystallized to yield dextrose monohydrate. The mother liquor from the second crystallization, dark brown in color, is called "hydrol" or "feeding corn-sugar molasses" and is widely used in animal feeding and in industrial processes.

The production of anhydrous dextrose (containing no combined water) involves still further processing. Anhydrous dextrose belongs to a different crystal system from that of dextrose monohydrate. Crystalline dextrose monohydrate from the centrifuges is redissolved in water to form a solution containing about 55% dry substance. Activated carbon is added to this solution and the mixture is heated to about 160°F for 30 min. The solution is then filtered until perfectly clear after which it is run into an evaporator called a "strike pan" for crystallization. Seeding with crystals from a previous batch is not required since graining conditions are effected through evaporation. Evaporation is continued until crystallization has reached the desired stage. Centrifuging separates the crystals from the remaining liquid; then the crystals are washed with a hot-water spray. Finally, the anhydrous dextrose is removed from the centrifuges, dried, pulverized, and packed.

Recent advances in corn hydrolysate treatment have made it possible to prepare syrups having a substantial amount of D-fructose (Table 3.8). The advantages of these syrups to the snack producer are the higher solids content which can be obtained, the greater sweetness, and the lower viscosity as compared to conventional corn syrups. It is possible to get sweetness equal to sucrose syrups at a substantially lower cost (Piekarz 1968).

TABLE 3.8

EXAMPLES OF CARBOHYDRATE COMPOSITION OF CORN SYRUPS CONTAINING FRUCTOSE

Syrup Identification	D-glucose	D-fructose	Maltose	Higher Saccharides
A	43	14	31	12
B	50	30	10	10
C	50	42	3	5
D	70	20	3	7

Source: Courtesy of Corn Industries Research Foundation.

OTHER SWEETENERS

Nutritive Sweeteners

Honey.—The composition of honey varies depending upon the source and other factors, but a typical analysis would be: water 17.2%, protein 0.3%, ash 0.2%, carbohydrates 82.3%, and no fat. The sugars are mostly D-glucose and D-fructose, with substantial percentages of sucrose and traces of several other sugars.

Honey is classified on the basis of the flower from which the nectar was obtained. There are probably more than a hundred honey types sold to consumers throughout the world, but only a few are of importance to bakers. The lighter colored and milder flavored honeys of commercial importance are clover and alfalfa, while sage and buckwheat are darker and more strongly flavored. An imported honey of considerable importance because of its low price but relatively strong pleasant flavor is Yucatan honey.

Maple Syrup.—This material is a very expensive source of sweetening and characteristic flavoring. It is made by boiling down the sap obtained from the sugar maple tree. The typical flavor is not present in the sap and develops only during processing. Maple syrup contains about 66% solids and the solid sugar cake about 93%. Most of the solids consist of sucrose but several percent of invert sugar are also present, and there is around 1% ash.

Lactose.—Lactose is derived from the fractionation of milk or some of its by-products such as whey. Prior to 1974, it was more expensive than sucrose and it is much less sweet (about 0.16 compared to sucrose as 1.00). For applications where a bland solid material is needed to fill out a formula, as in cases where it is needed as a carrier for a salty or savory flavor and starch or corn syrup solids have been found to be unsatisfactory diluents, lactose may be of special value. This sugar is offered in crude, edible, and pharmaceutical grades, in several particle sizes, and in at least two crystal forms.

Nonnutritive Sweeteners

Although several new artificial sweeteners are under investigation by government and private research groups here and abroad, this category is presently limited to saccharin and aspartame. Saccharin (2,3-dihydro-3-oxobenzisosulfonazole) is about 300 to 500 times sweeter than sugar in dilute aqueous solutions, although it has flavor notes which are not present in sugar, and vice versa. An exact match to the flavor contributed to a food product by sucrose may sometimes be unobtainable with saccharin. Furthermore, many people find that saccharin has a bitter taste, especially at higher concentrations. Saccharin can be used only in those foods and at those concentrations permitted by the U.S. Food and Drug Administration, and special labeling regulations must be followed.

Aspartame (the methyl ester of 1-aspartyl-1-phenylalanine) was cleared in July 1974 for certain food uses. These uses do not seem to include snack products, although cold breakfast cereals is one of the approved categories. On the average, it is considered to be about 180 times as sweet as sucrose. At present, the supply is limited and the substance is very expensive.

BIBLIOGRAPHY

ANON. 1952. Sugar—Its Types and Uses, Competitive Products. Sugar Information, New York.

ANON. 1965. Corn Syrups and Sugars, 3rd Edition. Corn Industries Research Foundation, Washington, D.C.

BECK, K., and ZIEMBA, J. V. 1966. Are you using sweeteners correctly? Food Eng. *38*. No. 12, 71–73.

BIESTER, A., WOOD, M. W., and WOHLIN, C. S. 1935. Relative sweetness of sugars. Am. J. Physiol. *73*, 387–400.

BOHN, R. T., and JUNK, W. R. 1960. Sugars. *In* Bakery Technology and Engineering, S. A. Matz (Editor). Avi Publishing Co., Westport, Conn.

CAMERON, A. T. 1947. The taste sense and the relative sweetness of sugars and other sweet substances. Sugar Res. Found. Sci. Rept. Serv. *9*.

CORSON, G. E. 1957. Critical Data Tables, 2nd Edition. Corn Refiners Assoc., Washington, D.C.

COTTER, W. P., LLOYD, N. E., and HINMAN, C. W. 1971. Method of isomerizing glucose syrups, U.S. Pat. 3,623,953. Nov. 30.

DAHLBERG, A. C., and PENCZEK, E. S. 1941. The relative sweetness of sugars as affected by concentration. N.Y. State Agr. Expt. Sta. Bull. *258*.

DEPPERMAN, L. O., and WITTENBERG, W. R. 1966. Sugar granulation as related to product uniformity. Biscuit Cracker Baker *55*, No. 11, 30–33.

ERICKSON, E. R., BERNTSEN, R. A., and ELIASON, M. A. 1966. Viscosity of corn syrup. J. Chem. Eng. Data *11*, 485–488.

FLICK, H. 1964. Fundamentals of cookie production, including soft type cookies. Proc. Am. Soc. Bakery Engrs. *1964*, 286–293.

HOOVER, W. J. 1964. Corn sweeteners. *In* Food Processing Operations, Vol. 3, M. A. Joslyn, and J. L. Heid (Editors). Avi Publishing Co., Westport, Conn.

JUNK, W. R., and PANCOAST, H. M. 1973. Handbook of Sugars. Avi Publishing Co., Westport, Conn.

KELLY, N. 1964. Sugar. *In* Food Processing Operations, Vol. 3, M. A. Joslyn, and J. L. Heid (Editors). Avi Publishing Co., Westport, Conn.

MATZ, S. A. 1965. Water in Foods. Avi Publishing Co., Westport, Conn.

MONEY, R. W., and BORN, R. 1951. Equilibrium humidity of sugar solutions. J. Sci. Food Agr. 2, 180–185.

NEWTON, J. M. 1970. Corn syrups. *In* Symposium Proceedings on Products of the Wet Corn Milling Industry in Food. Corn Refiners Assoc., Washington, D.C.

NIEMAN, C. 1960. Sweetness of glucose, dextrose, and sucrose. Mfg. Confectioner *40*, No. 8, 19–24, 43–46.

PANGBORN, R. M. 1963. Relative taste intensities of selected sugars and organic acids. J. Food Sci. *28*, 726–730.

PIEKARZ, E. R. 1968. Levulose-containing corn syrup. Bakers Dig. *42*, No. 5, 67–69.

REGER, J. V. 1958. New aspects of an old sugar—lactose. Cereal Sci. Today *3*, 270–272.

STONE, H., and OLIVER, S. M. 1969. Measurement of the relative sweetness of selected sweeteners and sweetener mixtures. J. Food Sci. *34*, 215–222.

WATSON, S. A. Manufacture of corn and milo starches. *In* Starch: Chemistry and Technology, Vol. 2, R. L. Whistler and E. F. Paschall (Editors). Academic Press, New York.

WATT, B. K., and MERRILL, A. L. 1963. Composition of Foods. U.S. Dept. Agr. Handbook *8*.

CHAPTER 4

Cheese and Other Dairy Products

Cheese has proven to be one of the most popular flavors for crackers, corn collets, and popcorn (Hess 1972). Potato chips are rarely flavored with cheese. In puffed or expanded pieces, dehydrated cheese must be used. Because of the expensive nature of this ingredient, it is often extended or fortified (and sometimes completely replaced) by dried buttermilk, acid or sweet whey powder, artificial flavors, etc. Nearly all dehydrated cheese flavors are artificially colored so that the finished snack has a medium dark orange-yellow color reminiscent of the usual type of cheddar cheese sold from dairy cases in the supermarket.

Cheese powders are generally made by forming an emulsion of the cheese and any additives that are required and spray drying it. A typical top grade snack dusting material would contain sharp aged cheddar cheese, emulsifier, salt, food color, and lactic acid. It would have a moisture content between 2 and 3%, a fat content between 46 and 50%, and granulation such that 100% would pass through a U.S. No. 25 mesh sieve. These powders are usually packed 50 or 100 lb to a multiwall Kraft paper bag with polyethylene liner or in fiber drums. About 10 oz of such dehydrated material is considered to be equal to a pound of fresh cheese. In applying these powders to puffed corn meal collets, the ratio of 60% collets, 25% oil, 10 to 12% cheese powder, and 2% salt would provide a high-quality product. Cheaper snack products would utilize 6% cheese powder, or even less, and the minimum amount of oil needed to get adhesion. The very cheapest products rely on fractional percentages of artificial flavors with the needed amount of oil and color, but with ample salt. Cheese-flavored popcorn might require 10 to 12% addition of cheese powder, or 22 to 27% of an oil-powder mixture, while potato chips would be dusted with 5 to 7% of the powder (Anon. 1969).

The less expensive versions of cheese powders would have compositions such as aged cheddar cheese, vegetable oil, corn syrup solids, buttermilk, salt, lactic acid, artificial food color, whey, disodium phosphate, and artificial flavor. The analytical data would be:

	Minimum	Maximum	Average
Moisture (%)	1.0	2.0	1.5
Fat (%)	41.0	47.0	44.0
Protein (%)	16.0	20.0	18.0
Carbohydrate (%)	20.0	24.0	22.0
Ash (%)	5.0	9.0	7.0
Chloride (as NaCl) (%)	2.0	5.0	3.0
pH	5.0	6.0	5.4

The bacteriological evaluations would show no coliform or *Salmonella*, 10 yeast and mold per gram, and 25,000 for the standard plate count.

Citric acid, monosodium glutamate, starch, hydrolyzed plant protein, and whey (acid or sweet) are other ingredients sometimes used in these cheese powders to modify flavor or act as less expensive fillers.

Cheese powders are frequently mixed with oil and other ingredients in the snack plant and spray- or dribble-coated on to the pieces in a tumbler-type machine.

The cheese applicators are revolving drums (inclined at an angle), through which the snack pieces pass after they have gone through an oven to reduce the moisture content. In the tumbler, a spray nozzle or drip spout introduces the hot oil and cheese mixture. The pieces become more or less uniformly coated with the suspension as a result of the rubbing and smearing contact as they tumble past one another. Plasticized cheese flavorings which consist, for example, of spray-dried cheddar cheese, hydrogenated vegetable oils, buttermilk powder, dried whey, mono- and diglycerides, salt, and lecithin, are offered by some suppliers. These mixtures eliminate the need for blending oil and powder in the plant. They are pumpable after heating to 90°F, and can be sprayed on to snacks after heating to 120 to 130°F. Higher temperatures can drive off some of the aromatics while lower temperatures lead to difficulties in spraying. Constant agitation is required after melting to prevent separation of the components. These coatings are packed in 55-lb pails or 55-gal. drums and are said to retain good quality during storage for 3 to 4 months at 75°F.

According to Zick (1969), the manufacture of dehydrated cheese flavor for snacks includes the following steps: (1) mechanical cleaning and trimming of block cheese, (2) random cutting of the blocks, (3) preparing an emulsion of the cheese solids, (4) mixing dry ingredients and vegetable oil with the cheese emulsion, (5) bringing the blend to pasteurization temperatures, (6) homogenizing, and (7) spray-drying. Alternative processes are to dry only the cheese and then blend in the remaining ingredients, or to dry the cheese and some of the other ingredients and add-mix the balance. A straight cheddar powder would contain, on a dry weight basis, 97.0% cheddar cheese, 2.5% emulsifier, and 0.5% artificial color. A typical cheddar cheese product contains 45.0% cheddar cheese, 14.0% nonfat dry milk, 24.0% dried whey, 9.0% vegetable oil, 3.0% salt, 1.5% autolyzed yeast, 0.8% artificial cheddar flavor, 0.7% hydrolyzed plant protein, 0.6% lactic acid, 0.5% citric acid, 0.4% monosodium glutamate, 0.3% emulsifier, and 0.2% artificial food color.

Specifications for raw materials, and labeling of the finished snack product must take into account the federal Standards of Identity for cheese, which describe in considerable detail every variety of cheese likely to be encountered by the snack manufacturer.

A listing of cheese varieties with their physical and organoleptic characteristics was compiled by Sanders (1953), and thorough descriptions of the production methods and quality factors of most kinds of cheeses have been given by Kosikowski (1966). Although it would not be appropriate to give a lengthy survey of these points in the present volume, some general information on the subject should be valuable to the snack technologist in the event that flavorings must be developed from the basic ingredients.

There are perhaps 18 different kinds of cheese, although hundreds of different names have been applied. The 18 basic types are represented by Brick, Camembert, Cheddar, Cottage, Cream, Edam, Gouda, Hand, Limburger, Neufchatel, Parmesan, Provolone, Romano, Roquefort, Sapsago, Swiss, Trappist, and whey cheese. Cheese can also be classified in other ways, as follows.

(1) Very hard (grating cheese) ripened by bacteria
    Example: Parmesan
(2) Hard
    (a) Ripened by bacteria, without eyes
        Example: Cheddar
    (b) Ripened by bacteria, with eyes
        Example: Swiss
(3) Semisoft
    (a) Ripened principally by bacteria
        Example: Brick
    (b) Ripened by bacterial and surface microorganisms
        Example: Limburger
    (c) Ripened principally by blue mold in the interior
        Example: Roquefort
(4) Soft
    (a) Ripened
        Example: Camembert
    (b) Unripened
        Example: Neufchatel

Typical analytical values for some of these cheeses are given in Table 4.1.

Cheddar cheese, the most popular variety sold in the United States and the most popular cheese flavoring for snack foods, has been described as possessing a clean flavor reminiscent of walnuts, with a waxy body which breaks down smoothly when portions of the cheese are kneaded between the fingers. It has a minimum of gas holes. Both the flavor and the texture are affected by the aging process which is an essential step in its manufacture. The aroma becomes more pungent, the taste more acidic and "full," and the texture crumblier. The natural hue of the cheese is often almost white, but various pigments are permissible additives to give the yellow-orange color preferred by the majority of consumers.

Most cheddar cheese is ripened at about 36°F and 86% RH for 4 to 12 months. It may be force-cured by withdrawing it from the cold room after two months and holding it at 50 to 60°F until the desired temperature has been obtained. In time, a peak of desirable flavor is attained, after which a subtle deterioration sets in. Optimum flavors in cheese made from pasteurized milk are usually achieved in 12 to 24 months at 36°F, 9 months at 50°F, or 5 months at 60°F. The flavor peak is reached sooner in raw milk Cheddar.

There are variants of Cheddar cheese, e.g., washed curd and Colby which are of little importance to the snack food business. Stirred curd cheese is used extensively in the manufacture of coating powders. It cannot be labeled as Cheddar, however.

Substances thought to be involved in the flavor and aroma of nonripened fermented dairy products include lactic and acetic acids, acetaldehyde, formic acid, ethanol, carbon dioxide, diacetyl, dimethyl sulfide and other sulfur compounds, methyl ketones, primary and secondary alcohols, methyl and ethyl esters of aliphatic acids, and lactones. Over 100 compounds were identified in the volatile fraction of Cheddar cheese. Those believed to be important include acetic, butyric, caproic, and caprylic acids, hydrogen sulfide, glutamic acid, methional, and carbonyl compounds. Compounds of microbial origin important to the flavor of blue cheese are free fatty acids, methyl ketones, and secondary alcohols. In Swiss cheese, propionic acid and proline are distinctive.

Fatty acids are produced during the ripening of cheese, both by the hydrolysis of milk fat and by fermentation, and these are largely responsible for the flavor changes. The rate of flavor production in cheese from pasteurized milk is slower than in that made from raw milk, probably due both to the natural enzymes and more varied flora, and in some cases, rancidity can develop (Dixon *et al.* 1970).

Nearly all flavor houses supplying snack producers have a range of artificial cheese flavors, some of them quite good. Cheddar of different degrees of ripeness, Swiss, and blue are common variants. Mozzarella, Parmesan, Romano, and Roquefort are also available. These are generally marketed either as oil soluble liquids or as encapsulated or absorbed liquids which can be applied as powders. They are high in aromatic components but low in the taste sensation we generally expect from fresh or dehydrated cheese. Fermentation based cheese flavors, derived from

TABLE 4.1

TYPICAL COMPOSITION OF CHEESE

Type	Cheese	Moisture %	Fat %	Protein %	Fat in Dry-matter %	Salt %	Ash %	Lactose %	Calcium %	Phosphorus %
Soft-unripened Low-fat	Cottage	79.0	0.4	16.9	1.9	1.0	0.8	2.7	0.09	0.05
	Creamed cottage	78.3	4.2	13.6	19.3			3.3	0.09	0.05
	Quarg	72.0	8.0	18.0	28.5	1.0	0.8	3.0	0.30	0.35
	Quarg (high fat)	59.0	18.0	19.0				3.0	0.30	0.35
Soft-unripened high	Cream	51	37	8.8	75.5	1.0	1.2	1.5–2.1	0.08	0.06
	Neufchatel	55	23	18.0	51.1	1.0	2.0			
Soft-ripened by surface bacteria	Limburger	46	27	21.5	50.0	2.0	3.6	0–2.2	0.5	0.4
	Liederkranz	52	28	16.5	58.3	1.5	3.5	0	0.3	0.25
		53	25.5	16.8	54.2	1.7	3.9			
Soft-ripened by external molds	Camembert	51	26	20.0	53.0	2.5	3.8	0–1.8	0.6	0.5
	Brie	45	30	21.6	54.5	2.0	4.0	0–2.0	0.6	0.4
Soft-ripened by bacteria, preserved by salt	Feta	57	24	20	55.8	5.0				
	Domiati	55	25	20.5	55.5	4.8				

Category	Cheese										
Semi-soft, ripened by bacteria with surface growth	Brick	42	31	21	53.4	2.0	4.2	0–1.9	0.6	0.4	
	Muenster	44	28	25	50.0	1.8			0.5	0.35	
Semi-soft, ripened by internal molds	Blue	41.5	30.5	21.5	52.1	4.0	6.0	0–2.0	0.7	0.5	
	Roquefort	40.0	31.0	22.0	50.1	4.2	6.0		0.65	0.45	
	Gorganzola	36.0	32.0	26.0	50.0	2.4	5.0			0.5	
Hard, ripened by bacteria	Cheddar	37.0	32.0	22	50.8	1.6	3.7	0–2.1	0.7	0.5	
	Colby	39.0	31.0	21	50.8	1.7	3.6	0	.7	.5	
Hard, ripened by eye-forming bacteria	Swiss	37	28	27.5	44.4	1.3	3.8	0–1.7	1.0	0.6	
	Edam	39	25	28.0	40.9	2.0	4.4	0–1.0	0.75	0.45	
	Gouda	36.5	29	25.0	45.6	1.7		0–1.0	0.60	0.38	
Very hard, ripened by bacteria	Parmesan	30.0	26.0	36.0	37.1	1.8	5.1	0–2.9	1.1	0.8	
	Romano	32.0	30.0		44.1	4.6	5.4	0			
Pasta filata (stretch cheese)	Provolone	38	28	28	45.1	3.0	4.0	0	0.7	0.6	
	Mozarella	53	18	22	38.3	1.0		0.3			
Low-fat or skim milk cheese (ripened)	Euda	56.5	6.5	30.0		2.6		1.0			
	Sapsago	37.0	7.4	41.0		4.5					
Whey cheese	Ricotta	72.0	10.0	12.5	35.7	1.2	3.6	3.0			
	Primost	13.8	30.2	10.9	35.0	—		36.6			
Process cheese	Process Cheddar	39.5	31.5	22.2	52.0	1.7	4.9	0	0.7	0.7	
	Process Cheese food	43.0	24.0	20.5	42.1	1.0		7.0	0.6	0.6	

Source: Hargrove and Alford (1974).

controlled bacterial action (e.g., "starter" cultures) on milk components, can overcome this deficiency to some extent. Some of the artificial cheese flavors undergo oxidative or other changes during storage which can result in unpleasant or atypical aromas or tastes. Of course, natural cheese products will also undergo deteriorative changes. For this reason, snacks incorporating artificial cheese mixtures should be taste tested after at least a week of high-temperature storage before a final selection is made.

Some representative formulas were given by Merory (1968).

Imitation Blue Cheese MF 79

	Gm
Phellandrene	2
Butyl butyryl lactate	44
Iso-valeric acid	44
Butyric acid	44
Caproic acid	22
Ethyl butyrate	44
Propylene glycol	800
Total	1000

Imitation Cheddar Cheese Flavor MF 80

	Gm
Butyl butyryl lactate	20
Iso-valeric acid	50
Ethyl butyrate	20
Butyric acid	120
Ammonium iso-valerate	30
Lactic acid	10
Caproic acid	60
Propylene glycol	690
Total	1000

Cheese Imitation Flavor MF 81

	Fl Oz
Butyl butyryl lactate	2.0
Iso-valeric acid	2.0
Ethyl butyrate	2.0
Butyric acid	2.0
Caproic acid	1.0
Methyl-n-amyl ketone	1.0
Alcohol, 95%	16.0
Propylene glycol	102.0
Total—or one gallon flavor	128.0

Imitation Cheese Flavor, Roquefort MF 82

	Fl Oz
Butyl butyryl lactate	2.0
Iso-valeric acid, anhydrous, pure	2.0
Ethyl butyrate	2.0
Butyric acid	2.0
Ammonium iso-valerate	4.0
Caproic acid	1.0
Methyl-n-amyl ketone (free of fusel)	1.0
Alcohol, 95%	16.0
Propylene glycol	98.0
Total	128.0

About 1 to 2% of autolyzed yeast or hydrolyzed plant proteins can reproduce some of the taste effects reasonably satisfactorily. To provide "bite" or tartness, small amounts of lactic or citric acid may be added. Monosodium glutamate is also helpful in flavor building.

Cheese powders can be blended with spices and other flavoring materials to provide pizza, chili con queso, or other ethnic type snacks.

BIBLIOGRAPHY

ANON. 1969. The cheese phenomenon in snack flavors. Snack Food 58, No. 6, 29–31.

DIXON, R. P., DE MAN, J. M., and WOOD, F. W. 1969. Production of volatile acids during Cheddar cheese ripening. Can. Inst. Food Technol. J. 2, 127–135.

HARGROVE, R. E., and ALFORD, J. A. 1974. Composition of milk products. In Fundamentals of Dairy Chemistry, 2nd Edition, B. H. Webb, A. H. Johnson, J. A. Alford (Editors). Avi Publishing Co., Westport, Conn.

HESS, J. 1972. More consumers say "cheese." Snack Food 61, No. 10, 37–39.

KOSIKOWSKI, F. 1966. Cheese and Fermented Milk Foods. Edwards Brothers, Ann Arbor, Mich.

MERORY, J. 1968. Food Flavorings: Composition, Manufacture and Use, 2nd Edition. Avi Publishing Co., Westport, Conn.

SANDERS, G. P. 1953. Cheese Varieties and Descriptions. U.S. Dept. Agr. Handbook 54.

SANDINE, W. E., and ELLIKER, P. R. 1969. Abstr. Papers. Am. Chem. Soc. 158, MICR 29.

VAN SLYKE, L. L., and PRICE, W. V. 1952. Cheese. Orange-Judd Publishing Co., New York.

WONG, N. P. 1974. Milk clotting enzymes and cheese chemistry, Part II. In Fundamentals of Dairy Chemistry, 2nd Edition. B. H. Webb, A. H. Johnson, and A. H. Alford (Editors). Avi Publishing Co., Westport, Conn.

ZICK, W. F. 1969. Lipid- and protein-derived flavors for snack foods application. Cereal Sci. Today 14, 205–206.

CHAPTER 5

Salt

INTRODUCTION

Salt is an indispensable flavoring for all snacks, sweet or savory. In many varieties of these foods it is the predominant flavor note. Hundreds of chemical compounds are classified as salts, and most of the water-soluble ones exhibit what we recognize as a salty taste, but only pure sodium chloride gives this flavor in a form not modified by sour, bitter, or sweet tastes. Other substances may provide a similar flavor but they generally also have off-notes which are described by the consumer as disagreeable or atypical. For example, sodium sulfate has a less pronounced salty flavor and potassium chloride has a bitter, cooling aftertaste.

In addition to having a pronounced and generally agreeable taste of its own, sodium chloride will modify other flavors. In most test situations, it has been found to enhance the sweetness of sugars and decrease the sourness of acids (Fabian and Blum 1943). In some liquid products, the addition of small quantities of salt, even below the threshold level, will increase the apparent sweetness of dissolved sucrose.

Although food grade salt is quite pure compared to many other ingredients, it always contains some other ionic species which may affect the taste. Calcium and magnesium ions lead to a stronger, more bitter flavor. These ions tend to concentrate near the surface of the particles and have an even more potent effect than they would have if evenly distributed, especially in the case of snack toppings.

Wherever practical, salt should be applied to snacks as a topping. This insures a quickly sensed saltiness which is a primary determinant of consumer acceptability. A sufficient level (superthreshold) should be applied to yield a distinct salty flavor, but gross oversalting should be avoided because it can mask or depress desirable flavor notes such as the mild sweetness of potatoes, or accentuate undesirable flavors.

MANUFACTURING SALT

The primary sources of commercial salt are saline water from the sea and certain lakes and wells and underground deposits of the crystalline mineral. Saline water is often trapped in shallow pools and allowed to evaporate, leaving a solid deposit which is scraped up and crushed. Underground deposits are either removed in solid form by conventional mining procedures or dissolved by circulating water through bore holes. Crude salt obtained from these sources is nearly always further purified by recrystallization and other procedures to yield a material suitable for food use. For some nonfood uses, crude rock salt may be acceptable. The physical form, which is a very important factor in determining the salt's utility for various snack food applications, is dependent upon the methods used to process the brine.

Granulated Salt

Granulated or common salt is made from brine that has been evaporated in a vacuum pan. The resultant cube-shaped salt crystal is extremely hard and has a medium solubility rate. Alberger process flake salt is made from brine that has been treated to remove much of the calcium and magnesium. The Alberger method of crystallization gives small flake-like crystals which dissolve more rapidly than granulated salt. Some of the physical and chemical properties of Alberger salt grades used for snack products are given in Tables 5.1, 5.2, and 5.3.

TABLE 5.1

CHEMICAL PROPERTIES OF ALBERGER SALT

Chemical Property	%
$NaCl$ (salt)	99.95 ± 0.01
$CaSO_4$ (calcium sulfate)	0.02–0.06
$CaCl_2$ (calcium chloride)	0.00–0.02
$MgCl_2$ (magnesium chloride)	0.00–0.01
Na_2SO_4 (sodium sulfate)	None
Cu (copper)	Less than 0.00015 (1.5 ppm) on the avg
Fe (iron)	Less than 0.00015 (1.5 ppm) on the avg
Moisture	Less than 0.1
Insolubles	Less than 0.002 (20 ppm)
Polysorbate 80	Less than 0.001 (10 ppm)

Source: Courtesy Diamond Crystal Salt Company.

Dendritic Salt

Dendritic salt is produced by vacuum pan evaporation under patented processes. A distinctive crystal formation is achieved by chemically pretreating the brine with yellow prussiate of soda (sodium ferricyanide decahydrate). A maximum of 13 ppm of the additive is permitted in the finished product. Unlike the cubical structure of vacuum pan (granulated) salt or the relatively flat crystal aggregates of flake type salt, dendritic crystals are branched or star-like in form and contain numerous mi-

TABLE 5.2

PHYSICAL PROPERTIES OF ALBERGER FINE FLAKE SALT

Tyler	On 28	On 35	On 48	On 65	On 80	On 100	Thru 100
U.S.S. equiv.	On 30	On 40	On 50	On 70	On 85	On 100	Thru 100
Minimum	0.0	0.0	7.5	37.5	14.5	15.5	2.8
Average	0.0	0.6	11.7	43.4	17.3	20.0	7.0
Maximum	0.0	1.3	15.9	49.1	20.0	24.4	11.3

Apparent density	0.833
Rate of solubility (Dunn Method)	7.6 sec
Caking resistance	2–3 months
Crystals per pound	54,000,000
Specific surface (sq ft/lb)	64.1

Source: Courtesy Diamond Crystal Salt Company.

croscopic cavities. It has a relatively low bulk density, high specific surface area, good resistance to caking, and a rapid rate of solution. Because it is porous, this form of salt is a fairly good absorbent for some nonaqueous liquids, taking up comparatively large amounts of oil, for example, before becoming wet or pasty. It also demonstrates excellent blending properties when mixed with powders. Some manufacturers offer dendritic or flake salt mixed with antioxidants as a convenient means of retarding rancidity development in snacks. Microphotographs of granulated, flake, and dendritic forms are shown in Fig. 5.1.

A commercial variety of granulated salt (e.g., Morton's Culinox 999) designed to have minimum prooxidant qualities, is made by introducing EDTA (ethylendiaminetetraacetic acid) during production to complex or sequester heavy metal ions. As a result of this treatment copper, iron, etc., ions remain in solution as the salt is precipitated. The solution containing the EDTA and the complexed cations is removed and traces washed away in subsequent processing steps. By this procedure, the copper content is reduced to a typical value of 0.1 and iron is reduced to a typical value of 0.4 ppm. This type of salt is too coarse to employ as a

TABLE 5.3

PHYSICAL PROPERTIES OF ALBERGER FINE PREPARED SALT

Tyler	On 65	On 80	On 100	On 115	On 150	On 200	Thru 200
U.S.S. equiv.	On 70	On 80	On 100	On 120	On 140	On 200	Thru 200
Minimum	0.1	0.5	15.1	16.3	18.7	17.1	9.6
Average	0.4	1.1	21.4	19.4	22.6	20.4	14.7
Maximum	0.8	1.7	27.8	22.4	26.4	23.7	19.8

Apparent density	0.968
Rate of solubility (Dunn Method)	4.6 sec
Crystals per pound	130,000,000
Specific surface (sq ft/lb)	125.4

Source: Courtesy Diamond Crystal Salt Company.

Courtesy of Morton Salt Company

FIG. 5.1. MICROPHOTOGRAPHS OF GRANULATED (TOP), FLAKE
(CENTER), AND DENDRITIC SALT (BOTTOM)

topping, but could be used in puffed or formed snacks where the salt can be incorporated into dough.

As an example of typical specifications for dendritic salt, the following description of Star Flake brand of the Morton Salt Company is of interest. Star Flake dendritic salt complies with the Food Chemicals Codex tolerances for sodium chloride. Guaranteed tolerances on a moisture-free basis are 99.9% sodium chloride, 0.01% maximum calcium and magnesium expressed as calcium, 1.0 ppm maximum copper, and 2.0 ppm maximum available iron. The moisture level should not exceed 0.1%. The level of residual yellow prussiate of soda (sodium ferrocyanide) does not exceed the Federal tolerance of 13 ppm. The No. 1 USDA standard specifies a limit for coarse, water insoluble sediment of 0.5 mg per 250 gm sample, which is equivalent to 2.0 ppm. A typical chemical analysis of dendritic salt on the moisture-free basis is 99.94% sodium chloride, 0.05% sodium sulfate, and 0.01% calcium salts. The pour bulk density is required to fall in the range of 53 to 58 lb per cubic foot (0.85 to 0.93 gm per ml). Particle size specifications are 5% maximum retained on U.S.S. 40-mesh screen, and 20% maximum passing through a 100-mesh screen.

According to Strietelmeier (1974), dendritic and fine-screened grades of flake salt with a particle size range of 4 to 140 U.S.S. mesh, are the preferred types for salting deep-fat fried potato and corn chips, nuts, and cheese-coated snacks. Perhaps the main advantage of dendritic salt in these applications is that it is nearly dustless when applied either by a mechanical salting roll or by forced air. It flows freely and more uniformly through mechanical salters, providing a more even distribution on the product. This type of salt also adheres well to snacks especially if the surface is tacky or oily at the time of application. The sharp-pointed edges of the crystals tend to penetrate the surface fat and become firmly attached to the snack. There is also greater oil absorption by dendritic salt, which contributes to the adhesion. Relatively dry snacks require a finer screened grade of "flour" salt to secure adequate adhesion. Flour salts have a particle size range of 70 to 200 U.S.S. mesh and must be treated with a free-flowing, anticaking agent, usually 1.0 to 1.5% tricalcium phosphate or calcium polysilicate. Flour salts may consist of the finest screenings of flake, granulated or ground granulated production. These grades are extremely dusty and difficult to dispense by conventional roll-type salters. Flake flour salt is usually the least dusty but is not universally available.

Pulverized Salt

For snack applications where the salt may be incorporated into an oil or oil and cheese topping, pulverized (minus 200 mesh) salt is often employed. Pulverized salt must contain 1.7 to 2.0% tricalcium phosphate to prevent caking. It is extremely dusty and cannot be dispensed as a topping because it will not free-flow.

Solubility

The dissolving rate of salt, which affects the flavor and some other factors of interest to the snack food manufacturer, can be estimated by standardized procedures.

Figure 5.2 illustrates an apparatus for determining the solubility index of salt. It provides conditions of maximum exposure to solvent. In a typical test, 17.3 gm of salt will be dropped through a 100-ml column of water 9.5 in. in height, and the volume collecting in the graduated bottom part of the tube will be recorded. Another procedure, the float

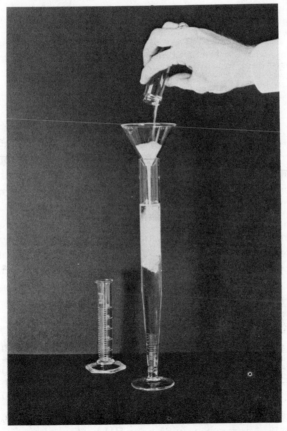

Courtesy of Morton Salt Company

Fig. 5.2. Apparatus for Solubility Index Determination

method, limits the exposure to solvent by suspending or floating the salt on the surface of the liquid. A ring-shaped cork float suspends a screen on the surface of the water. A weighed amount of salt is placed on the screen and the time required for complete solution is recorded. The solubilities of certain types of salt when determined by these methods are given in Table 5.4.

TABLE 5.4

TYPICAL SOLUBILITIES OF VARIOUS TYPES OF SALT BY TWO METHODS

Type	Particle Size Microns	Solubility Index Method[1]	Float Method[2]
Dendritic	150–420	98	380
Fine flake	150–420	95	240
Fine granulated	150–420	87	140
Medium flake	300–850	83	295
Coarse flake	420–1,200	78	350
Coarse granulated	200–600	74	180

Source: Courtesy of Morton Salt Company.
[1] Amount dissolving "instantly," as %.
[2] Time required to dissolve 20gm, in seconds.

SALT WITH ADDITIVES

Various additives have been applied to salt, usually as a means of obtaining good distribution of a micro-ingredient. Vitamins, mineral nutrients, and antioxidants and anticaking agents are some of the additives which are commercially available in mixtures with salt (Anon. 1972).

Free-flow Agents and Antioxidants

Two manufacturers can supply dendritic salt with free-flow agents or with antioxidants. The free-flow agent may be calcium polysilicate or tricalcium phosphate. Antioxidant-treated salt is designed for use on oil-fried snacks such as potato chips and corn chips. The formulation of R.G. dendritic salt of Morton Salt Company is:

	%
Dendritic salt	99.092
BHA	0.130
Propyl gallate	0.048
Citric acid	0.130
Calcium polysilicate	0.500
Propylene glycol	0.100

When this salt is added at the 2.0% level to snacks containing 40% oil, it will contribute approximately 65 ppm BHA, 65 ppm citric acid, and 25 ppm propyl gallate to the oil. Federal tolerances based on the fat content of the food are 100 ppm for any one antioxidant, 200 ppm for total antioxidants, and 100 ppm citric acid.

Fortification with Nutrients

From time to time, the statement is made by consumerists and others that snack foods should be fortified with various nutrients so the uneducated customer and especially children will not be filling up on "empty calories." A simple method for adding valuable nutrients is to use vitamin-fortified snack topping salt. Diamond Crystal and Morton Salt Co. offer a vitamin-fortified snack topping on either a 40-mesh salt base or on a flour salt base (70 to 200 U.S.S. mesh). The nutrients are agglomerated on to the salt to avoid segregation and dusting during application.

The standard formula is designed to furnish one RDA (Recommended Daily Dietary Allowance) of five vitamins per pound of potato chips when it is added at the rate of at least 1.5%. The vitamins in this product are Vitamin A, thiamine (B-1), riboflavin (B-2), niacin, and ascorbic acid (C). Custom formulations can be supplied if sufficiently large quantities are ordered.

Iodized salt contains approximately 0.01% of added potassium iodide. The purpose of this additive is to reduce the incidence of goiter in susceptible populations. Because potassium iodide tends to decompose forming free iodine, which may have adverse effects on the color and flavor of foods, stabilizers such as dextrose are usually added along with it. Food processing salt is not normally iodized although it can be obtained in this form at nominal additional cost. There is probably no real need for using iodized salt in snacks, unless it can be shown that there is a consumer demand for more of the supplement than is obtained from table salt and other dietary sources.

STORAGE AND PACKING

Food-processing salt is packed in 80- or 100-lb multiwall paper bags having a 1.5-mil polyethylene liner. These packages give good protection against moisture absorption and caking as long as the liner remains intact. For antioxidant or nutrient-containing salt, a special liner of glassine/wax laminate may also be included to prevent evaporative loss of the additives.

Bags of salt should be stored on pallets in a dry warehouse away from damp floors and walls and condensate drip. Bulk delivery of salt in trucks or rail cars can also be obtained if the rate of consumption justifies the installation of a bulk handling and storage facility. Bulk handling tends to reduce the average particle size and some specialty salts are considerably more susceptible to this type of damage than is regular vacuum grade salt.

Unlike most other food ingredients, salt *per se* does not undergo chemical deterioration in storage. Absorption of foreign odors is rarely a problem. Undesirable physical changes, such as caking, can occur, however. Caking of salt seriously interferes with its application to snacks. Caking is a response to fluctuations in relative humidity around a critical point. Above about 75% RH, salt becomes deliquescent. Moisture from the atmosphere condenses on the surface of the crystals and forms a layer of saturated brine. These layers will coalesce where the crystals touch and, when the water evaporates as the ambient relative humidity falls below 75%, form a recrystallized bridge or weld between the crystals. Obviously, then, salt should either be protected from fluctuations in atmospheric humidity by packaging in impervious containers, or it should be stored in rooms where the relative humidity can be kept below 75%. Most fine grades of snack topping salt are treated with anti-caking agents such as tricalcium phosphate which adsorb moisture vapor. Dendritic salt is very resistant to caking in the absence of such treatment because of its moisture absorption characteristics. However, these grades will cake if exposed to severe condensation.

SALT REQUIREMENTS FOR SNACKS

Potato Chips

Most potato chip manufacturers will add from 1.5 to 2.5% salt (basis total weight). The actual amount used in a specific case will depend on several factors, such as size of the salt crystals, type of salt, amount of oil, and sweetness of the chip (related to maturity and storage history of the potato). The salt level should be increased about 0.25% when the potatoes contain more than the normal amount of reducing sugars. Since the ingredient is cheap, the manufacturer will be inclined to use as much as he can without causing undue saltiness or masking other desirable flavor notes.

Chip salt should be of high purity. Since all chip salters feed volumetrically, the salt should flow freely and have a uniform bulk density. Particle size should be fine enough to insure 80 to 90% adherence, but not excessively fine or exceptionally fragile since these characteristics can lead to excessive dusting or drifting during application.

Since most potato chips will age a few days to a few weeks before being consumed, it is important to take measures to retard development of rancidity. Salt with a minimum content of copper and iron should be specified because these metals tend to accelerate the oxidative deterioration of fats. Salt coated with antioxidants can also be obtained, and this will improve storage life although it will not prevent the ultimate development of rancidity. The effects of antioxidants are discussed in Chapter 2.

Popcorn Salt

The main problem in salting popcorn is securing uniform distribution. Uniformity is a function of the type of salt, the method of application, the amount and state of the oil which has been applied to the corn, and the handling the product receives after the salt has been applied. Small, uniform particles are essential.

The very fine particle size of popcorn salt causes it to cake very quickly unless a free-flowing agent is blended with it. Tricalcium phosphate, magnesium carborate, and calcium carbonate have been used.

Traces of copper and iron accelerate the development of rancidity. If the popcorn is to be eaten shortly after it is salted, the presence of these trace elements is of little importance, but popcorn which is to be held for some time before consumption should be seasoned with salt having these contaminants reduced to a minimum.

Nuts

Securing good adherence of salt is a major concern of nut processors. The use of oils, rather than fats that are solid at room temperature, as the frying medium accentuates the difficulties. The surface coating of nuts fried in cottonseed oil, peanut oil, or other oils that are liquid at room temperature exerts very little binding action on the salt crystals. If the nuts are fried in a shortening that is solid at room temperature and the salt is applied while the nuts are cooling, at least some of the crystals will be trapped by the solidifying fat. Dendritic and fine flake (40 mesh) salt are usually employed with oil-roasted nuts.

If the nuts are dry roasted there is only a film of natural nut oil to cause adherence of particles to the nut surface, and this is not very effective. Flour salt should be used to improve adherence.

Failure of salt to adhere to the nuts so that it falls to the bottom of the bag, jar, or can during shipping, can be alleviated by the following procedure. After the warm nuts are removed from the fryer or roaster, let them cool slightly and then coat them with about 2% warm (80 to 85°F) coconut oil. After the fat has been uniformly distributed, add the salt, and allow the nuts to cool down to room temperature.

Corn Chips

Tortilla-style corn chips have a low fat content and a slick, smooth surface so that very fine or flour salt is usually required in order to obtain satisfactory adherence. Fine granulated (all through 50 mesh) may be suitable where flour salt or dendritic varieties are not available.

Courtesy of Morton Salt Company

FIG. 5.3. COARSE ROCK PRETZEL SALT, 16-30 MESH (2X MAG-
NIFICATION)

Note crystalline appearance and planed surface of particles.

Extruded and Puffed Products

Snacks such as corn curls ordinarily contain 25 to 45% by weight of oil in their finished form. If the applied fatty ingredient is of the right type, these amounts are sufficient to cause adequate adherence of salt. If the amount of oil is reduced to about the 18% level in order to reduce calories or ingredient costs, difficulties are encountered in applying the salt. When the salt is added as part of the mixture fed to the extruder, it is

TABLE 5.5

PRETZEL SALT SCREEN ANALYSIS

U.S.S. Sieve Number	Percentage Salt Retained on Screen	
	Coarse	Fine
14	0	5
16	23	5
20	72	75
30	5	16
40	Trace	4

found that the process is interfered with and variable results are obtained from batch to batch.

Cheese-coated snacks (collets) can be salted in the rotating drum employed to apply the coating. A dustless grade such as dendritic is pneumatically dispensed over the freshly coated snacks near the exit of the drum.

Another procedure for collets is to mix pulverized salt with the hot cheese/oil slurry prior to application. Since the salt will not dissolve in the slurry, it is important that ultra fine 200-mesh particles be employed to ensure a uniform dispersion of salt and to avoid damage to positive displacement pumps typically used to apply the coating. The disadvantage of this approach is that the oil enrobes the salt resulting in a blander flavor unless exceptional concentrations of salt are added.

Corbin *et al.* (1972) patented a method for applying most of the salt to these low fat extruded snacks in the form of an aqueous salt solution. The collet is then dried. It is apparent that some shrinkage and textural changes will occur during the application of any aqueous solution to a puffed collet, but this may be acceptable in some cases.

Pretzel Salt

There are various kinds of pretzel salt. The Morton Salt Company offers selected screenings of crushed rock salt obtained from a dome deposit on the Gulf Coast of Louisiana. It is a peculiar characteristic of this material that it breaks up into uniformly flat, rectangular-shaped particles upon crushing (see Fig. 5.3). These granules adhere particularly well to pretzels. They have a mild saline flavor with a note of "huskiness" provided by traces of mineral impurities. The average bulk density is 72 lb per cubic foot. Pretzel salt is offered in coarse and fine varieties having the screen analyses given in Table 5.5.

BIBLIOGRAPHY

ANON. 1972. Vitamin fortified toppings for fat fried snacks. Morton Salt Co. Tech. Serv. Bull. 72-5 A.

BRIGHTON, T. B., and DICE, C. M. 1931. Increasing the purity of common salt. Ind. Eng. Chem. *23*, 336–339.

CORBIN, D. D., MARQUARDT, R. F., and GABBY, J. L. 1972. Process for preparing a low calorie snack. U.S. Pat. 3,682,652. Aug. 8.

FABIAN, F. W., and BLUM, H. B. 1943. Relative taste potency of some basic food constituents and their competitive and compensatory action. Food Res. *8*, 179–193.

HESTER, A. S., and DIAMOND, H. W. 1955. Salt manufacture. Ind. Eng. Chem. *47*, 672–683.

KAUFMANN, D. W. 1960. Sodium Chloride. The Production and Properties of Salt and Brine. Van Nostrand Reinhold Co., New York.

LOCKER, C. D. 1954. Salt. *In* Encyclopedia of Chemical Technology, Vol. *12*, R. E. Kirk, and D. F. Othmer (Editors). Interscience Encyclopedia, New York.

MOSS, H. T. 1933. Tricalcium phosphate as a caking inhibitor in salt and sugar. Ind. Eng. Chem. *25*, 142–147.

SIMPSON, R. A. 1954. Tailored salts provide many plus values. Food Eng. *26*, No. 7, 78, 79, 130, 133.

STRIETELMEIER, D. M. 1969. On electrostatic salting. Snack Food *58*, No. 9, 46.

STRIETELMEIER, D. M. 1972. Personal communication. Chicago, Ill.

STRIETELMEIER, D. M. 1974. A new incentive for controlling salt content. Snack Food *63*, No. 10, 36–38.

STRIETELMEIER, D. M., and BAIRSTOW, M. 1968. Survey of antioxidant salt and MSG usage on potato chips. Potato Chipper *27*, No. 12, 38, 42.

CHAPTER 6

Water

INTRODUCTION

Although ingredient water is less of a quality factor in snacks than in most other foods because only small percentages of water are normally added to snack ingredient mixtures, it can affect flavor, appearance, and texture under certain circumstances and for those reasons is worthy of consideration.

Water is an ingredient which must meet Federal and local regulations the same as any other component. In addition, the amounts and types of dissolved minerals and organic substances in the water can affect the functional qualities of the process mixture and the organoleptic characteristics of the finished product. The average temperature of the water and fluctuations around the average can be important factors in the performance of a given formula. The differences in bacterial flora and other suspended contaminants are of obvious interest to any food producer.

The initial requirement for any water used as a food ingredient is that it comply to applicable legal standards for drinking water. These standards have been set forth clearly and in minute detail by Federal agencies.

Congress has passed (in 1967) a Water Quality Act requiring each State to establish water quality criteria for all interstate waters and to develop a plan for the implementation and enforcement of the criteria. In general, the standards must be such as to enhance the quality of natural waters for their ". . . use and value for public water supplies, propagation of fish and wildlife, recreational purposes, agricultural, industrial, and other legitimate uses. Numerical values should be stated for quality characteristics where available and applicable. Biological or bioassay parameters may be used, where appropriate." Until these laws are promulgated, other guidelines must be used and, of these, the U.S. Public Health Service Drinking Water Standards are the most familiar.

PUBLIC HEALTH SERVICE DRINKING WATER STANDARDS

The current standards for drinking water to be used on common carriers engaged in interstate commerce were prepared by the U.S. Public Health Service in 1946 and issued in revised form in 1962 by the U.S. Department of Health, Education, and Welfare. These Federal Standards are used as guidelines by many States and municipalities in setting up limits for potable water. The American Water Works Association regards the standards as being minimum requirements for the protection and well-being of individuals and communities, and advocates the establishment of the standards as the criteria of quality for all public water supplies in the United States. Water which does not conform to the standards should not be used as an ingredient in foods regardless of the legal status of the standards in the locality of the plant.

The standards specify that the water supply shall be obtained from the most desirable source which is feasible, and efforts should be made to prevent or control pollution at the source. If the source is not protected by natural means, the supply shall be adequately protected by treatment. Adequate protection by natural means is defined as involving one or more of the following processes of nature that produces water consistently meeting the requirements of the standards: dilution, storage, sedimentation, sunlight, aeration, and the associated physical and biological processes which tend to accomplish natural purification of water by infiltration through soil and percolation through underlying material and storage below the ground water table. Adequate protection by treatment is defined as any one or combination of the controlled processes of coagulation, sedimentation, absorption, filtration, disinfection, or other processes which will produce a water consistently meeting the requirements of the standards. This protection also includes: processes which are appropriate to the sources of supply; works which are of adequate capacity to meet maximum demands without creating health hazards, and which are located, designed, and constructed to eliminate or prevent pollution; and conscientious operation by well-trained and competent personnel whose qualifications are commensurate with the position and acceptable to the reporting agency (the official State health agency or its designated representative) and the certifying authority (the Surgeon General of the U.S. Public Health Service or his duly authorized representative).

The standards direct that frequent sanitary surveys be made of the water supply system in order to locate and identify health hazards which might exist in the system. According to definition, the water supply system includes the works and auxiliaries for collection, treatment, storage, and distribution of water from the sources of supply to the free-flowing outlet of the ultimate consumer.

Approval of water supplies is to be depended in part upon (1) enforcement of rules and regulations to prevent development of health hazards, (2) adequate protection of water quality throughout all parts of the system, as demonstrated by frequent surveys, (3) proper operation of the water supply system under the responsible charge of personnel whose qualifications are acceptable to the reporting agency and the certifying authority, (4) adequate capacity to meet peak demands without development of low pressures or other health hazards, and (5) records of laboratory examinations showing consistent compliance with water quality requirements of these standards.

Responsibility for conditions existing in a water supply system is held by (1) the water purveyor from the source of supply to the connection to the customer's service piping, and (2) the owner of the property served, and the municipal, county, or other authority having legal jurisdiction from the point of connection in the customer's service piping to the free-flowing outlet of the ultimate consumer. As can be seen from the preceding requirements, the food manufacturer would be considered responsible for the condition and quality of the water in his service piping before and after any treatment in his plant.

TABLE 6.1

DESIRABLE LIMITS FOR CHEMICAL SUBSTANCES IN POTABLE WATER

Substance	Maximum Concentration,[1] mg/l
Alkyl benzene sulfonate (ABS)	0.5
Arsenic (As)	0.01
Chloride (Cl)	250
Copper (Cu)	1
Carbon chloroform extract (CCE)	0.2
Cyanide (CN)	0.01
Iron (Fe)	0.3
Manganese (Mn)	0.05
Nitrate[2] (NO_3^-)	45
Phenols	0.001
Sulfate ($SO_4^=$)	250
Total dissolved solids	500
Zinc (Zn)	5

Source: Anon. (1962).
[1]These concentrations should not be exceeded if, in the judgment of the reporting agency and Certifying Authority, other more suitable supplies are or can be made available.
[2]In areas in which the nitrate content of water is known to be in excess of the listed concentration, the public should be warned of the potential dangers of using the water for infant feeding.

Tables 6.1 and 6.2 list desirable and essential limits for chemical substances in water supplies. Semiannual analyses for these substances are usually considered adequate. Furthermore, drinking water should contain no impurities which would cause offense to the senses of sight, taste, or smell. Turbidity should not exceed 5 units, color should not exceed 15 units, and the threshold odor number should be 3 or less for general use water (see references for definition).

TABLE 6.2

ESSENTIAL LIMITS FOR CHEMICAL SUBSTANCES IN WATER SUPPLIES

Substance	Maximum Concentration,[1] mg/l
Arsenic (As)	0.05
Barium (Ba)	1.0
Cadmium (Cd)	0.01
Chromium, hexavalent (Cr^{+6})	0.05
Cyanide (CN^-)	0.2
Lead (Pb)	0.05
Selenium (Se)	0.01
Silver (Ag)	0.05

Source: Anon. (1962).
[1]Presence of the substances in excess of the maximum concentration constitutes grounds for rejection of the supply.

EPA STANDARDS

On March 14, 1975, the Environmental Protection Agency published in the *Federal Register* a document entitled "Interim Primary Drinking Water Standards" which set forth some proposed changes in the limits of contaminants allowed in potable water supplies. The regulations were to take effect 18 months after promulgation and would apply with relatively few exceptions to every public water system. So far as inorganic contaminants are concerned, the permissible levels of arsenic, barium, cadmium, hexavalent chromium, cyanide, lead, selenium, fluoride, and silver remain the same as those given in Table 6.2, but maximum limits of mercury and nitrate (as N) are newly set at 0.002 and 10 mg per liter, respectively.

Maximum contaminant levels for certain pesticides (chlordane, endrin, heptachlor, heptachlor epoxide, lindane, methoxychlor, toxaphene, and chlorophenoxys) are to be established, as well as for total organic chemicals, turbidity, and microorganisms.

ANALYSES OF WATER

Standardized techniques are available for determining nearly all of the water quality factors thought to be important. Some of the procedures have attained official, or quasi-official, status through inclusion in *Standard Methods for the Examination of Water and Waste-water* (Anon. 1971). Nearly all of the tests referred to in *Drinking Water Standards* (Anon. 1962) are included in that book.

Some of the analyses which have been standardized are: acidity, alkalinity, aluminum, arsenic, boron, calcium, carbon dioxide, chloride, chlorine (residual chlorine), chlorine demand, chromium, color, specific conductance, copper, cyanide, fluoride, grease, hardness, iodide, iron, lead, lignin, magnesium, manganese, albuminoid nitrogen, ammonia nitrogen, nitrate nitrogen, nitrite nitrogen, oxygen, oxygen consumed, pH, phenol, phosphate, potassium, residue, selenium, silica, sodium, sulfate, sulfite, tannin, taste and odor, temperature, turbidity, and zinc. There are alternate or tentative methods for some of these factors. Since the standard methods are readily available, there is no point in reproducing them here.

Three more or less traditional bacteriological tests are applied to water: (1) counts of growth on gelatin plates incubated at 68°F, (2) counts on agar plates incubated at 99°F for 24 hr, and (3) the coliform count. The latter criterion has been superseded as a standard by the "Most Probably Number (MPN)" of *Escherichia coli*, which has been defined as "the bacterial density, which, if it had actually been present in the sample under observation, would more frequently than any other have given the observed analytical results."

Plate counts at 68 or 99°F are not particularly important as indicators of water safety, but they are useful as routine quality control tests in the various water treatment procedures and as means for determining the sanitary conditions of basins, filters, etc. The coliform test is the ultimate measure of bacteriological safety of water supplies.

It is sometimes necessary to make an estimate of the planktonic organisms (e.g., algae, diatoms) in a reservoir or other water source. This is usually done by direct microscopic count on a sample concentrated by filtration or other means.

Determination of biological oxygen demand (BOD) is important as an indicator of the extent of pollution of water sources and the potential for spontaneous recovery of these supplies. BOD is the total amount of oxygen taken up by those microorganisms which are present and which act on the available nonliving organic matter. The BOD has been defined by the American Public Health Association as "the oxygen in parts per million required during stabilization of the organic matter by aerobic bacterial action." It has also been stated that "Complete stabilization requires more than 100 days at 68°F but such long periods of incubation are impractical in any but research investigations, consequently a much shorter period of incubation is used."

The bacterial consumption of oxygen may be partially offset by the photosynthetic processes of microscopic plant life which produce oxygen and consume carbon dioxide.

Turbidity, color, and odor are routinely determined when a complete water analysis is performed. Turbidity is a measure of the extent to which light passing through the water is reduced in intensity by suspended material such as clay, organic debris, or certain industrial wastes. It can be determined by a Jackson candle turbidimeter or by continuously measuring and recording devices. For most snack food uses, a rather high level of turbidity can be tolerated. Color of water is defined as the difference in hue caused only by those substances actually in solution. It can be the results of mineral or vegetable pigments of natural origin or it can be caused by soluble organic or inorganic materials from sewage or industrial effluents. To obtain a quantitative estimation of "color" the analyst visually compares a sample with standard platinum-cobalt solutions or with colored glass tubes or discs under standard conditions.

WATER TREATMENT

Nearly all water obtained from surface sources (rivers, lakes, impounding reservoirs) must be treated to remove suspended material, dissolved color and flavor vectors, and microbiological contaminants. The details of these treatments are too complex to be set forth here, and the reader who is interested in such information as it applies to food is referred to *Water in Foods* (Matz 1965) and similar texts. In general, sedimentation with or without added coagulants, filtration, and chlorination is needed for most water supplies.

A thorough course of treatment for making potable a raw water of low quality might involve (1) sedimentation in a large impounding reservoir, (2) coagulation (the formation of a voluminous flocculent precipitate by adding iron or aluminum salts under the proper conditions), (3) allowing the bulk of the precipitate to settle out in a reservoir or tank, (4) filtering the decantate from the preceding step through specially prepared beds of sand or gravel to remove residual precipitate and other impurities, (5) filtering through activated charcoal which will adsorb many dissolved substances affecting the color, odor, and taste, and (6) chlorination.

It is often desirable to further treat municipal or well water in the plant to eliminate off-flavors or other undesirable characteristics. Some firms offer modular systems which can be used in any required combination to furnish ingredient water of the needed purity. The available operations include (1) prefiltration using filter media which eliminates the necessity for expensive, space-consuming flocculation and settling, and which can be cleaned by either automatic or manual back-flushing, (2) organic compound adsorption by activated charcoal or by special weak-base organic scavenger resin which can be automatically regenerated in place, (3) reverse osmosis pretreatment to reduce the salt content of saline water, (4) cation or anion exchange for removal of specific ions in waste treatment, and (5) ultrafiltration for final particulate re-

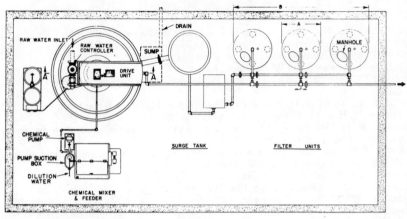

PLAN

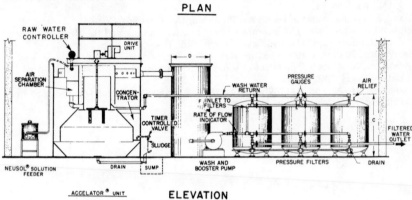

ELEVATION

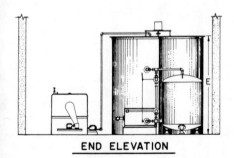

END ELEVATION

Courtesy of Infilco, Inc.

FIG. 6.1. A PACKAGED TREATING PLANT USING PRESSURE FILTERS

moval. Several manufacturers offer combinations of equipment or "package plants" for treatment of raw well or surface water or further treatment of municipal water by food manufacturers (Fig. 6.1).

BIBLIOGRAPHY

ANON. 1962. Drinking Water Standards. U.S. Public Health Serv. Publ. *956.*

ANON. 1963. Water Analysis Procedures. Hack Chemical Co., Ames, Iowa.

ANON. 1971. Standard Methods for the Examination of Water and Waste Water, 13th Edition. American Water Works Assoc., New York.

ANON. 1975. Interim Primary Drinking Water Standards. Federal Register *40,* 11990–11998.

BETZ, W. H., and BETZ, L. D. 1970. Betz Handbook of Industrial Water Conditioning, 3rd Edition. W. H. and L. D. Betz, Philadelphia.

BROOKE, M. M. 1939. Survey of basic work on the effects of various minerals on fermentation. Proc. Am. Soc. Bakery Engrs. *1939,* 84–86.

BROWN, E. B. 1939. The effect of minerals on baking, using sponge and straight doughs. Proc. Am. Soc. Bakery Engrs. *1939,* 88–91.

DUNN, J. A. 1941. Salt and water. Proc. Am. Soc. Bakery Engrs. *1941,* 70–71.

HAAS, L. W. 1927. Water in baking. Proc. Am. Soc. Bakery Engrs. *1927,* 80–81.

JAMES, G. V. 1965. Water Treatment, 3rd Edition. Chemical Rubber Co., Cleveland.

KIRK, D. J. 1951. Effects of hardness and acidity of water and fermentation. Bakers Dig. *25,* No. 6, 28–30, 34.

MATZ, S. A. 1965. Water in Foods. Avi Publishing Co., Westport, Conn.

PICKERING, C. S. 1936. The interpretation of a water analysis, and the effect of water on dough fermentation. Am. Soc. Bakery Engrs. Bull. 104.

SKOVHOLT, O. 1948. Water in baking. Bakers Dig. *22,* No. 4, 65–66, 81.

SWANSON, A. T. 1966. The effect of water on the fermentation process. Proc. Am. Soc. Bakery Engrs. *1966,* 48–57.

SWENSON, H. A., and BALDWIN, H. L. 1965. A Primer on Water Quality. U.S. Govt. Printing Office, Washington, D.C.

CHAPTER 7

Nuts

INTRODUCTION

Nuts, particularly peanuts, are consumed in large quantities as snacks, and they are also important as ingredients in combination snacks such as "Cracker Jack." Being natural agricultural products, their inherent characteristics as they come from the field are susceptible to only limited modification by processing, so that careful selection of stocks available from suppliers and judicious timing of purchases and deliveries are the most important factors in quality assurance.

Most authorities agree that nuts should be stored in the shell and at low temperatures whenever possible, since these conditions will minimize staling. Shelled nutmeats are very susceptible to the development of off-flavors. Rancidity is the main problem, but the kernels also tend to absorb off-odors from their environment, such as those from tobacco and paint. It is generally the practice, however, for nuts to be shelled immediately after the producer delivers them to the processor. The nutmeats are then stored for shipment throughout the year until the next crop comes to market. Refrigerated storage is strongly indicated for these materials.

Although it is certainly easy to identify rancid nuts, the detection of advanced storage deterioration which has brought the material to the brink of detectable rancidity is not so easy. Tests for peroxides and free fatty acids may be helpful, but they do not always give a clear indication of the remaining storage life.

The presence of aflatoxin in nuts (particularly peanuts) has received considerable governmental attention in the last few years and doubtlessly will attract increased regulatory action as detection and enforcement methods are perfected. Aflatoxin refers to a series of compounds all of which are carcinogens. They are produced by certain fungi which are widely distributed in nature. Contamination by aflatoxin can be found in lots of nuts which are not obviously moldy. The U.S. Food and Drug Administration has been increasing its surveillance of commercial shipments of peanuts, and lots are occasionally condemned because of the presence of excessive amounts of aflatoxin. All reliable suppliers will have had their stocks analyzed for the contaminant, but it is advisable for purchasers to specify that shipments must conform to U.S. FDA standards in this as in all other respects.

TABLE 7.1

TYPICAL COMPOSITION OF NUTS

Nuts	Water (%)	Protein (%)	Fat or Oil (%)	Total Carbohydrate (%)
Almonds	4.7	18.6	54.1	19.6
Brazil nuts	5.3	14.4	65.9	11.0
Cashew nuts, roasted	4.1	19.6	47.2	26.4
Peanuts, Virginia roasted in shell	2.6	26.9	44.2	23.6
Pecans	3.0	9.4	73.0	13.0
Pistachio nuts	5.6	19.6	53.2	18.6
Walnuts, black	2.7	18.3	58.2	18.7
Walnuts, English	3.3	15.0	64.4	15.6
Chestnuts	53.2	2.8	1.5	41.5
Coconut, fresh	46.9	3.4	34.7	14.0

PEANUTS AND PEANUT BUTTER

Peanuts

Many people regard the peanut as the first snack food, although others regard popcorn as occupying this niche in history. Roasted peanuts in the shell were the first form of this food, introduced into New York in 1870 by Barnum's circus; then around 1906 a Wilkes-Barre storekeeper conceived the idea of selling salted, roasted, shelled peanuts in small bags. This was the origin of the "nickel lunch." Per capita consumption of peanuts has been increasing since the mid-1950's and is now at the highest level since World War II.

Use of peanuts in the United States for edible purposes, i.e., use as candy, salted nuts, peanut butter, and peanut butter sandwiches, amounted to about 1,065 million pounds for the 12 month period ending July 31, 1971. These peanuts reflect the full price support given by the Federal government. Additional large amounts of nuts are crushed to produce cooking oil and meal, the latter serving mostly as a protein supplement for animal foods.

According to Woodroof (1966), the classification of types and varieties of peanuts is loose and many intermediate forms are found. The relatively small, somewhat spherical Spanish type is grown mainly in South Africa and in the Southwestern and Southeastern United States. The Virginia type is considerably larger and the kernel is longer or football-shaped.

The runner type is approximately intermediate in shape and size. There are also Valencia and Tennessee red or white types but they are currently of relatively minor economic importance. Some of the types of peanuts are covered by U.S. Standards (Anon. 1948, 1956, 1959, 1965).

TABLE 7.2

COMPOSITION OF PEANUT KERNELS

Constituent	Range (%)	Average (%)
Moisture	3.9–13.2	5.0
Protein	21.0–36.4	28.5
Lipids	35.8–54.2	47.5
Crude fiber	1.2–4.3	2.8
Nitrogen-free extract	6.0–24.9	13.3
Ash	1.8–3.1	2.9
Reducing sugars	0.1–0.3	0.2
Disaccharide sugar	1.9–5.2	4.5
Starch	1.0–5.3	4.0
Pentosans	2.2–2.7	2.5

Source: Freeman *et al.* (1954).

The number of kernels per pound of Spanish peanuts will vary from about 1,100 to 2,000, of runners from 900 to 1,100, and of Virginia from 512 to 864.

Peanuts when freshly dug have a moisture content of 25 to 60%. Upon air-drying in the shell, this drops to 5 to 10%. Roasted peanuts will have a moisture content of about 0.5 to 1.5%, the same as peanut butter. The average composition of peanut kernels is shown in Table 7.2.

Dried in shell peanuts retain good organoleptic properties for six months, more if kept free from infestation and protected from foreign odors which they might absorb. At 47°F storage, the life is nearer nine months. When shelled the storage life is reduced by about ⅓; if blanched and split, another ⅔ of the potential stability is lost. Most warehousemen store shelled peanuts in refrigerator rooms, which substantially increases the shelf-life. Chopping and roasting can further reduce stability.

Spoilage is the result of development of oxidative rancidity, absorption of foreign odors, and color changes. Woodroof (1966) states that these changes occur three times faster in shelled peanuts and nine times faster in hulled and blanched nuts than in raw peanuts in the shell. Roasted nuts have a short storage life due to the acceleration of the reactions leading to oxidative rancidity. Chopping the nuts further speeds the development of off-flavors because of the greater surface area exposed to oxygen. Thus, cured peanuts in the shell might have a useful storage life of nine months or more if held under cool and dry conditions, while roasted and ground peanuts can be expected to exhibit off-flavors in a couple of weeks.

For most commercial applications, the roaster buys cured shelled peanuts packaged in burlap bags. Depending on their intended application, whole peanuts or split peanuts are procured. Some roasters are installing systems for the receiving and storage of bulk (unbagged) peanuts, and the amount of nuts shipped in this form can be expected to increase in the future.

Shelled peanuts are cleaned by feeding them over a vibrating screen through which air is blown. The amplitude of vibration and the angle of the screen are controlled so that heavy foreign objects move toward the back of the screen where they fall off and are discarded. Light trash such as loose skins and dust are sucked out through an exhaust system. The peanuts move forward onto conveyors that deliver them to the oven hoppers.

Peanuts are either oil roasted for salted peanuts and some confections, or dry roasted for peanut butter, flavored dry roasted peanuts, and some candies. Either process develops the characteristic nut flavor and crunchy texture of the nuts. The nut also becomes darker in color and its moisture content is reduced from about 5 to 8% to about 0.5 to 1.5%. Skins become brittle and more easily separated from the kernel.

About 300 to 400 lb of peanuts are roasted in direct-fired rotary batch roasters at 280 to 320°F, the temperature depending on the degree of roast desired.

After roasting, the nuts are rapidly cooled either in bins having slotted bottoms through which air is circulated, or on slotted conveyor belts.

Blanching, or skin removal, follows the cooling step. The peanuts are passed through narrow passages between brushes or ribbed belts. The friction gently rubs off the skins, which are then blown off and collected in a cyclone separator.

Various methods have been suggested to improve stability of roasted nuts. Among these are patented procedures involving adding of antioxidants or coating the nuts with a mixture of zein and acetylated monoglycerides. These methods are sometimes useful in extending the life of fresh nuts a few weeks or months, but they will not reverse deterioration which has already occurred.

Peanut Butter

Peanut butter is an important constituent of certain snacks. A standard of identity has been issued by the Federal government. This allows as much as 10% of certain seasoning and stabilizing ingredients including up to 3% of hydrogenated vegetable oils. Salt, dextrose, and emulsifiers are other ingredients commonly used in peanut butter. Either blanched or unblanched kernels may be used, but in the latter case, a label declaration of unblanched peanuts must be made. Extent of roast is classified from USDA No. 1 (lightest) to USDA No. 4 (dark).

Peanut butter is sold in bulk in steel drums, in fiber drums with plastic liners, and in fiber boxes with liners. There are three basic grinds: (1) smooth, in which the particle size has been reduced below the perceptible limit; (2) regular, having a definite grainy texture; and (3) chunky, similar to regular with the addition of a substantial amount of large particles (about $\frac{1}{16}$ in. in the greatest dimension).

Peanut butter to be used as an ingredient in snacks is sometimes of a special high-roast type to yield the most flavoring power per dollar of cost. The peanuts should be roasted to a point just short of the development of bitterness or scorched flavor. Blanched peanuts are highly desirable as the raw material. Specification of the variety of peanut is generally not feasible. The question of whether or not to use additives depends to some extent on the method of dispensing the material. If it is to be scooped from partially filled barrels, the presence of some stabilizer is essential to prevent stratification in the drum, i.e., oil separation. When fiber boxes with plastic liners are used, stabilizers are also desirable in order to prevent leakage and other handling problems.

There are three grades of peanut butter: (1) U.S. Grade A is practically free from defects and has good color, flavor, texture, and aroma; (2) U.S. Standard Grade is fairly free of defects and has fairly good flavor, aroma, color, and consistency; and (3) Substandard includes products failing to meet the requirements of the other two grades.

Peanut butter is fairly resistant to the development of rancidity if stored in light-proof containers and protected from oxygen. Temperatures in the 50°F range are preferred. If the oils are allowed to soak into fiberboard, rancidity occurs quickly. Stability of the butter depends to a considerable extent on the conditions to which the peanuts were subjected before being ground.

Peanut Meal

Peanut meal, made by pressing most of the oil from ground peanuts, is useful as a flavoring material for fillings. Much of the peanut flavor is lost with the oil, however.

Defatted Peanuts

A process for so-called "Defatted Peanuts" was developed by the Southern Utilization Research and Development Division of the U.S. Agricultural Research Service. As much as 80% of the oil is mechanically expressed from the whole raw peanuts. The peanuts are considerably distorted, but regain their original shape and increase in moisture content when they are placed in water. When roasted, the defatted nuts have a harder, tougher texture and less peanut flavor than the original would have had if normally roasted. They are, however, lower in calories and might be useful for the dry roasted, flavored type of peanut snack. It is clear that the defatted peanuts cannot be declared as "peanuts," without modifying objectives, on the consumer package.

ALMONDS

There are two species of almonds, the bitter and the sweet or edible almond. Of the latter, there are two classes, the hard shell and the soft shell. The almonds of commerce are principally soft-shelled.

In world trade, the major producers of almonds are Spain, Italy, Portugal, Morocco, Iran, and the United States. Some almonds are imported into the United States, mostly from Spain, but most of the almonds consumed here are domestically grown.

California is the only important almond-growing state in this country. Seven varieties that dominate the U.S. market are Nonpareil, IXL, Ne Plus, Peerless, Drake, Mission, and Jordanolo.

The U.S. Dept. of Agr. has established Standards for Grades of Shelled almonds including U.S. Fancy, U.S. Extra No. 1, U.S. No. 1, U.S. Select Sheller Run, U.S. Standard Sheller Run, U.S. No. 1 Whole and Broken, and U.S. No. 1 Pieces for almonds of similar varietal characteristics, as well as Mixed Varieties, and Unclassified. There are also Standards for grades of almonds in the shell, which are not of much significance to the snack industry.

Standard Sheller Run, the lowest quality class, consists of kernels just as they come from the cracking machines, of mixed sizes, free of dust and shell but containing some broken pieces. Select Sheller Run contains few broken pieces and is offered in different sizes. Fancy consists of almonds closely sized and without defects, as regular in shape, size, and color as is practicable. They are free of double-nuts, dust, shell, and broken pieces.

Standard Sheller Run is satisfactory for making almond paste, chopped or sliced nuts, and the like. For garnishing, where appearance is of primary importance, Fancy should be specified. The size, which should also be specified, is defined as the number of almonds to the ounce.

In the United States most almonds are marketed through growers' federations, such as the California Almond Growers Exchange, which enforces quality standards and markets cooperatively.

For many applications, it is desirable that the kernels be blanched, i.e., the brown skin removed. A typical treatment involves contacting the almonds with 180°F water for 3 min, and then skinning by hand or by special machines. If they are to be kept any length of time, the excess moisture must be removed. California almonds can be obtained already blanched and with the moisture content brought back to safe levels.

Nuts should always be stored in air-tight containers to prevent moisture pickup.

Almonds are relatively resistant to rancidity development as compared to pecans or peanuts, but they will deteriorate with time, and the decrease in acceptability is accelerated by blanching or roasting. It is desirable to store the nuts at freezing temperatures if they are to be held for several months. Moisture-proof and insect-resistant packaging should be used for room temperature storage, of course. If the nuts must be roasted and held for several days or weeks, it is helpful to apply the antioxidants TBHQ, BHA, and BHT in spray form or in the roasting oil.

CASHEWS

Cashew nuts are produced in India, Brazil, and East Africa. Virtually all of the cashew consumption in the United States, amounting to about 70 to 80 million pounds annually, is imported from India, but may include also some East African nuts processed there. Brazil is assuming increasing importance as an exporter of cashew nuts. There is no commercial production of cashew nuts in the United States.

The United States has not established standards for cashew nuts, but the Cashew Export Promotion Council of India has established grading and marketing rules which apply to cashews exported from that country. These rules define grades for whole cashews, scorched wholes, dessert wholes, white pieces, scorched pieces, and dessert pieces.

Whole cashew kernels are classified according to size.

Identification Symbol	Size Range, Kernels per Pound
W210	200 to 210
W240	220 to 240
W280	260 to 280
W320	300 to 320
W400	350 to 400
W450	400 to 450
W500	450 to 500

Cashews are packed in square metal tins which are hermetically sealed. Generally, a few inches of vacuum is drawn before sealing. The tins of approximately 4-gal. capacity contain 28 lb of Baby Bits and 25 lb of other sizes. The nuts are comparatively stable when stored in the unopened tin at room temperature, and a year of useful shelf-life can be expected when they are refrigerated.

PECANS

Pecans are highly regarded as components of nut mixtures or as an ingredient in confections and certain snack foods. Because of their higher cost relative to most other nuts and the fluctuations in supply, they are usually reserved for premium goods.

The pecan is a native American tree. It will not survive severe winters and so is restricted to the milder climatic areas. Nuts are harvested from both wild and cultivated trees. There is more variability in the pecan crop than that of most other nuts. Because of the fairly recent trend toward cold or frozen storage, the fluctuations in supply and prices have been somewhat reduced.

After harvesting, pecans are put in dry storage to undergo curing. During this period of about 3 weeks at room temperature, the moisture of the entire nut decreases to about 8.5 to 9.0% and that of the meat to about 4.5%. Free fatty acids and peroxide value of the lipids increase, and the tannins of the seed coat oxidize with a resultant color change from pale to medium brown. The overall effect of these changes is to develop in the kernel the characteristic pecan appearance, aroma, flavor, and texture. The nuts will gradually develop staleness and rancidity at a rate dependent upon the temperature as storage continues. Conditions of 40°F with 70 to 80% RH are necessary if the fresh flavor is to be retained more than 3 months. At 0°F the in-shell kernels will retain good quality for more than 5 yr.

The following describes the procedure used by one large pecan sheller for storing these nuts under freezing conditions:

When pecans arrive from the orchard, they are immediately cleaned and sized. They are stored in wood pallet boxes, each containing 2,000 lb of nuts, in rooms held at 0°F. When an order is received, the requisite amount of pecans is taken out of the freezer, brought to room temperature and shelled, all of which takes place within a period of not more than three days. The kernels pass through inspection stations at which dark colored halves, chipped and imperfect pecans, etc., are removed. The shelled pecans are placed in cartons lined inside with a waterproof and greaseproof plastic coating.

Heating pecan meats to an internal temperature of 176°F in dry air or oil doubles the shelf-life by inactivating oxidative enzymes. Higher temperatures produce a partially cooked flavor, while roasting to 365°F for 15 min by dry roasting or infrared rays destroys natural antioxidants (accelerates rancidity development) but increases the aroma and flavor many times. This is a favorable situation for nuts which are to be consumed soon after preparation but has unfavorable implications for snacks and similar items entering a rather long distribution system. Antioxidants such as BHA or BHT can be added to increase shelf-life.

If pecan meats are dried below about 3.5 to 4.0% moisture content, they become too brittle and suffer excessive breakage during handling.

U.S. Standards for shelled pecans were promulgated in 1956. The four grades for shelled pecans are U.S. No. 1 Halves, U.S. Commercial Halves, U.S. No. 1 Pieces, and U.S. Commercial Pieces. There is also an "Unclassified" category which is not regarded as a grade. The size of halves are specified according to the number of halves per pound while pieces are classified in accordance with a sieve test. The following list illustrates the halves grading system.

	Halves per Pound
Mammoth	200 to 250
Junior Mammoth	251 to 300
Jumbo	301 to 350
Extra Large	351 to 450
Large	451 to 550
Medium	551 to 650
Topper	651 to 750
Large Amber	400 or less
Regular Amber	more than 400

Development of the so-called amber color is a kind of storage deterioration probably related to relatively high temperature and high moisture content.

COCONUT

Some coconut is used in mixed snack food products. Most of the coconut preparations available are based on the dried meat of the ripe coconut. This may be further processed by cooking the coconut with sugar and glycerol or by toasting. Principal countries of origin are the Philippines and Ceylon.

In unsweetened coconut, the only variables available to the purchaser are particle size and particle shape. This material will consist of about 65% fat, 2 to 3% moisture, fiber, carbohydrate, protein, etc. The fat is comparatively stable.

Sweetened coconut shreds contain approximately 39% fat, 53% carbohydrate, and 3% water. It is considerably more expensive than the unsweetened product. Plastic coconut is the finely ground dried meat. Liberation of the oil through rupture of the cell walls causes the product to feel softer than the granular dried material. Mixed products such as coconut syrup and coconut honey are sometimes offered. They generally consist of some form of sugar ingredient cooked with the freshly ground coconut meat. Colored coconut shreds are available from a few manufacturers.

Toasted coconut can be added to snacks to give a nut-like flavor and crisp and chewy texture. Other ingredients are added before the coconut particles are oven toasted. The piece sizes offered by one dealer are described as thin flake, extra fine, macaroon, medium, rice cut, nuggets, shred, and chip.

Specifications for coconut products should include a limitation on moisture of about 3%, a fat minimum of not less than 60%, and a description of the particle size and shape. Flavor and appearance specifications, always difficult to quantitate, should include at least a requirement that the product be free of visible foreign material and discolored particles, and have a typical taste and aroma. Specifications for yeast and mold count, lipase activity, and *Salmonella* are desirable if they can be enforced.

WALNUTS

The American black walnut (*Juglans nigra*) is used in relatively small amounts in snack foods where its distinctive flavor can make a positive contribution to acceptability. It is more common to use English walnuts (*J. regia*) which are milder in flavor and lower in cost.

The English walnut is widely cultivated today, with centers of production in middle and southern Europe, as well as in California and some other parts of the southern United States. Some of the popular varieties in the United States are Payne, Eureka, Hartley, and Franquettes. These all yield nuts satisfactory for consumption in mixed nuts, with the Franquettes being somewhat lighter in color than the others.

When harvested, the kernel moisture content may be as high as 35%. This is reduced by mechanical drying, and the in-shell nuts are stored at about 4.0% kernel moisture until needed for shelling.

Shelled walnuts quickly darken and develop rancidity under unfavorable exposure to moisture, heat, light, and air. Maximum stability is at about 3.1% moisture. There is an almost linear increase in stability with decreasing temperature of storage. Antioxidants can be applied to the kernels to retard rancidity development.

U.S. Standards for grades of shelled walnuts were issued in 1959. The grades consist of U.S. No. 1, and Commercial. There is also an "Unclassified" category for portions of walnut kernels which have not been classified in accordance with either of the foregoing grades.

OTHER NUTS AND NUT-LIKE PRODUCTS

Because the usual varieties of nuts are expensive relative to other snack food ingredients, there is a constant search for cheap substitutes. None have been particularly successful. One approach is the nut piece, such as provided by the invention of Maloney *et al.* (1970). Pillsbury's Bitsyn® is a rather good simulation of pecan pieces, but is fairly expensive and could create labeling problems in certain situations. Tamanuts® and other versions of glandless cottonseeds have reasonably good texture and flavor (Lawhon *et al.* 1970), but are small in size and rather unattractive in color. Also, the supply situation is unclear. Great strides have been made in processing soybeans so that they resemble nuts in appearance, texture, and flavor. If progress continues at the same rate, we may soon see the development of an excellent nut replacement from soybeans within the next few years. Soybeans have the great advantage for a raw material of always being in plentiful supply at low prices compared to nuts.

BIBLIOGRAPHY

ANON. 1948. U.S. Standards for Cleaned Virginia Type Peanuts in the Shell. U.S. Dept. Agr. Consumer Mktg. Serv., Washington, D.C.

ANON. 1956. U.S. Standards for Shelled Runner Type Peanuts. U.S. Dept. Agr. Agr. Mktg. Serv., Washington, D.C.

ANON. 1959. U.S. Standards for Shelled Virginia Type Peanuts. U.S. Dept. Agr. Agr. Mktg. Serv., Washington, D.C.

ANON. 1965. U.S. Standards for Grades of Shelled Spanish Type Peanuts. U.S. Dept. Agr. Consumer Mktg. Serv., Washington, D.C.

DELA CRUZ, A., CAVALETTO, C., YAMAMOTO, H. Y., and ROSS, E. 1966. Factors affecting macadamia nut stability. 2. Roasted kernels. J. Food Sci. 20, 1217–1221.

FREEMAN, A. F., MORRIS, N. J., and WILLICH, R. K. 1954. Peanut Butter. U.S. Dept. Agr. A1C-370.

HARRIS, N. E., WESTCOTT, D. E., and HENICK, A. S. 1972. Rancidity in almonds—shelf life studies. J. Food Sci., 824–827.

HOWES, F. N. 1953. Nuts—Their Production and Everyday Uses, 2nd Edition. Faber and Faber, London.

LAWHON, J. T., CATER, C. M., and MATTIL, K. F. 1970. Preparation of a high-protein low-cost nut-like food product from glandless cottonseed kernels. Food Technol. 24, No. 6, 77–80.

LAWLER, F. K. 1965. Presses calories from nuts. Food Eng. 1965, No. 9, 138–139.

MALONEY, J. F., SANDER, E. H., and WILHELM, A. 1970. Edible food product and process. U.S. Pat. 3,505,076. Apr. 7.

MASELLI, J. A. 1969. Peanut butter's qualities can be assets to candies. Mfg. Confectioner 1969, No. 10, 3–8.

MCWATTERS, K., HEATON, E. K., and CECIL, S. R. 1971. Storage stability of peanut pie mixes. Food Prod. Develop. 5, No. 1, 69, 71, 73, 74.

STUCKEY, B. N. 1954. Rancidity prevention in candy containing almonds. Mfg. Confectioner 34, No. 6, 47–49.

WATT, B. K., and MERRILL, A. L. 1963. Composition of Foods. U.S. Dept. Agr., Agr. Res. Serv. Handbook 8.

WILSON, C. T. 1973. Peanuts—Culture and Uses. American Peanut Research and Education Assoc., Stillwater, Okla.

WOODROOF, J. G. 1966. Peanuts: Production, Processing, Products, 2nd Edition. Avi Publishing Co., Westport, Conn.

WOODROOF, J. G. 1967A. Coconuts: Production, Processing, Products. Avi Publishing Co., Westport, Conn.

WOODROOF, J. G. 1967B. Tree Nuts: Production, Processing, Products, Vol. 1 and 2. Avi Publishing Co., Westport, Conn.

YOUNG, C. T. 1973. Influence of drying temperature at harvest on major volatiles released during roasting of peanuts. J. Food Sci. 38, 123–125.

CHAPTER 8

Potatoes and Other Plant Products

POTATOES

The member of this category which is most used in snack foods is potatoes. The potato chip can probably be considered the archetype of snacks, although potatoes in this form are also used in great quantities as a vegetable course with regular meals. In this chapter most of the text will be devoted to potato quality, with a concentration on suitability for chipping. Minor attention will be given to other vegetables.

In selecting potatoes for chipping, the processor is concerned with economic factors, such as those in the following list.

(1) Yield of chips per dollar of potato cost
(2) Uptake of oil by the potatoes during frying
 And consumer acceptance factors, the following for example, must be considered.
(1) Appearance, flavor, and texture of the finished chips
(2) Storage stability

The yield of chips from a given quantity of potatoes is highly correlated with the specific gravity of the tubers. Specific gravity is affected by variety, climatic conditions (temperature, solar radiation, and precipitation) during the growing season, nutrients available to the plant, soil characteristics, effects of predators (e.g., insects) and disease, and cultivation practices. Specific gravity is closely related to the dry matter content. Smith (1961) showed there is an increase of 1% in yield of chips for each increase in specific gravity of 0.005, within a variety. For example if a 100-lb lot of peeled potatoes of 1.060 specific gravity yields 27 lb of chips, a similar lot of 1.085 specific gravity would produce about 32 lb.

The direct method of measuring specific gravity requires a comparison of the weights of a sample of tubers in air and immersed in water. The specific gravity is then calculated as follows:

$$\frac{\text{weight in air}}{\text{weight in air minus weight in water}} = \text{specific gravity}$$

Scales are available which have specific gravity units on the dial so the desired units can be read directly.

The Smith (1967) potato hydrometer simplifies this determination with some loss of accuracy. A weighed sample (e.g., 8 lb) of potatoes is placed in a wire basket suspended from the bulb of the hydrometer. When the whole assembly is placed in a container of water, the level reached by the water surface on the stem of the hydrometer indicates the specific gravity of the potatoes.

Potato variety affects yields. Each variety will have an inherent tendency to develop a certain percentage of total solids, but this will be affected, of course, by conditions prevailing during the growing season. Different varieties grown under similar conditions will tend to have different total solids content.

Common varieties of potato particularly suitable for chipping are Katahdin, Kennebec, Irish Cobbler, Norchip, Monona, La Chipper, Superior, and (in some cases) Russet Burbank. Of these varieties, Norchip and Kennebec are the most popular.

Stage of maturity at harvesting affects yield and quality of chips as well as the keeping quality and conditioning of the stored potatoes. Fully mature tubers are considered most desirable for chipping, partly because they have a higher specific gravity and give a higher percentage of finished product than less mature potatoes. According to Smith (1975), it is usually necessary to process immature potatoes from the South and West in the April to June period. Choice of the proper variety of early harvested immature tubers is very important, and Sebago and Irish Cobbler from the South or Kennebec from California and Arizona are the most readily available even though they are, in general, relatively low in specific gravity.

There are various cultural factors which affect specific gravity of potatoes and the yield of chips made from them. Potatoes grown in irrigated fields are often of lower specific gravity than those grown without irrigation. High soil moisture from any source late in the growing season tends to decrease specific gravity. Heavy fertilization, particularly with nitrogen, usually leads to lower specific gravity. It appears that some insecticides have a similar effect. Vine killing, light intensity, and probably many other factors also influence specific gravity.

The color of potato chips has an important effect on consumer acceptance. Dark brown colors are disliked. The development of brown color in chips depends upon reactions between reducing sugars and amino acids (Habib and Brown 1957). Large scale screening studies revealed that only reducing sugars gave consistent evidence of correlation with chip color. Brown (1960) states, however, that constituents other than reducing sugar must be present before any color forms under the conditions which are used to fry potato chips. Certain other components, among them ascorbic acid, basic amino acids, and specific enzyme systems, may at times have an effect on color (Hoover and Xander 1961). Potatoes containing about 0.2% reducing sugar usually develop the desired golden brown color when fried, but with higher levels of reducing sugars, the product tends to be dark and unattractive. The chips may also have a bitter or burnt flavor.

The reducing sugar content usually depends upon variety, cultural practices, and storage conditions. Variety and cultural practices have a relatively minor influence on the sugar content in most cases. Freshly harvested tubers will normally contain only traces of these carbohydrates. The usual cause of high sugar concentration is storage of the tubers at low temperatures. When they are stored at 69 to 86°F, there is a slight rise in total sugars to about 1.5%, but storage at temperatures lower than about 50°F leads to an accumulation of 3 to 10% sugar in 4 to 8 weeks, the rate of development and final level depending somewhat on the variety. The change in concentration of sugars is accompanied by an opposite change in starch content.

The ideal temperature for storage of tubers intended to be used for french fries is between 50 and 55°F. Some sugars may accumulate and some sprouting may occur, but, on balance, these temperatures cause the least total damage.

The lower the temperature (down to freezing), the more rapid the rate of sugar accumulation.

Methods have been developed for reconditioning potatoes to prevent excess browning of the chips made from them. The reducing sugar content of potatoes which have been stored at low temperature can be decreased by holding them at higher temperatures for 3 to 6 weeks. Gradual cooling is preferred to sudden cooling. Presence of 5% carbon dioxide in the atmosphere of the storage space will slow the rate of reducing sugar development. The reducing sugar concentration can be different at the apical and basal portion of potatoes (Iritani et al. 1973), but this difference is usually slight and of no practical importance.

Peeled and sliced potatoes will also increase in reducing sugars if held for several days at room temperature. A dip in bisulfite solution saturated with carbon dioxide prior to cooling will prevent sugar accumulation for some time or, at least, darkening of the fried chips.

Genetic constitution affects response to cold storage. Lauer and Shaw (1970) identified an interspecific hybrid which produced acceptable potato chips when processed directly from 40°F storage during 3 yr of testing. A group of 600 seedlings from S. tuberosum parents treated comparably produced black colored chips without exception.

In addition to low storage temperatures, the following conditions are conducive to higher than normal sugar content.

(1) Low specific gravity (there is some disagreement on this point).
(2) Harvesting prior to maturity.

(3) Small size of tubers (if due to disease or immaturity).

(4) Microbial or mechanical injury to the potato.

(5) Irradiation with gamma rays.

(6) High temperature after harvesting (Timm *et al.* 1968); however, storage at 60 to 65°F for two weeks is recommended for wound healing of tubers.

(7) High soil temperatures (Motes and Greig 1970).

Some potato varieties normally accumulate too much reducing sugar for good chip color. Varieties such as Green Mountain, Pontiac, Bliss Triumph, Ontario, Houma, Erie, Essex, Mohawk, and Warba produce unacceptably dark chips when the tubers are grown and stored under usual conditions. Varieties which will produce light-colored chips under similar conditions are Norchip, Kennebec, Superior, Irish Cobbler, Katahdin, and Sebago.

A number of chemical or physical treatments have been suggested for improving potatoes which yield dark chips even after attempted reconditioning by holding in an optimum temperature range. Some of these depend upon extraction by various solutions of the reactants from potato slices. Hot dilute aqueous solutions of calcium chloride or certain other salts are said to be effective. A dip in sodium bisulfite solutions or exposure to sulfur dioxide gas is definitely helpful but at the risk of possible flavor effects. Hot water alone is also of some value, although washing does remove some flavor constituents.

ONIONS

Onions are important constituents of imitation onion rings made by the extrusion process and are used as a flavoring material in many other kinds of snacks. The dehydrated form is used almost exclusively, because it is easier to store and handle, and is more uniform than fresh material.

The characteristic odor of onions, and other members of the Allium genus, arise from enzymatic reactions initiated by cutting or otherwise damaging the tissue.

Because onions are used almost entirely for their flavor, the exceptions being when some textural contribution is desired, the predominant quality factors are determined organoleptically or by some chemical tests related to sensory effects. It should be obvious that sanitation and microbiological requirements must also appear in the specification.

Processing for dehydration begins by grading the onions for size. The onions may have been stored for periods of several weeks to a few months before processing, although fresh onions are also used during the harvesting season. Length and temperature of storage of the bulbs do have effects on the color and flavor of the dried product. The bulbs are flame-peeled at high temperatures and washed to remove the organic debris and other undesirable material. High-velocity air and then high-pressure water jets are used to wash away the charred skins and dirt. After inspection and removal of damaged material, the tops and roots are trimmed off. The cleaned and trimmed bulbs are sliced or chipped by automatic equipment. There is no blanching step in onion dehydration since enzymic activity is needed to develop the typical flavor.

Dehydration occurs in hot air tunnels. The older type of dehydrator used stacks of wooden trays to convey the onion pieces through the tunnel. In the newer dehydrators, the onion slices or dices are placed on stainless steel belts. In either case, hot air is blown in alternate directions in successive compartments of the dryer, entering first at the exit end. The total drying period of about 3 hr is divided into three temperature stages: 165°F, 145°F, and 130 to 140 °F. The onions will have not reached the necessary level of moisture content upon leaving the dryer, and are put into finishing bins where warm air currents complete the drying and facilitate equilibration of the remaining water. The resulting pieces may be milled to give products of varying particle size.

Dehydrated onions are very hygroscopic. Moisture absorption leads to physical and chemical instability, and causes the powdered forms to cake severely. If kept cool and dry, dehydrated onions can be stored for several years (Jones and Mann 1963).

OTHER PLANT PRODUCTS

Shallots, chives, leeks, and garlic are also available in dehydrated form. Shallots and leeks are much more expensive than onions and, although much appreciated by gourmets, probably do not contribute distinctive flavor notes which could be recognized by the average consumer in a snack product. Chopped chives may be useful for the appearance factor. Garlic has obvious value in many kinds of snack products when used as a flavoring ingredient.

BIBLIOGRAPHY

ANON. 1969. Official Standards and Methods of the American Dehydrated Onion and Garlic Association for Dehydrated Onion and Garlic Products. American Dehydrated Onion and Garlic Assoc., San Francisco, Calif.

BROWN, H. D. 1960. Problems of the potato chip industry. In Advances in Food Research, Vol. 10, C. O. Chichester, E. M. Mrak, and G. F. Stewart (Editors). Academic Press, New York.

COMIN, D. 1946. Onion Production. Orange Judd Publishing Co., Chicago.

COX, A. E. 1967. The Potato: A Practical and Scientific Guide. W. H. & L. Collingridge, London.

FITZPATRICK, T., and PORTER, W. 1966. Changes in the sugars and amino acids in chips made from fresh, stored, and reconditioned potatoes. Am. Potato J. 43, 238–248.

HABIB, A., and BROWN, H. D. 1957. Role of reducing sugars and amino acids in browning of potato chips. Food Technol. *11*, 85–89.

HESS, J. 1974. Will chippers have to "beachcomb" for potatoes again? Snack Food *63*, No. 1, 33, 36–37.

HOOVER, E., and XANDER, P. 1961. Potato composition and chipping quality. Am. Potato J. *38*, 163–170.

IRITANI, W. M., WELLER, L., and RUSSELL, T. S. 1973. Relative differences in sugar content of basal and apical portions of Russet Burbank potatoes. Am. Potato J. *50*, 24–31.

JONES, H. A., and MANN, L. K. 1963. Onions and Their Allies: Botany, Cultivation, and Utilization. John Wiley & Sons, New York.

KIRKPATRICK, M. ET AL. 1956. French-frying quality of potatoes as influenced by cooking methods, storage conditions, and specific gravity. U. S. Dept. Agr. Tech. Bull. *1142*.

KUNKEL, R., and HOLSTAD, N. 1972. Potato chip color, specific gravity, and fertilization of potatoes with N-P-K. Am. Potato J. *49*, 43–62.

LAUER, F., and SHAW, R. 1970. A possible genetic source for chipping potatoes from 40°F storage. Am. Potato J. *47*, 275–278.

LYMAN, S., and MACKEY, A. 1961. Effect of specific gravity, storage, and conditioning on potato chip color. Am. Potato J. *38*, 51–57.

MOTES, J. E., and GREIG, J. K. 1970. Specific gravity, potato chip color, and tuber mineral content as affected by soil moisture and harvest data. Am. Potato J. *47*, 413–418.

PRESSEY, R., and SHAW, R. 1966. Effect of temperature on invertase, invertase inhibitor, and sugar in potato tubers. Plant Physiol. *41*, 1657–1661.

SAMOTUS, B., and SCHWIMMER, S. 1962. Predominance of fructose accumulation in cold-stored immature potatoes. J. Food Sci. *27*, 1–4.

SCHAPER, L. A., YAEGER, E. C., FLICKKE, A. M., and JUNNILLA, W. A. 1967. Storage temperature maintenance and its effect on processing potatoes. Am. Potato J. *44*, 159–164.

SCHWIMMER, S., and BURR, H. K. 1967. Structure and chemical composition of the potato tuber. *In*, Potato Processing, 3rd Edition. W. F. Talburt, and O. Smith (Editors). Avi Publishing Co., Westport, Conn.

SMITH, O. 1961. Factors affecting and methods of determining potato chip quality. Am. Potato J. *38*, 265–272.

SMITH, O. 1967. Potato chips. *In* Potato Processing, 3rd Edition. W. F. Talburt, and O. Smith (Editors). Avi Publishing Co., Westport, Conn.

THOMPSON, N. R. 1967. Potato varieties. *In* Potato Processing, 2nd Edition. W. F. Talburt, and O. Smith (Editors). Avi Publishing Co., Westport, Conn.

TIMM, H., YAMAGUCHI, M., CLEGG, M.D., and BISHOP, J. C. 1968. Influence of high-temperature exposure on sugar content and chipping quality of potatoes. Am. Potato J. *45*, 359–365.

CHAPTER 9

Flavors and Colors

The use of flavoring materials, both natural and artificial, and of added coloring to make snacks more attractive or merely to differentiate them from competition obviously is becoming more prevalent. Although many if not most of the ingredients described in previous chapters have some effect on the appearance and taste of the finished products, this chapter will contain discussions of those substances which are added in relatively small amounts for the specific purpose of favorably influencing color and flavor.

FLAVORS

Sjöström (1970), who is widely recognized for his pioneering work in the flavor profile method of evaluating flavors, defined a flavoring ingredient as "any substance added to food, drugs, or other products taken into the mouth, the clearly predominant purpose and effect of which is to provide all or part of the unique and identifying flavor in the final product."

Flavoring materials which have been added to snack foods include but have not been restricted to spices and other natural products such as chocolate, vanilla, and essential oils, artificial flavors such as vanillin, and modified natural materials such as cultured butter and starter distillate. The amounts and kinds of flavors which can be used in foods have come under increasingly restrictive Federal and local regulations in recent years. This trend appears to be accelerating. In considering the use of any flavor in a snack food, the first question the formulator must ask himself is: Is it permitted by applicable laws and guidelines?

The fact that a flavor has a natural origin does not assure its compliance with existing regulations. Examples of natural flavors which cannot be used because they contain allegedly toxic constituents are tonka bean extract containing coumarin and sassafras root containing safrol.

Spices

Spices include seeds, bark, leaves, fruits, and other plant derived materials which can contribute substantial amounts of desirable flavoring principles to foods. The flavor principle is frequently found to be an essential oil extractable by fat solvents. Many spices can be obtained commercially as whole spice, ground spice, essential oil or oleoresin, or standardized fluid extract. Only the spices most important to the snack industry will be described in the following pages.

Capsicum.—Includes red pepper, paprika, and allied species. Various types of capsicum spices are grown throughout the world. Africa, Mexico, and the United States are the largest suppliers of these materials. Each kind of capsicum spice will have a different color and pungency. The pungent principle is a chemical compound called capsaicin, which can be quantitated as scoville heat units by a standard method, and the color is due to carotenoid-type substances. There is no measurable amount of volatile oil.

Caraway.—This spice is the fruit of a biennial of the parsley family, native to Europe. It is imported mostly from The Netherlands and Poland. Netherlands caraway is recognized as top grade. For the bakery trade, caraway is available ground as well as whole. This true caraway has no relationship to the charcoal caraway or black caraway, a small black seed that is used mainly for toppings on breads of central European and Russian origin.

Caraway yields from 3 to 7% of volatile oil. One part of the oil will dissolve in 10 parts of 80% alcohol.

Caraway is almost indispensable in the making of rye bread, and it is also good in corn muffins, cheese rolls, and certain kinds of cakes and cookies. It is compatible with the flavors of sugar, apple, peach, apricot, and rye.

Celery.—This plant is a native of Eurasia, but is now found throughout the world. Dehydrated leaves, stalks, and seeds are available for flavoring purposes. Extractives (oleoresins) are made from the seed. Most of the seed is produced in India and France from a plant *Apium graveolens,* a member of the parsley family not identical with the parsley we serve as a vegetable.

Cinnamon.—The dried bark of evergreen trees of the *Cinnamomum* species, a division of the laurel group.

Cinnamomum cassia, known in the spice trade merely as "Cassia," is the bark most commonly sold in this country as cinnamon. In its different subvarieties, Cassia cinnamon is native to China, IndoChina, and Indonesia.

Cinnamomum zeylanicum, sometimes called "true cinnamon," is a member of the *Cinnamomum* species that is native to and grows mostly on the island of Ceylon. This type of cinnamon is popular in other parts of the world, particularly Mexico, but is rarely used in the United States. It is almost buff-colored and quite mild. It would scarcely be recognized as cinnamon to Americans, who have always been used to Cassia cinnamon.

Cassia cinnamon, or just cinnamon, is a reddish-brown bark. It has an agreeable, aromatic odor and a pungently sweet taste.

The young shoots of the cinnamon trees are cut and the bark is peeled twice a year. New shoots then grow on the stump. The bark is rolled into quills (a trade term for what we think of as cinnamon sticks), to minimize breakage. Broken quills are sold as "chips" (Jones 1956).

Cumin.—Cumin seed is the small, dried fruit of an annual plant of the parsley family. It has a strong taste, somewhat reminiscent of caraway. This is an important flavoring constituent for Mexican style foods (chili, etc.) as well as curry powder.

Dill.—Dill seed is the dried fruit of an annual plant of the parsley family. Although the plant is native to the Orient and Mediterranean countries, it is now cultivated throughout the world. Most of the domestic consumption is imported from India. Oils are produced from the entire herb and from the seed, and they differ in character. The pungent, characteristic flavor of the spice probably has more potential in snack foods than is generally realized.

Garlic.—The garlic plant is a member of the lily family. The botanical nomenclature is *Allium sativum.* The dehydrated, ground cloves (or bulb segments) are most commonly used in snacks flavoring, but oleoresins from both the cooked and fresh vegetable are also available. Garlic salt is a mixture of ordinary salt with ground dehydrated garlic. Generally, an anticaking agent has been added.

Marjoram.—This spice is available as the dried, gray-green leaves of a perennial herb of the mint family. Most of the supply for the United States is imported from Mediterranean countries and from South America. The flavor of wild marjoram from Spain is harsh and somewhat reminiscent of eucalyptus, while sweet or garden marjoram has a mild, pleasant herb-like character. It is often used as a component of the mixed spices in some lunch meats (bologna, liverwurst, etc.), and could be used in savory snacks. Marjoram is commercially available whole, ground, or as the oleoresin.

Nutmeg and Mace.—Nutmeg is the seed of a peach-like fruit of an evergreen tree (*Myristica fragrans* Hutt) that grows in the tropics. The fleshy aril or skin that covers the seed is mace. Nutmeg has a fairly pungent or biting flavor and a warm, spicy aroma, while mace has a softer, better rounded, and less intense flavor. They are available as moderately coarse oily powders, primarily in spice blends for the food manufacturing industries. Both nutmeg and mace are widely used in cookery, especially in baking, and should be compatible with the flavors of many kinds of sweet or savory snacks, although they are seldom found in commercial varieties today.

Onion.—Onion is probably the most commonly used plant-derived flavor in food products. Many variations of the basic material are available, including different physical forms, different cooking treatments, extractives, etc. Because it is relatively inexpensive and very widely accepted, onion flavor has many current and potential applications in savory snacks.

Pepper.—This is certainly one of the most important spices used in foods, as attested by the ubiquitous presence of the pepper shaker and the almost universal listing of the ingredient in main-dish recipes. There are two principal types of pepper, the black and the white. Both are derived from the same plant, a climbing or trailing vine-like shrub native to southern India but now mainly grown commercially in India and the East Indian archipelago. The most important commercial types of white pepper are the so-called Muntok white pepper from Sumatra, and Sarawak white pepper. Whole and ground peppercorns and various kinds of extractives are available (Anon. 1970).

Rosemary.—Rosemary is a small evergreen shrub native to countries bordering on the Mediterranean. It has a pleasant, slightly camphoraceous odor and taste and is used as a condiment in various foods (Anon. 1970).

Sage.—Sage is found abundantly on the Dalmatian coast and the adjacent areas of the Adriatic Sea. There are many other kinds of sage produced commercially from various types of *Salvia* species some of which do not have the same flavor characteristics as the above material. It has a warm spicy taste very important in many sausage flavoring compositions (Anon. 1970).

Tarragon.—Tarragon, also known as Estragon, enjoys use in all types of connoisseur food products. The plant grows throughout southern Europe and the United States. Its flavor is reminiscent of basil and anise but has a peculiarly characteristic aromatic odor (Anon. 1970).

Thyme.—Thyme grows primarily in the Mediterranean area, Spain being the principal producing region. There has been some confusion with oils from other *Thymus* species, many of which do not contain the principal constituent, Thymol. This phenolic substance definitely characterizes the variety (Anon. 1970).

Turmeric.—This plant is a native of southern Asia. The rhizomes are ovate or pear-shaped having a dark yellow-orange color internally. In its ground form it is used extensively in curry-type flavors. The coloring principle is due to curcumin which is bis(4-hydroxy-3-methoxy cinnamoyl) methane (Anon. 1970).

Synthetic Flavors

The formulating of imitation flavors from pure chemical compounds is an art and a science much surrounded by secrecy. One of the few books giving any insight into modern practices in this field is Merory's *Food Flavorings* (1968). The identification of the chemical substances present in natural flavor materials, as a preliminary to simulating the natural flavor with mixtures of synthetic compounds, requires the most sophisticated methods of instrumental analysis, including the use of gas chromatography, mass spectrometry, infrared spectrophotometry, etc. Since only the largest food firms can support an independent activity of such complexity, it is necessary for most snack food manufacturers to rely on flavor suppliers for identification and formulation work. It is generally not feasible for a food producer to attempt to manufacture flavor materials from the basic ingredients.

Unfortunately, flavor suppliers vary greatly in their competence and reliability. Some merely repackage materials obtained from other firms, or, at best, formulate simple mixtures such as artificial vanilla. Other firms have extensive analytical laboratories, either in this country or in Europe, and manufacture many of the required chemical intermediates themselves. A few of these large firms obtain the bulk of their sales from the cosmetic and soap industries, and the food flavoring business is on the order of a sideline, but they are still willing to provide much useful technical service if they can foresee a reasonable sales volume.

Flavor manufacturers are restricted by FDA regulations in the substances they can use in products intended for the food industry. There is a rather lengthy list of pure chemical compounds to draw from (the GRAS list) as well as naturally derived ingredients. Federal approval of substances can be withdrawn at any time, with unpleasant consequences if substantial amounts of product containing the suspect ingredient are in retail channels at the time.

Pure and Artificial Vanilla

Vanilla is a most common flavoring added to other foods, but not much is consumed by the snack industry. It is not only a very desirable flavor by itself but also is an essential note in chocolate goods and many other compound flavors.

Vanilla is the fruit of an orchid cultivated in tropical and semitropical countries. The pods, much like long string beans in appearance, are cured in special ways to bring out the characteristic aroma and taste of vanilla.

Although the ground pod can be added to foods as an ingredient, it is more common to use alcohol extracts, which incorporate nearly all of the desirable flavor notes of the pod. These extracts are aged in glass or stainless steel containers until the reactions which lead to improved aroma have occurred.

The vanilla beans of commerce are described by type or region of origin, such as Mexican (heavy or coarse taste and aroma), Bourbon (floral and smooth tasting), and Tahitian (spicy, due to the natural paramethoxybenzyl alcohol content). Blends of Mexican and Bourbon are mostly used. It is of little value for the buyer to prescribe the bean blend for the extracts he purchases, since there is no reliable way to police the deliveries.

Pure vanilla extract is defined under U.S. FDA Standards in part as follows: "Pure vanilla extract is the solution in aqueous ethyl alcohol of the sapid odorous principles extractable from vanilla beans. In pure vanilla extract the content of ethyl alcohol is not less than 35% by volume and the content of vanilla constituents is not less than one unit weight of vanilla beans per gallon." The term "unit weight of vanilla beans per gallon" is defined as 13.35 oz of beans having a moisture content of 25% or less. Ingredients such as glycerine, propylene glycol, sucrose, dextrose, and invert sugar are listed as optional ingredients.

Since all manufacturers make vanilla extracts by very similar processes, and with concentrations of flavorants limited by the Standards, the flavor quality of different extracts will depend upon the types of vanilla beans used.

It is unfortunate that vanilla extracts sold to manufacturers have the reputation of being the most frequently adulterated of all food ingredients. The food manufacturer will find it advisable to buy only from reliable suppliers and to be prepared to pay more than the lowest quoted price. Careful evaluation of incoming shipments is important for keeping the supplier honest. Chemical analyses alone cannot determine the quality of the extract or its purity, but the lead number, the ash content, and the vanillin content are helpful in establishing a baseline. Considerable advances have been made in the use of gas chromatography to detect adulteration.

Powdered vanilla or vanilla sugar is made by mixing ground beans with sugar or by coating sugar granules with an extract.

Oleoresin of vanilla is prepared by evaporating under vacuum the filtered extract of comminuted vanilla beans. It consists mostly of the resinous substances of the vanilla beans with very little of the balsamic flavor of the original extract. Oleoresin can be diluted with solvents to give the 10-fold extracts sometimes sold. No more than 4-fold concentration is possible by extraction methods.

It can be shown by panel tests that consumers prefer products which are made with pure vanilla extracts as opposed to artificial flavors when two products are compared side by side (Bowden 1968). The difference in acceptability ratings is usually slight, on the average, and tends to disappear when there is not a comparison sample immediately available. Many very successful food products such as the most popular chocolate bar rely entirely on artificial vanilla flavor.

Although it is often said that artificial flavors cannot give the fine aroma and delicate taste contributed by good quality vanilla extracts, most consumers find vanillin or mixtures of vanillin and ethyl vanillin to be perfectly acceptable substitutes for the natural product. In very delicately flavored foods (snacks are seldom in this category) the differences between vanilla and vanillin are fairly obvious, but it is considerably harder to distinguish between chocolate products, for example, which are flavored with the two substances. Since the synthetic materials are so much cheaper on an equivalent strength basis than the vanilla extract, most food manufacturers are willing to forego the fine nuances of flavor, which usually are not appreciated by their customers, in favor of a more competitive ingredient cost.

The aromatic components of both artificial and natural vanilla extracts tend to be lost in a heating or extruding process. The minor constituents of vanilla distill off quicker than vanillin or ethyl vanillin, so the change during heat treatment is more apparent in products containing true vanilla extracts. The loss is greatest at elevated temperatures and in products brought almost to dryness, such as puffed snacks.

Ethyl vanillin is considerably more expensive than vanillin, but it does have a stronger flavor. The flavor quality of the two compounds is different, however, so a meaningful comparison of ingredient cost is difficult. Frequently the best results are obtained by using mixtures of the two compounds.

Cacao Products

Chocolate is prepared from the seeds of an evergreen tree of the genus *Theobroma*. Several of these seeds are borne in a pod resembling somewhat a large acorn squash. After removing the seeds from the pod and fermenting them to facilitate the removal of the surrounding pulp (and to cause certain desirable flavor changes), they are dried or "cured." Next follows a roasting step which further develops the flavor and loosens the outer covering of the seeds. After the hull and the germ are removed by milling, the remainder of the seed, called the nib, is ready for processing into chocolate. The nibs are subjected to grinding and milling procedures which reduce the nonfatty substances to colloidal size and develop the texture which is characteristic of chocolate liquor, bitter chocolate, and bakers' chocolate (these are synonyms).

Addition of milk solids and/or sugar together with other flavoring materials such as vanilla, and sometimes emulsifiers, to chocolate liquor produces the "eating" chocolate varieties milk chocolate, sweet chocolate, and bittersweet chocolate. These products contain a minimum of 10, 15, and 35% chocolate liquor, respectively. In Table 9.1 the composition of several cacao products is compared.

Cocoa is chocolate from which a substantial proportion of the fatty

TABLE 9.1

COMPOSITION AND pH OF PRODUCTS DERIVED FROM CACAO

	Nibs, Roasted (%)	Chocolate Liquor (%)	Breakfast Cocoa (%)	Cocoa, Medium Fat (%)	Dutched Breakfast Cocoa (%)	Sweet Chocolate (%)	Milk Chocolate (%)	Chocolate Syrup (%)
Moisture	3.0	2.3	3.9	4.0	5.0	1.4	1.1	31.0
Ash	2.8	3.2	5.0	5.5	8.0	1.4	1.7	1.9
Protein	10.5	8.0	8.0	8.7	7.7	4.0	6.0	3.5
Fat	55.0	55.0	23.8	16.0	21.5	33.0	33.5	1.1
Fiber	2.6	2.6	4.6	5.0	4.3	1.4	0.5	0.6
Alkaloids	1.45	1.4	1.9	2.1	2.0	0.4	0.4	—
Carbohydrate[1]	25.2	26.6	44.3	48.7	40.2	61.3	55.2	56.0
pH	4.9	5.2–6.0	5.2–6.0	5.2–6.0	6.1–8.8	5.2–6.0	5.2–6.0	—

Source: Watt and Merrill (1963).
[1]Other than fiber.

material has been removed. This is usually accomplished by pressing, but sometimes solvent extraction is used to remove a very large percentage of the cocoa butter. It is usually agreed that the best chocolate quality and flavor are contained in the liquor, especially when it has been produced from good quality Accra, Bahia, or Arriba beans. Off-grade and poorer beans such as Sanchez mid-crop, Lagos, and mid-crop Accra are frequently used to prepare the low priced cocoas of commerce. Cocoa from superior beans can be obtained readily, but it sells for a premium.

Even though chocolate liquor undoubtedly has a superior taste and aroma, cocoa performs quite adequately in many foods. The advantages of chocolate liquor are more evident in eating chocolate and in coatings.

As stated above, the flavor quality of chocolate liquor is related to the type and origin of the bean, and the desirability of cocoas is related to the type of chocolate from which they were produced and to the percentage of fat which they contain. By U.S. FDA Standards (Anon. 1958), breakfast cocoa must contain a minimum of 22% cocoa butter. Bakers frequently use a less expensive grade which contains from about 8 to about 15% cocoa butter. The Standards identify cocoa having less than 10% fat as "low-fat cocoa."

Treatment of cocoa nibs with alkalies (such as potassium carbonate) at some stage during the roasting process profoundly changes their characteristics and gives the "Dutched" chocolates and cocoas of commerce. Some of the changes observed are: the color of the cocoa becomes darker; the flavor changes and becomes stronger; some of the cocoa butter is saponified; the starch is partially gelatinized; cellulose materials swell; the natural cocoa acids are neutralized and the slightly acidic pH changes to a nearly neutral or slightly alkaline pH; and, some of the tissues are disintegrated. The net changes of importance are the changes in color and flavor, which sometimes make it possible to use less of the flavoring material, and an increase in the solution stability of the product.

Federal specifications are frequently useful guides for establishing commercial requirements. The U.S. Dept. of Agr. specification for cocoa (Anon. 1952) stipulates that breakfast-type cocoa must contain at least 22% cocoa fat and medium-fat cocoa must contain from 10 to 22% of cacao fat. Dutch process cocoas, according to this specification, shall have added alkaline materials (selected from among those permitted to be used for this purpose by the FDA) equivalent in neutralizing power to not more than 3 parts of anhydrous potassium carbonate per 100 parts of cocoa (sic) nibs from which the cocoa was made. All cocoas shall have a particle size that will allow 99% of the cocoa to pass through a No. 140 U.S. Standard Sieve after it (the cocoa) has been disintegrated and washed with petroleum benzine.

The U.S. Dept. of Agr. specification for chocolate and sweet chocolate (Anon. 1949) requires that bitter chocolate contain not less than 50 and not more than 58% of cacao fat. Vanilla-flavored sweet chocolate shall contain not less than 25% of chocolate liquor together with the addition of such amounts of cacao butter as may be needed to give a smooth consistency with the amount of sugar used, the total fat being not less than 30%. The total sucrose content shall not exceed 57% and the product shall be flavored with vanilla, vanillin, or ethyl vanillin, and it may contain lecithin, phosphatides, and other stabilizers.

Tests applied to cacao products to determine their quality include (1) fat content, (2) particle size, (3) color—both intrinsic and potential, (4) pH, (5) moisture content, (6) flavor, and (7) microbiological contamination. The color and flavor are best evaluated in the food in which the ingredient will be used, since both of these characteristics can change markedly depending upon the processing conditions and associated materials. Color, especially, is much affected by the pH of the finished product.

Because of the method of manufacture, it is difficult for the cocoa processor to hold the fat content closer than within 1% of the specification figure. For this reason, it is customary to specify at least a 2% range, such as 10 to 12% fat. There are significant differences between fat content determined by the Soxhlet extraction technique and the acid hydrolysis method, the former giving considerably lower values, so it is necessary to state in the specification the analytical procedure by which the fat will be determined. In this country, the Soxhlet extraction technique is the preferred method.

The pH of a water slurry of natural (nonalkalized) cocoas will vary depending on many factors in the history of the material, and some of these factors are not controllable by the manufacturer. Alkalized cocoa can generally be held within a range of ±0.2 units by controlling the processing conditions.

Particle size is critical for some applications such as chocolate milk, confectioner's coatings, etc. A cocoa of very fine particle size might be described as 0.5% maximum on a 325-mesh screen. A reasonably satisfactory fineness for cocoa to be used in batters would be 0.5% maximum on a 200-mesh screen. Although there are several procedures being used for wet-screening cocoas, a method developed by technologists at the Robert A. Johnston Company seems to have several advantages over all others. It avoids the usual rubbing and spraying techniques needed to disperse agglomerates and facilitate the passing of fine particles through the sieve. This method involves suspending a wire mesh, which has been fitted with sidewalls more than an inch high, in a bath of solvent that can be ultrasonically vibrated. The solvent enters through the screen and fills the sieve about halfway, so that there are two "chambers," the ultrasonic bath and the sieve compartment, communicating with each other through the screen. The rapid and continuous agitation afforded by the ultrasonic vibration keeps the particles suspended in a continually renewed bath of solvent. A few seconds usually suffices to wash out all of the soluble fat.

COLORS

Artificial colors are not widely used in the snack food industry. Probably their principal application today is in the powders and oils which are applied to expanded products, including popcorn. All food colors must be selected from a series approved by government agencies. There are two categories of acceptable color additives, the certified list and the uncertified group. The certified colors include the chemically synthesized dyes and their lakes. Lakes are made by adsorbing or chemically combining water-soluble dyes on to insoluble substances. Uncertified colors are derived from plant or mineral raw materials by various processing techniques and have been approved for food use in many cases because of a long history of satisfactory prior usage. At the present time, the legal status of many substances given in Table 9.2 is in a state of flux.

Caramel Color

Caramel color is an important ingredient for the food industry because it can give rich brown shades difficult to achieve with dyestuffs and it is relatively inexpensive.

TABLE 9.2

UNCERTIFIED COLOR ADDITIVES FOR FOOD[1]

Color Additive	Restrictions for Use
Annatto extract	—
Beta-carotene	—
Beet powder	—
B-apo-8'-carotenal	15 mg per lb or pt
Canthaxanthin	30 mg per lb or pt
Caramel	—
Carmine	—
Carrot oil	—
Cochineal extract	—
Cottonseed flour, toasted partially defatted cooked	—
Ferrous gluconate	Black olives only
Fruit and vegetable juices	—
Grape skin extract (enocianina)	Beverages
Paprika and paprika oleoresin	—
Riboflavin	—
Saffron	—
Titanium dioxide	1% or less
Turmeric and turmeric oleoresin	—
Ultramarine blue	Salt only
Carbon black (Impingement)[2]	—

[1] Feed (animal) additives not included.
[2] Provisionally listed.

Caramel coloring has no relationship to caramel candy. It is often manufactured by heat treating high DE corn syrups in the presence of reactants such as certain alkalies. It is a very dark, almost black syrup of about 11 lb-per-gal. density. The pH will be near 4.0 to 4.2, the solids content will be near 65%, and it will be fairly low in viscosity. The flavor will be mild.

Although the hues or spectral distributions of colors from the different kinds and brands of caramel are similar, this ingredient imparts varying shades depending upon the amounts used and the background colors. The color of the finished product may range from a light tannish yellow to a very dark brown.

The common containers for caramel color are 5-gal. steel pails, 30-gal. steel drums, and 56-gal. steel drums. Bulk deliveries in rail tankers and tank trucks are available. Color changes continue in storage, but even so, high-quality caramel colors can be stored a year or longer at room temperature. Under certain adverse conditions, caramel can polymerize into an amorphous irreversible gel.

Dry, powdered forms are sold by some manufacturers, but they are usually considerably more expensive than the syrups.

BIBLIOGRAPHY

ANON. 1949. Chocolate and sweet chocolate. U.S. Dept. Agr. Spec. JJJ-C-271a, Feb. 7.

ANON. 1952. Cocoa. U.S. Dept. Agr. Spec. JJJ-C-501b, Oct. 7.

ANON. 1958. Definitions and standards for chocolate and cocoa products. Food, Drug and Cosmetic Act No. 2, Par. 14, U.S. Food and Drug Admin., Washington, D.C.

ANON. 1970. The Oleoresin Handbook. Fritzsche Bros., New York.

ANON. 1971. Spices and snacks: seasoned to profit? Snack Food 60, No. 2, 52–54.

BOWDEN, G. L. 1968. Judicious use of vanilla. Proc. Am. Soc. Bakery Engrs. 1968, 299–304.

FRANCIS, F. J., and CLYDESDALE, F. M. 1972. Color measurement of foods. XXXIII. Miscellaneous: Part III. Food Prod. Devlop. 6, No. 3, 86–89.

FRANCIS, F. J., and CLYDESDALE, F. M. 1975. Food Colorimetry: Theory and Applications. Avi Publishing Co., Westport, Conn.

JONES, L. W. 1956. A Treasury of Spices. American Spice Trade Association, New York.

KINNISON, J. W., and CHAPMAN, R. S. 1972. Extrusion effects on colors and flavors. Snack Food 61, No. 10, 40, 42.

MERORY, J. 1968. Food Flavorings—Composition, Manufacture, and Use. Avi Publishing Co., Westport, Conn.

SCHUEMANN, H. W. 1972. How to use chocolate power wisely. Snack Food 61, No. 10, 50–51, 81.

SJÖSTRÖM, L. B. 1970. Functionality of flavorings in cereal-based prepared mixes and convenience foods. 11th Ann. Symp. of the Central States Sect., Am. Assoc. Cereal Chem., Feb. 14.

WATT, B. K., and MERRILL, A. L. 1963. Composition of Foods. U.S. Dept. Agr., Agr. Res. Serv. Handbook 8.

Products and Processes

CHAPTER 10

Potato Chips

The discussion in this chapter will deal exclusively with chips made from fresh sliced potatoes. Processing of potato dough to form simulated potato chips is discussed in Chapter 14. Many of the points concerning potato chips are applicable to other low moisture fried forms of potatoes, such as the ridged or corrugated slices and the shelf-stable shoestring potatoes.

POTATO CHIP PROCESSING

The most important steps involved in potato chip processing are in the following list.

(1) Selecting, procuring, and receiving potatoes
(2) Storage of potato stock under optimum conditions
(3) Peeling and trimming the tubers
(4) Slicing
(5) Frying in oil
(6) Salting or applying flavored powders
(7) Packaging

Selection and Storage

All authorities agree that it is important to select potatoes of high specific gravity since this characteristic indicates superior yield and lower oil absorption.

It is even more important to select potatoes with low reducing sugar contents or to store them at temperatures conducive to the minimizing of these substances. See Chapter 8 containing a discussion of potatoes for further details on the subject of reducing sugar content. Sprouting and fungal damage must also be minimized by the storage conditions.

Peeling

The ideal peeling operation would remove only a very thin outer layer of the potato leaving no eyes, blemishes, or other material for later removal by hand trimming. It should not significantly change the physical or chemical characteristics of the remaining tissue. Preferably, it should use small amounts of water and result in minimal effluent, to satisfy the regulatory authorities who control the watercourses. As a practical matter, compromises will have to be made in all of these aspects of peeling.

Harrington and Shaw (1967) describe the peeling process in great detail. First, the potatoes are thoroughly washed, not only for sanitary reasons, but also to prevent dirt or grit from abrading the equipment the tubers will later contact. Washing may take place in flumes, as the potatoes are being conveyed by water streams, or in equipment provided with means for scrubbing the potato with brushes or rubber rolls. In barrel-type washers, potatoes are cleaned by being tumbled and rubbed against each other and against the sides of the barrel while they are immersed in, or sprayed with, water.

After washing, the potatoes are allowed to drain, usually on mesh conveyors, and then they travel over an inspection belt where foreign material and defective tubers are removed.

The more common peeling methods are abrasion, lye immersion, and steam. Abrasion peelers, which may be either batch or continuous, use disks or rollers coated with grit to grind away the potato surface. An important design feature is to ensure that all surfaces of the tuber are equally exposed to the rasping action. The peel fragments are flushed out of the unit by water sprays. Such systems work best with uniform, round, undamaged potatoes. Varieties with deep set eyes require much additional hand trimming, which is uneconomical. Some of the advantages of abrasion peelers are their simplicity, compactness, low cost, and convenience. They are particularly suitable for peeling potatoes intended for chipping since they do not chemically alter the surface layers. About 10% of the original tuber weight is lost through abrasion peeling prior to chipping. The loss may be greater when potatoes are peeled for other uses because more trimming is required.

In lye peeling, a hot caustic solution loosens and softens the surface skin blemishes and eyes so that they are readily dislodged by pressure spray washers, which also remove the residual chemical. High-temperature lye peeling uses solutions (15 to 25% concentration) heated above the starch gelation point of about 160°F, leading to development of a cooked surface ring or heat layer on the peeled tuber. Low-temperature lye peeling uses caustic in the temperature range of 120 to 160°F to decompose, hydrate, and soften the surface layers of tissue without cooking or denaturing the material that will be further processed. Because the lye solution is more viscous, less reactive, and does not penetrate as rapidly when it is below 160°F, the low-temperature method is slower and somewhat more difficult to control.

Steam peeling is an efficient processing method but has the disadvantage of leaving a cooked layer at the potato surface which adversely affects the appearance of potato chips. Because the high-pressure steam used in this method of peeling affects all surfaces of the potato uniformly, contour has little effect on its efficiency. Removal of about $\frac{3}{16}$ of an inch depth is required for adequate peeling. Some trim is required.

Hot brine, flame, and hot oil have also been used to peel potatoes in experimental installations.

Although it seems intuitively obvious that potatoes should be peeled before being made into chips, there have been suggestions that acceptable product can be prepared from sliced, unpeeled tubers. In one study (Shaw *et al.* 1973) it was reported that chips from peeled and unpeeled potatoes are similar in flavor, appearance, and shelf-life. On close inspection, the difference in appearance was noticed by consumers. The chips from unpeeled potatoes had a higher oil content. There was a 7% increase in potato solids and a reduction in waste as a result of omitting the peeling step.

Slicing

The peeled potatoes are cut into slices from $\frac{1}{15}$ to $\frac{1}{25}$ in. thick by rotary slicers. Centrifugal force presses the tuber against stationary gauging shoes and knives. Thickness is varied, not only to meet customer preferences, but also to fit the condition of the tubers and the frying temperature and time. Slices produced at any one time must be very uniform in thickness, however, in order to obtain uniformly colored chips. Slices with rough or torn surfaces lose excess solubles from ruptured cells and absorb larger amounts of fat.

It is necessary to remove from the surface of slices the starch and other material released from the cut cells so that the slices will separate readily and completely during frying. The slices are washed in stainless steel wire mesh cylinders or drums rotating in a rectangular stainless steel tank. High-pressure water sprays flush away the loose material from all surfaces of the tumbling slices. After washing and an additional rinse in similar equipment, the potato piece may or may not be dried in centrifugal extractors, by high-velocity air currents (heated or unheated), pressure rolls of sponge rubber, or vibrating mesh belts, etc. Surface drying, by helping to shorten the frying time, increases the capacity of the cooking unit.

Reeve (1971) showed that expected losses of solids due to extrusion of the contents of cut cells range from 8 to 12% at 2-mm slice thickness and from 16 to 24% at 1-mm slice thickness based on average parenchyma cell diameters of 160 to 240 μ. It was assumed that clean, sharp cuts were made without tearing of additional cells, which would, of course, occur in practice and result in even higher losses.

In some frying systems, potatoes are sliced directly into the frying fat, i.e., they are not washed after slicing. As a result, they are said to be more flavorful. Some sticking together of the slices, due to adhering starch, is usually observed.

Frying

From the dryer or rinser the sliced potatoes are conveyed directly to the fryer. The capacity of this piece of equipment is generally the limiting factor in the process line. Most manufacturers currently use continuous fryers but some batch equipment is still employed. Modern continuous fryers process from 2 to 4 tons of raw potatoes per hour. The essential elements of these units are (1) a tank of hot oil in which the chips are cooked, (2) a means for heating and circulating the oil, (3) a filter for removing particles from oil, (4) a conveyor to carry chips out of the tank, (5) a reservoir in which oil is heated for adding to the circulating frying oil, and (6) vapor-collecting hoods above the tank. In most frying units, rotating cylinders or wheels near the receiving end of the tank push any floating slices under the surface of the oil and also slow their progress so that they receive sufficient heat treatment. Near the discharge end of the tank is a series of perforated baskets or "rakes" suspended on camshafts above the oil surface. Their function is to turn the slices and submerge them as they finish cooking. The chips are removed from the tank and drained on mesh belts. The temperature of the oil in new or well maintained fryers can be held within a range of about $\pm 2°F$. Temperatures normally used are from 350 to 375°F at the receiving end and 320 to 345°F at the exit end. Normally the oil heater settings are not varied. If a temperature change is desired, it is obtained by varying the rate at which slices are fed into the tank.

The oil used for deep-fat frying of potato chips has two functions: (1) it serves as a medium for transfering heat from a thermal source to the tuber slices, and (2) it becomes an ingredient of the finished product. Use of highly refined oil is of great importance in flavor and stability of chips. Flavor, texture, and appearance are affected both by the amount of oil absorbed and its characteristics as it exists in the chip (i.e., not necessarily its initial chemical and physical parameters). There seem to be some geographical differences in consumer preferences for oil flavor, some regions preferring peanut oil, etc. (King et al. 1936).

Oils change continuously during the frying process, but the heat abuse resulting from chip cooking is relatively mild. Temperatures rarely rise above 385°F at any point.

The steam evolved from the potatoes forms a blanket of nonoxidizing gas over the oil. Approximately 1 lb of steam is generated each hour for each 3 lb of heated oil. This not only furnishes the steam blanket, but also provides a continuous deodorization of the oil. Undesirable products do not accumulate because there is a relatively rapid turnover of the oil, a result of constant replenishment with fresh fat to compensate for that absorbed by the chips.

Chips fried at 350°F have a lower oil content than chips fried at 300°F, the lower viscosity of the hotter oil favoring more complete drainage of oil from the cooked chip (Shaw and Lukes 1968). As the chips emerge from the frying oil, they are still giving up moisture and tend to cool. This rapid cooling increases oil viscosity and inhibits draining. By increasing the temperature of chips after they emerge from the fryer, a significant reduction in oil content can be achieved. Dousing with super-heated oil, running chips through a hot air tunnel, or use of a radiant heater are some of the means suggested for decreasing oil retention.

Better control over chip color could be obtained if the final stage of moisture removal could be achieved without the browning reactions which always accompany it in the frying process. Smith (1967) described the successful use in the laboratory of hot air tunnels, infrared heat, and microwave drying for this purpose. Although some increase in browning always accompanied additional drying it was less than when all moisture was removed by the frying process. It appears that microwave drying, described in detail by Smith et al. (1965), is particularly suitable for drying with minimal discoloration. Experiments conducted on a large scale in a chip plant showed that microwave drying could be conducted under conditions leading to no additional color development. Furthermore, the oil content of microwave-finished chips was lower than those conventionally fried and the texture was slightly different. According to Smith et al. (1965) the use of microwave ovens for finishing allows processors to make salable chips from raw stock that would produce dark brown chips when fried conventionally. Microwave finishing makes it possible for potato chips to be removed from the oil with 6 to 10% moisture remaining in them after which they are dried to a desirable level of 2% with no additional coloring. It is also claimed that rancidity development is slower in these chips. In commercial installations, it was found that microwave finishing required closer control of moisture content than was practicable, and so the initial enthusiasm for this theoretically superior method has abated considerably.

Chips may be sorted for size at this point, with larger chips being diverted to the bulk packs and larger pouches, while the smaller pieces are used for vend pack and other individual service containers. Potato chip sizing is also accomplished by separating the peeled potatoes into large and small sizes, which are then sliced and fried separately.

The chips are salted immediately after they leave the fryer. It is important that the fat be liquid at this point to cause maximum adherence of the granules. Powders containing barbecue spices, cheese, or other specialty flavoring materials may be added at this point. The salt may contain added enrichment materials or antioxidants. Equipment used for applying salt is described in Chapter 23 of this book.

After salting, the chips pass on to a conveyor where they are visually inspected and off-color material removed. If the chips are allowed to cool before packaging, better adherence of salt and flavor powders is obtained.

Factors Affecting Oil Content

Under most market conditions, vegetable oils cost more per pound than potatoes, on a dry matter basis. Therefore, processors wish to keep the oil content of potato chips at the lowest level consistent with consumer satisfaction. Some of the factors affecting the amount of oil absorbed by the potato slice are (1) solids content of the tuber, (2) fat temperature, (3) duration of frying time, and (4) thickness of slices.

Fat pickup by potato chips is said to be 10 to 15% lower when the chip is fried in liquid oil rather than a fat which is solid at room temperature. This has been attributed to low absorption of the oil as well as the marked resistance to hydrolysis of hydrogenated fat, but better draining properties of the oil probably contributes a great deal to the phenomenon.

As long as steam is being rapidly evolved from the cooking slices, fat absorption will be at a low level. As the protective layer of water vapor begins to dissipate in the final stage of frying, fat can enter the voids left in the dehydrated cells. This condition will exist at different times in different parts of the chip because of their varying rates of dehydration. Excess fat, the amount being related partly to fat viscosity and the surface unevenness of the chip, adheres to the chip as it is removed from the frying. A large part of this fluid material is drawn into the unfilled cavities in the cells as the chip cools, and the remainder runs off. As would be expected, reducing slice thickness increases the oil content of chips. The violence with which the chips are agitated on the draining conveyor must have a considerable effect on the fat content of properly fried chips. A hot air blast at the point where the slices merge from the fat should also reduce fat absorption somewhat.

Partial drying of raw sliced potatoes before frying reduces the oil content of chips. Leaching raw slices with hot water (e.g., for removal of excess reducing sugars) results in an increased uptake of oil.

QUALITY FACTORS

The principal factors affecting potato chip acceptability are piece size, color, and, of course, flavor. These factors are controllable primarily by selecting the raw material, adjustment of processing conditions, and packaging.

Flavor of Potato Chips

The components comprising potato chip flavor have been extensively investigated. As in most natural foodstuffs which have been subjected to high-temperature processing, hundreds of flavored compounds probably exist but relatively few of them make a meaningful contribution to the flavor perceived by the consumer.

Deck *et al.* (1973) reported on a systematic characterization of the volatile compounds in potato chips. Pleasant, desirable flavors were found in 53 substances. These included 8 nitrogen compounds, 2 sulfur compounds, 14 hydrocarbons, 13 aldehydes, 2 ketones, 1 alcohol, 1 phenol, 3 esters, 1 ether, and 8 acids. The aromas of the alkyl-substituted pyrazines, the 2–4 dienals, phenyl acetaldehyde, and furyl methyl ketone indicated they might play important roles in contributing to the desirable flavor of potato chips. The aromas of the fractions 2,5-dimethyl pyrazine and 2-ethyl pyrazine were described by organoleptic panelists as either "strong potato" or "roasted peanut."

Maga (1973) claimed that blindfolded panelists could not detect the difference between dark and light chips selected from the same batch of fresh product. As the duration of the storage period increased, a preference developed for the dark chips, apparently because they developed rancid odors at a slower rate.

STORAGE STABILITY

If the frying oil is stabilized and has not deteriorated through use, and if the packaging is opaque and has a low moisture vapor transmittance rate, a shelf-life of 4 to 6 weeks should be achieved when stored at temperatures of about 70°F. This is the longest shelf-life available without vacuum packaging, freezing, or other special treatments, and recognizes that some decline in acceptability will occur, but the average consumer will accept the product without complaint under these conditions.

Once potato chips are in the bag, the three forms of quality degeneration having the greatest effect on consumer acceptance are breakage, absorption of moisture with loss of crispiness, and fat oxidation leading to development of rancid odors.

The mechanical abuse causing breaking of the chips can be partially prevented by using stiff packaging material, making the package "plump" with contained air, and avoiding crushing in the shipping case.

Absorption of moisture is prevented largely by proper choice of packaging material. Fortunately, potato chips are not as hygroscopic as popcorn and most other puffed snacks, but they will eventually absorb enough moisture to become soft or chewy if exposed to a sufficiently challenging environment. Cellophanes coated with various moisture barriers have proved to be satisfactory pouch films for the relatively short shelf-life expected (generally stated to be 4 to 6 weeks). This subject is discussed at greater length in Chapters 21 and 22.

Light (especially fluorescent light) accelerates oxidation (Fuller *et al.* 1971), so that opaque packaging material must be used to obtain maximum shelf-life.

BIBLIOGRAPHY

ANON. 1968. Rutgers scientist discovers the compounds that give potato chips their flavor. Am. Potato J. *45*, 36–37.

ANON. 1974. Unpeeled potato boost chip yield. Snack Food *63*, No. 1, 39.

BROWN, H. D. 1960. Problems of the potato chip industry—processing and technology. Advan. Food Res. *10*, 181–232.

CARLIN, G. 1974. Chip lines. Snack Food *63*, No. 2, 45.

CHANG, S. S. 1967. Chemistry and technology of deep fat frying. An introduction. Food Technol. *21*, 33–34.

DECK, R. E., POKORNY, J., and CHANG, S. S. 1973. Isolation and identification of volatile compounds from potato chips. J. Food Sci. *38*, 345–349.

FULLER, G. *et al.* 1971. Evaluation of oleic safflower oil in frying of potato chips. J. Food Sci. *36*, 43–47.

HARRINGTON, W. O., and SHAW, R. 1967. Peeling potatoes for processing. *In* Potato Processing, 2nd Edition, W. F. Talburt, and O. Smith (Editors). Avi Publishing Co., Westport, Conn.

JACOBSON, G. A. 1967. Quality control of commercial deep fat frying. Food Technol. *21*, 147–152.

KING, F. B., LOUGHLIN, R., RIEMENSCHNEIDER, R. W., and ELLIS, N. R. 1936. The relative value of various lards and other fats for the deep fat frying of potato chips. J. Agr. Res. *53*, 369–381.

MAGA, J. A. 1973. Influence of freshness and color on potato chip sensory preferences. J. Food Sci. *38*, 1251–1252.

MELNICK, D., LUCKMAN, F. H., and GOODING, C. M. 1958. Composition and control of potato chip frying oils in continuing commercial use. J. Am. Oil Chem. Soc. *35*, 271–277.

MITCHELL, R. S., and RUTLEDGE, P. J. 1973. Control of colour in potato chips by water treatment before frying. J. Food Technol. *8*, 133–137.

QUAST, D. G., and KAREL, M. 1972. Effects of environmental factors on the oxidation of potato chips. J. Food Sci. *37*, 584–588.

QUAST, D. G., KAREL, M., and RAND, W. M. 1972. Development of a mathematical model for oxidation of potato chips as a function of oxygen pressure, extent of oxidation, and equilibrium relative humidity. J. Food Sci. *37*, 673–678.

REEVE, R. M. 1971. One cell: the thin slice of profit. Am. Potato J. *48*, 47–52.

ROBERTSON, C. J. 1967. The practice of deep fat frying. Food Technol. *21*, 34–36.

SHAW, R., and LUKES, A. C. 1968. Reducing the oil content of potato chips by controlling their temperature after frying. U.S. Dept. Agr., Agr. Res. Serv. *ARS73-58.*

SMITH, O. 1961. Factors affecting and methods of determining potato chip quality. Am. Potato J. *38*, 265–271.

SMITH, O. 1967. Potato chips. *In* Potato Processing, 2nd Edition, W. F. Talburt and O. Smith (Editors). Avi Publishing Co., Westport, Conn.

SMITH, O., DAVIS, C. O., and OLANDER, J. 1965. Microwave processing of potato chips. Part 2. Potato Chipper 25, No. 3, 72, 74, 79–82, 86–88, 90, 92.

TALBURT, W. F., and SMITH, O. 1975. Potato Processing, 3rd Edition. Avi Publishing Co., Westport, Conn.

YEATMAN, J. N., and AULENBACH, B. B. 1973. Relationship of instrumental measurements to visual impressions of potato chip color. *In* Sensory Evaluation of Appearance of Materials. P. N. Martin, and R. S. Hunter (Editors). American Society for Testing and Materials, Philadelphia, Pa.

CHAPTER 11

Meat-based Snacks

INTRODUCTION

Although most snack products are based on cereals, there are a few that are composed primarily of animal-derived raw materials. As discussed briefly in the Preface, the definition of "snacks" can be broad or narrow, and the looseness of the definition will govern the number of meat products which can be included in the present category. Almost every consumer would agree that fried bacon rinds are snacks, since they have texture, appearance, and flavor very similar to puffed or fried cereal pieces and are sold in portion size pouches for eating mostly between meals. Jerky departs somewhat from the typical snack in form and flavor but its sales presentation and mode of consumption fit the familiar pattern. Small individual sticks of fermented cured sausage of the Slim-Jim® type would generally be recognized as snack products because of the way they are offered for sale and consumed, even though their predecessor forms are used in sandwiches or in other ways as meal components. I have also included a discussion of pickled pigs' feet, even though many readers will surely object that the form of packaging and the physical characteristics of the food set it apart from the usual snack. My reason for discussing it in this book is that it is representative of the pickled snacks (sausages, eggs) which are sold in taverns from glass jar multipacks or sometimes in individual pouches.

Lunch meat, sliced ham, and the like are, of course, used by many persons for between meal consumption, but their perishability and more common usage as meal components led me to exclude them from the present discussion.

POPPED PORK RINDS[1]

Popped pork rinds, sometimes called bacon skins or "skeens," have been popular as a between meal snack in the south for many years.

Distribution has been increasing in other parts of the country. Total annual volume of sales for 1972 was estimated at $27 million (Scales 1973). There have been reports that the retail demand is outrunning the maximum available supply of raw material. In simplest terms, these products are pieces of pork skins which have been processed so that they puff to many times their original volume. The flavor is rather bland

---

[1] This section based on information supplied by John Forsyth (1974).

and reflects the fat in which the product was processed. The texture is very crisp and friable, giving the impression of dryness, almost powderiness. These materials do not seem to be as hygroscopic as many other puffed snack products, and they also differ in being composed primarily of protein, although the nutritional quality, the Protein Efficiency Ratio (PER), of the protein is very low.

Perhaps 6 or 7 manufacturers supply processed pellets to retailers, who do the final popping. The raw materials are green belly skins, green fat hog skins, and green ham skins from any type of hog. Belly skins are said to deliver the best finished product. Prior to about the mid-1950's, rinds from cured and smoked bacon were used, but this raw material is no longer available due to changes in the method of processing bacon (i.e., the rind is removed before the side is smoked and cured). At least one processor claims to use a secret process for smoking and curing the green rinds. After removal from the carcass by a Townsend machine, the skin is dipped for about 30 sec in an air-agitated brine solution held at 212°F. The brine might be composed of 26 lb dextrose, 25 lb sucrose, and 150 lb salt in 200 gal. of water. After removal from the hot dip tank, the skin is drained and cooled to room temperature before being diced. The dicer will cut the skin into pieces ½-in. square, if small pieces of the finished product are desired for 10¢ bags, or into pieces about 1-in. square for large retail containers. Considerable unusable scrap is generated in the cutting process.

The next step is "rendering." Typically a 250 gal. steam-heated kettle is used. Violent agitation is absolutely essential. Two 1½ hp Lightnin' mixers with 11 or 12 in.-diameter propeller-type stirrers are adequate. Temperature of the prime steam lard in the kettle must be maintained between 230 and 240°F. Above 240°F, cottony puffs will develop when the pieces are eventually popped. Silicone antifoam agents and antioxidants must be added to the lard.

After about 4 hr dwell time in the kettle, the pellets will start to float to the surface. They are then removed from the hot lard and allowed to cool and drain. The cooled pieces are "scalped," or sized by screens, and undersized material rejected. Pellets are customarily packed 50 lb to a poly bag in a corrugated shipper. The pellets are stable and can be stored at room or refrigerator temperature for about six months. Freezing has an adverse effect on their performance.

The retailer who performs the final processing steps may test the pellets for peroxide value before accepting them. A value of 10 is grounds for rejection, and in the range of 5 or 10 decision must be made as to the desirability of accepting the risk involved. They may also perform a simple popping test in which the contents of a 100-ml beaker filled level full with the pellets, is puffed in 425°F lard. The cooled material is dumped into a 600-ml beaker, and judgment of its quality made as follows: (1) beaker excessively overflows—excellent quality, (2) beaker slightly over-flows—good quality, (3) beaker full but not overflowing—fair, and (4) beaker less than full—poor.

Acceptable pellets are processed as needed in 400 to 425°F fat. The puffs are sprinkled with flake flour salt or a seasoning mixture.

The shelf-life will average 3 to 4 weeks in the south and 6 to 8 weeks in the north. Bags must be coded, of course, and the route man must check the stock at every delivery and remove stales.

Immediately after popping, the total solids content will be over 99%. Other typical analytical values are protein, 57%, fat, 34%, carbohydrate, 5%, and ash, 4%.

Anderson and Smith (1958) addressed themselves to the problem of the nonpuffing pieces which, they say, constitute about 15 to 40% of the total amount obtained from the treated skins. Their invention involves treating the rinds with certain acid solutions prior to the puffing step. Specifically, if the cured pork skin pieces are immersed in acetic acid so-lutions for at least 15 sec, substantially all of them will puff when subse-quently cooked in hot oil. These authors claim that most of the moisture needed for puffing is taken up and held in gelatinous material which forms as a result of the breakdown of collagenous connective tissue found in the dermal layer of skin. A second type of connective tissue (re-ticular) strengthens the rinds and often prevents them from expanding as desired. The reticular tissue is weakened by the acetic acid treatment and does not interfere with puffing. Most other acids investigated by the inventors did not have as favorable an effect as acetic.

PICKLED SNACKS

Sausages, tongue, eggs, fish chunks, and similar food pieces are of-fered preserved in vinegar pickle and packed in large glass jars for dis-pensing in individual portions in taverns, etc. Pickled sausage pieces are also being packed in sealed pouches for sale over the counter. This dis-cussion will deal primarily with pickled pigs' feet, which undergo a more complex manufacturing technique and are sold mostly in retail size jars.

Pickled Pigs' Feet[2]

The fore feet of the pig are usually employed, since the hind feet are not as meaty and are frequently deformed by the shackles from which the carcass is suspended. Sometimes meaty portions of the hind feet are pickled to make what is known as "tid-bits." The largest feet, of 16 oz or more in weight, are generally used in preparing "semi-boneless feet,"

---

[2] This discussion is based on a patent by Forsyth (1952).

after being partially boned and sliced. Feet weighing less than 16 oz may be slit in half to form "split feet." Both fresh and frozen feet have been used as raw materials for these products.

In the conventional method of pickling, the feet are placed in large vats capable of holding 1,000-lb batches. To the vat is added about 110 gal. of a brine or pickle saturated with salt and containing about 175 ppm of sodium nitrate. It is essential to circulate the brine, and this can be accomplished either by daily hand agitation of the vat contents or by using a pump to circulate the liquid through a series of vats. When the older method of hand agitation was used, curing took 11 to 14 days, but with pumped circulation the minimum curing time can be reduced to 7 days. Temperature of the brine is held at about 40°F during this process.

The cured pigs' feet are transferred to large stainless steel tanks containing boiling water where 1,000 to 1,500 lb are cooked at a time. The water should be acidulated by the addition of about 3 gal. of 90 grain vinegar per 1,000 lb of feet. Live steam is injected into the mass to maintain a slow boil for 3 to 3½ hr or until the bones start to break through the skin and the feet are just short of falling to pieces. When the cooking period is terminated, the product is chilled with cold running water until it reaches an internal temperature of 55 to 60°F. The feet are then boned, leaving in place only a few bones which will not interfere with slicing, and washed in cold water. Loose fat and meat particles rise to the surface of the wash water and are skimmed off.

After slicing into 3 or 4 pieces, a weighed amount of the product is packed into jars which are then filled with vinegar of 45 to 55 grain strength containing spices and other condiments.

The jars are sealed under vacuum and held at 55 to 60°F for two weeks or longer. During this time, the meat absorbs about 20 to 25% by weight of the pickling liquid. The product is then distributed. As shown from the preceding, total time of processing including the holding period will be from 3 to 4 weeks.

In a satisfactorily processed material, the lean meat should be uniformly reddish or reddish-pink in color and free from dark streaks resulting from the retention of blood in the capillaries.

The skin and fat should also be uniformly colored, with the skin having a slightly pinkish cast. The color should not change to grayish tan upon exposure to moderate amount of oxygen, as when a jar is opened and held for a time before the contents are consumed. The liquid should be clear and free of turbidity.

The color changes in the meat which result in the desirable reddish hue are due to reactions of the nitrite from the brine with the hemoglobin remaining in the tissues.

The patent of Forsyth (1952), describes a method for reducing the total processing time prior to filling into jars to about 8 hr. Pickling in brine-filled vats is eliminated in his procedure. The raw meat is subjected to a progressive heating and cooking operation in the presence of dilute (2 to 10% by weight) aqueous salt solution containing sodium nitrite. Although fairly good results can be obtained without using any salt in the solution, the pigments are less stable and have a tendency to become lighter on storage. On the other hand, salt concentrations in excess of 10% lead to an excessively reddish skin color and interfere with the laking of entrapped blood, the latter leading to black streaks in the finished product. From 50 to 180 ppm of sodium nitrite are used in the brine. The preferred pH is from 5.5 to 6.0. A typical solution for use in jacketed kettles will contain 4% salt, 125 ppm sodium nitrite, and sufficient vinegar to give a pH of 6.0 to 5.5. If steam injection is to be used for cooking, the concentration of the ingredients should be increased slightly to compensate for the dilution.

Heating or cooking is controlled so that the temperature in the deepest portion of the feet increases gradually from their initial temperature to 130 to 140°F in 1.5 to 2 hr. The gradual temperature rise facilitates laking, i.e., the extraction of blood remaining in the small blood vessels. The temperature increase in the first stage should approximate a straight line curve as nearly as possible.

After completion of the first heating stage, during which the skin color may become somewhat grayish, the temperature is brought to cooking temperature as rapidly as the equipment permits, but not longer than 50 min in any event. A reddish color begins to develop during this transition or "come-up" heating phase, brighter colors resulting from faster heating rates. Finally, heating is continued for about 3.5 hr at a temperature of 200°F to boiling for completion of cooking. The fat is skimmed off, the cooking liquor is removed, and the feet are rapidly chilled with cold running water to an internal temperature of 55 to 60°F.

The cooled pieces are boned and sliced, care being taken to split the heavy cartilage tube extending down the center of each foot so as to expose the lean meat inside to the solution employed in the next step of the process. Some of the inner lean meat portions will have a cooked rather than cured appearance at this time. To bring these parts to the proper color, the cooked, cooled, and sliced feet are placed in stainless steel tanks or vats which are then filled with a solution containing about 0.75% acetic acid and about 125 ppm of sodium nitrite. The pieces are allowed to remain in the solution (55 to 100°F) for about 20 to 30 min, after which the liquid is drained off, and the meat is packed in jars. Large jars should be filled with 45 grain vinegar while the smaller jars should be filled with 55 grain vinegar. In either case, it is desirable to use a salt concentration of about 2.5 to 3.0%.

The jars are sealed under vacuum and held at about 55 to 60°F for at least 14 days. The curing continues during the storage period. At the beginning, the liquid is somewhat cloudy, the skin is grayish to light tan, and the lean meat has a dull red-gray cast, but during storage the lean gradually assumes the desirable pink color, the skin brightens, and the liquid clarifies.

The pickling of sausage and the like does not involve a curing stage (although the sausage itself may be cured before pickling) but generally requires only the filling of the product pieces into the final container and covering them with a hot vinegar and salt solution containing appropriate condiments. A few days storage suffice for allowing the pickle to penetrate throughout the product.

Jerky

Meat jerky was prepared originally by the American Indian who salted strips of muscle tissue from deer, buffalo, and other game animals, and cured the strips in the sun or over smoky fires for long periods of time.

Commercial beef jerky has been made by marinating strips of beef, drying the marinated meat, and finally cutting the dried strips to desired dimensions. The process takes a great deal of time and involves substantial loss of product as scraps from the cutting operation. Beef jerky can also be made by subjecting a mixture of ground and chunk beef to saline treatment for about 12 hr and then to freezing temperatures for 1 to 3 weeks. The meat preparation is then sliced into strips and dried.

Worden (1974) patented an improved method consisting of (1) cutting frozen lean meat into chunks, (2) partially thawing the frozen chunks to release natural juices containing soluble proteins, and (3) mixing the partially thawed chunks with seasonings and curing agents for a time sufficient to form a cohesive film of soluble proteins over the chunks. Subsequent processing steps follow along conventional lines. The chunks and seasoning mixture is formed into blocks by pressing it into molds, and these blocks are held at 45 to 55°F for 2 to 5 hr to cure the meat. In preparation for slicing, the "cured" blocks are frozen and tempered to about 18 to 20°F. Slices of about $\frac{1}{16}$ to $\frac{3}{32}$ in. in thickness are heated so as to congeal the surface film of soluble protein, to develop the red color of cured meat, and to partially dry the materials. The heat treatment is performed by placing the frozen strips on a supporting grid and subjecting them to temperatures starting at about 160°F and gradually reduced to about 90°F. Time is chosen to be sufficient to coagulate the surface film and reduce the moisture content to a water:protein ratio of at least 0.75 to 1.

BIBLIOGRAPHY

ANDERSON, M. G., and SMITH, C. F. 1958. Method of puffing bacon rinds. U.S. Pat. 2,855,309. Oct. 7.

FORSYTH, J. S. 1952. Methods of Processing meat products. U.S. Pat. 2,613,151. Oct. 7.

FORSYTH, J. S. 1974. Personal communication. Western Springs, Ill.

HALPERN, P. P. 1968. Method of preparing pork rink food products. U.S. Pat. 3,401,045. Sept. 10.

SCALES, H. 1973. Meat snacks gain respect. Snack Food 62, No. 1, 64–66, 68.

VAN MIDDLESWORTH, R. Q., BUCK, M. E., and SEEMANN, B. R. 1973. Method for producing an expanded meat food product and product produced thereby. U.S. Pat. 3,745,021. July 10.

WORDEN, G. E. 1974. Method of making meat jerky. Can. Pat. 950,262. July 2.

CHAPTER 12

Snacks Based on Popcorn

INTRODUCTION

It would be redundant to emphasize the commercial importance of popcorn and the snacks derived from it, since everyday observation confirms the fact. What may not be so obvious is the firm bases on which the popularity of these snacks rests. The crisp texture, fluffy white appearance, and convenient piece size of popped corn provide an almost unique combination of properties that can be utilized to advantage in many different kinds of snack products. When the ease of processing and the relative cheapness of the raw material are also considered, the widespread use and acceptance of popcorn snacks can be readily understood.

Popcorn does have a number of disadvantages, among them its fragility, its nonuniformity, and its response to adverse environmental influences.

FACTORS AFFECTING THE QUALITY OF POPCORN

The Bases of Quality

The consumer is primarily interested in price, flavor, appearance, and texture. Flavor is strongly influenced by the butter or oil used as a topping and by the salt. In discussions of popcorn quality in the literature, flavor is rarely mentioned. Yet, freshly popped corn does have a distinctive and appealing flavor. This soon dissipates, and most commercial corn provides a very bland base for the oil and salt. Texture and appearance are the main quality factors arising from the corn itself.

Texture is strongly related to the intrinsic specific volume of the popped kernel, which may be quite different from the apparent specific volume, the latter being much affected by the shape of the kernel. Texture also reflects in part the presence of hard particles, the remnant of the hulls.

The hull is the outer covering or pericarp of the corn kernel. It varies considerably in thickness between the different varieties, but no variety is completely hull-less even though this description has been applied to some of the better ones. Larger kernels generally have thicker pericarps than do smaller ones. The hull is torn and fragmented as the kernel violently expands, and some of it is dislodged from the corn but most of it remains. Light-colored pericarp is much less noticeable and, if it is also very thin, the hull-less condition may appear to be achieved on casual inspection. In addition to the undesirable appearance contributed by residual hulls, the texture of popped corn is adversely affected and the hulls "get between the teeth" (a common complaint).

Shape of the kernel is affected by variety, moisture content, and popping conditions. Popped corn with a round or ball shape is called the mushroom type while kernels yielding a highly irregular pronged shape is known as butterfly corn. Kernels having the mushroom configuration are preferred by manufacturers of coated or flavored corn because they break up less during the mixing operation and accept a more even coating of syrup. The same resistance to rough handling makes it more acceptable to vending machine operators and by central poppers. Butterfly corn has a lower apparent bulk density and retains salt well. In most cases, its texture is also considered superior to mushroom-type corn. For these reasons, it is selected by the majority of on-site poppers, such as theaters.

White popcorn is rarely used for commercial production of snacks. This category includes varieties ranging in size from small kernels up to the size normally associated with the large yellow varieties. Small kernel size generally leads to the highest volume popped corn. Fragility of the kernels leads to excessive crumbling when they are processed on a large scale.

Popcorn is unique among grains in that a high degree of expansion can be achieved when it is heated at atmospheric pressure. Other grains must be superheated in pressurized vessels and then suddenly passed to a region of lower temperature if much expansion is to be obtained. Popcorn evidently behaves as it does because of the physical structure of the entire kernel and the microscopic structure of the endosperm.

The endosperms of different types of grains show different degrees of starch granule gelatinization when the kernels are expanded by popping. In barley and wheat, which do not expand greatly, some starch granules undergo complete gelatinization without apparent expansion while other gelatinized granules expand and fuse. Localized cell-wall rupturing occurs when the kernels split open, and a few intracellular voids or enlarged bubbles can be seen in the gelatinized starch granules as a result of the explosion. Ungelatinized and partly gelatinized granules predominate immediately below the aleurone layer and near the scutellum. Localized cell-wall rupturing also occurs in the expanded endosperms of popped grain sorghum, popcorn, and dent corn, but the spongy expanded endosperms consist of intact cells within which the gelatinized starch granules form a characteristic structure of "soap bubble" appearance, each bubble representing a starch granule. The cell walls are not destroyed and remain clearly identifiable except where wall-rupturing contributes to both expansion and formation of voids. The starch granules are not exploded, but are gelatinized and dried into a three-dimensional network or reticulum.

Expansion is much less pronounced in dent corn than in popcorn and sorghum, in which more cell rupturing occurs to form voids. The soap-bubble type of structure is less extensively developed in all poorly popped kernels. Some unaltered and partly gelatinized starch granules are also present immediately beneath the aleurone and near the scutellum even in fully popped kernels.

Differences in the distribution of horny and floury endosperm, and differences in their protein content, influence the capacity to expand (Reeve and Walker 1969). A major difference between popcorn and other types of corn is that popcorn has a denser and harder endosperm throughout. Apparently, this structure causes steam pressure to build up under certain conditions of heating, until a sudden rupture of the pericarp allows the endosperm to expand into a porous starch network which then "sets up" as the water evaporates. The temperature just before the explosion must be sufficient to create a high enough water vapor pressure without burning the pericarp and the rate of temperature increase must be fast enough to build up the required pressure before the water evaporates. Damaged kernels may pop poorly or not at all because the water tends to boil away without the development of high pressure.

Moisture content of the kernel has a pronounced effect on the popping behavior. Kernels that are too dry pop feebly, with a somewhat muffled sound, the kernel often partly splits open, and the unpopped area appears rather dark or scorched. Corn that is too moist pops with a rather loud explosion, but the kernels are small, rough, jagged, and tough (Lyerly 1940).

Measurement of Expansion

The apparent specific volume developed on popping is an important quality factor because it determines the weight (cost) of raw corn which must be processed to fill a given size container.

Testing Methods.[1]—The first official testing method was perfected by the Popcorn Processors' Association shortly after the war in cooperation with Cretor's Corporation in Chicago. This piece of equipment was called the Official Volume Tester and it measured popping expansion in multiples of the original size of the grain. A cup measure holding approximately 6 oz of grain was filled and then struck off with a straight edge rule. Approximately 35% oil was added and popping was done

---

[1] This discussion is based on Brown (1973).

under control of a pyrometer with reading dial to insure consistent results on temperature.

By the early 1950's, it was felt that a more perfect system of running popping tests was desirable. The weight volume tester was perfected in which 150 gm of popcorn was weighed and popped in the same equipment used for the Official Volume Tester. The equipment had been modified for higher expansion popcorns by the addition of a taller shield to keep the popped kernels inside the kettle during the popping operation. Also, the calibrated tube was made of clear plastic and calibrated in cubic inches of popcorn per pound. This is a strange mixture of both the metric and English systems. It had definite advantages in that the grains of popcorn could be measured within one kernel of popcorn quite accurately to 150 gm. Furthermore, popcorn is purchased by weight and usually sold by volume so the two standards were realistic when compared with the industry and its method of doing business.

A further improvement in volume testing equipment is felt necessary at this time. The Popcorn Institute, the successor to the Popcorn Processors Association is again working to develop a larger machine with not only temperature control but voltage control as well.

The very rapid change from ear harvesting of popcorn to shelled harvesting of popcorn has necessitated an improvement in grades and standards by the industry in judging what is good quality combined popcorn. In determining this, they have employed the bushel test weight method used in field corn grading. Moisture is also a factor. Foreign material and damaged kernels on good quality combine popcorn can usually be kept below 1½%. The Federal government has not yet established grades and standards but a pattern of acceptable standards is emerging among those processors that are heavily involved in combining and interested in top quality.

Brown (1966) described the major differences between the Official Volume Test (OVT) and the Weight Volume Test (WVT) as follows:

OVT is a comparison of unpopped volume to popped volume. WVT is a comparison of pounds of raw corn to cubic inches of popped corn.

WVT Testing
(1) Two ounces of oil are accurately measured.
(2) Oil is heated to 440°F and corn is added.
(3) 150 gm of raw popcorn is weighed within one kernel accuracy.
(4) The plastic WVT tube measures popped volume in cubic inches per pound.

OVT Testing
(1) Two ounces of oil are accurately measured.
(2) Oil is heated to 470°F and corn is added.

(3) A cup designed to hold 6 oz of raw popcorn is filled to overflowing. A straight edge is zig-zagged in a vertical position across the cup to remove the excess corn.

(4) The OVT tube measures popped volume in multiples of the original 6-oz cup.

Advantages of WVT

(1) Complete accuracy of corn measure.

(2) Pounds convert to boxes of popped corn since results are in cubic inches per pound.

Disadvantages of OVT

(1) Human error prevents accurate measurement by striking off a straight edge over the corn cup.

The OVT equipment was originally designed during during World War II when expansion of 26 or 27 to 1 was quite good. Now with 37 or 40 to 1, the 6 oz-kettle is overloaded. The OVT cup holds 195 to 205 gm, while the 150 gm of the WVT compensates for the higher expansion.

It can be concluded that the Weight Volume Test is the more logical of the two, since raw popcorn is bought by weight and sold popped by volume.

Most processors agree that the treatment of popcorn immediately subsequent to harvesting and the storage conditions prior to popping have a considerable effect on the quality of the finished product. Damage to the kernels as a result of harvesting varies according to the method used, whether combined or gathered in the ear, and affects popping response. The moisture level reached during storage and the equilibration of moisture between the kernels can have a pronounced influence on expansion and other characteristics of the finished product. These aspects of popcorn quality will be discussed in the following paragraphs.

Harvesting Practices

Popcorn can be shelled at the time of harvest or "cob-cured." Most of it is combined and thus shelled when harvested, according to current practice. Cob-cured corn is not shelled at the time of harvesting but is placed into bins with the kernels still on the ears. According to one prominent advocate of cob-curing, specially built air-drying storage bins are used to slowly cure the corn with warm air from gas or oil heaters. After 2 or 3 weeks of this slow curing, the popcorn is shelled out at about 14% moisture. The shelled grain is given another curing and equilibration period of at least 30 days before final processing. Drying conditions that are too extreme, as with the use of hot air for 24 to 48 hr, may cause cracks to develop in the kernels as the result of uneven or excessively rapid contraction.

According to Brown (1973), popcorn harvested at 17 or 18% moisture and exposed to temperatures of about 90°F for rapid drying (i.e., in a matter of hours rather than days) can have its popping potential seriously reduced. Best results are obtained when the drying is gradual, the ideal being natural air curing on the ear in a crib. Even on combined popcorn, relatively good results can be obtained from corn harvested at 16 to 17% moisture if the drying is slow and complete equilibration of the moisture content is attained.

Popping expansion seems to be improved by aging, at least up to a point. With shelled corn, improvement in expansion may occur up to 6 or 9 months after harvest. The older the corn, the more moisture it can carry and still give maximum expansion, up to about 14 or 14.5%.

Storage of Popcorn

The principal points to observe in storage of shelled popcorn in bags were summarized by Brown (1968).

I. Moisture Control
  A. Processed popcorn is protected in the package with a moisture barrier. Patch any broken or torn bags with masking tape when they are received. An undamaged bag can usually be kept at proper popping moisture for at least 90 days.
  B. Store popcorn in an unheated or cool room. Rotate stocks of popcorn by using up the old corn first.
  C. Dump the open bag of popcorn in a can or corn bin with a tight lid. Place in the popcorn machine corn storage bin only that amount of popcorn that can be used during that day.
  D. Storage space required
     1. Bag storage—1 cu ft = 50 lb of popcorn.
     2. Bulk storage bin—1 cu ft = approximately 60 lb.
  E. Under normal storage conditions, moisture content will slowly decline. As a rule of thumb, the moisture content can be considered to drop by about 1% in 4 months. This decrease is probably acceptable in most cases. Longer storage periods will generally require re-humidification of the kernels for best popping performance.

     Corn used in wet popping should have a moisture content of between 13 and 13.5% for maximum expansion and a minimum amount of unpopped kernels. Immediately after popping, the moisture will be about 1.5 to 3%. For best results, dry popping calls for corn with more moisture, generally 14 to 15%. Popping reduced this to about 1 to 1.5%.

     There is a definite risk of fungal attack and development of musty odors for corn stored at moisture levels of above 14.1% for a week or more. Refrigeration can greatly retard mold growth, but is impractical in most cases. When a moisture content in excess of this danger level is required for optimum popping performance, the kernels should be humidified in the 24 hr prior to processing.

II. Rodent Control
  A. Keep storage area clean and orderly. Seal any cracks or crevices allowing entry from basement or outside.
  B. Store popcorn on platform or pallet off the floor.
  C. Trap the area for mice and set out D-Con or other warfarin type poisons. If this does not give you complete rodent protection, call in a professional control firm.
III. Insect Control
  A. What to look for
    1. Small moths that fly in the evening. Small white worms on the bags and in the popcorn.
    2. Clusters of popcorn kernels held together by webbing.
    3. Small weevils with pointed snouts. Insects are best identified by placing a few in a tight medicine bottle and sending them to your sanitation consultant for identification. Kill them with cleaning fluid.
  B. Fumigation
  Popcorn can usually be kept free of insect infestation in storage and processing through regular fogging and spraying with approved insecticides. Many suppliers will fumigate each shipment before it is bagged and shipped to their customer. This treatment will kill existing eggs, larvae, pupae, and adult forms. Grain insects are attracted to the corn, however, and may contaminate it at any time during transporting, storage, or use. No type of bag is completely proof against entry by every kind of insect.
  C. Precautions against insect damage
    1. Rotate stocks. In temperate climates or in temperature controlled storage, it takes 30 to 60 days for a noticeable insect population to develop from a few insects that might get into a fresh shipment of popcorn. You should know how to read the code dates printed on each bag of corn.
    2. Fog your storage. When the warehouse temperature is above 65°F (18°C), fog warehouse weekly with an approved pyrethrum chemical fog. The foggers cost between $60 and $100 and can be bought with timers so they will turn themselves off. Fog at the end of the week when closing down for the weekend. The last person to leave activates the fogging device in your popcorn storage area so by the time you open up the next working day, the fog will have disappeared. If you use this method, you will eliminate insects before they become a problem.
    3. Refrigeration. If the popcorn storage area is kept at 50°F (10°C), it will not be necessary to fog as insects are not active at this temperature.
    4. If you find insects. In case insects are discovered, contact your sanitation consultant for recommendations.
      a. Paper popcorn bags have a polyethylene sheet as one of the layers in their construction. If there is a heavy insect population in the popcorn, it is not easy to fumigate and get a kill because the gas generally does not penetrate the polyethylene liner. The U.S. distributor of Phostoxin claims that their material will give a good kill in bagged popcorn. Your sanitation

firm can secure Phostoxin and do the fumigation. Methyl bromide and ethylene dibromide have not been successful on bagged popcorn covered by poly tarps.

b. The Food and Drug Administration does not allow any tolerance for insect contamination in a food product. Therefore, the first goal is to kill the infestation so it does not contaminate other merchandise in your warehouse. Then you will have to talk with your supplier to determine whether it is better to destroy the insect contaminated product or send it back for reprocessing, or some other action.

c. Thoroughly clean the floors and walls in the old location. Destroy or fumigate all food and candy in the room. Fog with pyrethrum fogging equipment listed above. If fogging is not available, spray with household type aerosol bug spray daily for one week and weekly whenever the temperature is over 65°F, which will kill all types of insects as they become adult.

d. Bring in a fresh stock of popcorn and store it in a clean, new location, preferably 100 ft away from the old location so that any residual insects you have missed in the old location will not contaminate the fresh stock. After 60 days, if no other insects are noticed, you are probably safe in returning the popcorn to the old storage area.

POPPING PROCEDURES

Most home-prepared popcorn is popped in oil. Commercial popping can be accomplished either by heating with currents of very hot air or by using oil as a heat transfer medium. The former method is suitable for both batch and continuous procedures, while the latter is almost always a batch operation.

Wet popping can be automated by assembling a group of batch kettles having automated feed mechanisms which dump corn, oil, and salt as required. These units should have automatic cut-offs for ingredient feeds as well as on the heat supply. The dry popping process is easily adapted to continuous equipment and 3 or 4 companies now offer devices of this type.

In a continuous popper such as that made by Cretors, raw corn is automatically fed from a supply bin into a 24-in. diam, 78-in. long perforated metal tube containing a large auger screw. A large volume of hot (e.g., 420°F) air suspends the kernels and uniformly heats them until they pop.

It is necessary to maintain close control of conditions in continuous poppers in order to minimize scorching and secure maximum popping performance (expansion, kernel shape, percentage of unpopped kernels). Automatic controls on the Cretors equipment include (1) a pyrometer to control the temperature within a 10° range, i.e., ±5°F, (2) variable speed drive and tachometer for the fan, (3) variable speed drive for the auger screw in the perforated tube to provide different popping rates, and (4) feed rate of the raw popcorn. Recommended conditions include (1) a popping temperature between 410 and 430°F, depending upon the type of corn being used, (2) dwell time of corn in the machine— about 90 sec, and (3) feed rate of 400 to 450 lb per hour.

Brown (1966) summarized the advantages and disadvantages of wet and dry popping.

I. Wet Popping—Popping corn in oil using kettles from 6-oz to 5-lb capacity.
   A. Advantages
      1. Animation of popping.
      2. Tantalizing aroma.
      3. French-fried flavor.
      4. Adapted to popping at point-of-sale
         a. Inexpensive equipment.
         b. Compact operation.
         c. Minimum fire hazard.
         d. Simplicity of operation.
         e. Excellent efficiency on popping 25 to 120 lb per hour.
   B. Disadvantages
      1. Irregular oil coverage.
      2. Some of oil color is destroyed by the heat of popping.
      3. Waste of oil from smoke and screenings.
      4. Wet popping requires about 25% more oil than dry popping.
      5. Cleaning problem because oil gums and carbonizes on the kettles.
      6. Recommended corn:oil ratio is 3 to 1.
II. Dry Popping—Popping without oil, screening, and then adding oil and salt in a batch tumbler or continuous coating drum.
   A. Advantages
      1. Dry popping requires about 25% less oil than wet popping.
      2. Recommended corn:oil ratio is minimum 5 to 1 and maximum 4 to 1.
      3. Well adapted to popping over 120 lb of popcorn per hour.
      4. More even distribution of oil.
      5. Good oil color retention since it is not exposed to popping heat.
      6. Low labor cost. One operator for 3 to 5 (10-lb) dry poppers.
      7. Needed to manufacture other products such as, cheese popcorn, caramel corn, sugar corn, popcorn balls and molds.
   C. Disadvantages
      1. Fire hazard.
      2. Operator needs careful training.
      3. Between popping cycles a few kernels tend to catch in the popper screen and scorch.
      4. Expensive equipment.
      5. A well-managed processing technique is necessary for an evenly coated product, and good expansion.

Loss During Popping

Commercial poppers find that a considerable amount of the grain apparently disappears during processing. There are three principal causes.

About 11 to 13% moisture is lost to the atmosphere as the kernel expands. There is no way this loss can be significantly reduced without adversely affecting popping expansion, percentage of unpopped kernels, or finished product texture. In addition, some kernels are defective or will receive insufficient heat to cause expansion. The amount of unpopped kernels considered the minimum achievable in commercial processing is 2%. Finally, loss of hulls in the screening operation and of small pieces broken off of the popped kernels will amount to 1% or more in total. The sum of these losses indicate a minimum shrinkage in weight through the popping operations of about 14%, and probably 15 or 16% is more representative of reasonably efficient commercial plants.

Freshly popped popcorn has extraordinary moisture absorbing properties. It will take up water vapor from air having a relative humidity of 20% or more. As the moisture content increases, crispness is lost and the kernels shrink. At about 2.5% moisture, the corn starts to become somewhat elastic and chewy. To preserve the best textural quality, the corn must either be kept hot or packed immediately after popping in containers having very low water vapor transmission rates. Because most centrally processed popcorn is packed in pouches not containing a foil layer, it becomes tough or soggy after a few days, and generally reaches the consumer in a condition that would be considered unacceptable by many people. Only the most uncritical consumers will continue to buy such foods. Sales could undoubtedly be increased substantially if precautions were taken to preserve optimum texture until the corn is eaten. These precautions include packaging the product within a few minutes of popping in well-sealed pouches containing a layer of aluminum foil (or in other moisture impervious containers such as glass jars). Cost of the pouch material would increase somewhat over existing transparent films, of course.

CARAMEL CORN AND OTHER FORMULATED POPCORN SNACKS

Popcorn has been used as the basis for numerous combination snacks because it furnishes low bulk density and pleasing textural effects at relatively low cost. The basic popcorn snack is simply popcorn coated with melted butter and salt. In commercial versions, butter is replaced by flavored and colored vegetable oil. A typical ratio is 73% popped corn, 22% oil, and 5% salt. As an example of the more elaborate confections, Dame et al. (1972) describe a snack food product consisting of popped corn in a dough made of tapioca flour, corn flour, and potato starch. Sucrose and alkali metal bicarbonate are added to improve stability and enhance flavor. The pieces are deep fat fried.

Caramel corn is an old and popular confection that has given rise to a number of variants now in national distribution. Many include nuts, usually peanuts. The most popular nationally sold brand is Cracker Jack, although locally made caramel corn probably outsells all national brands by a considerable margin.

The coatings employed are made principally of sucrose and are, in effect, hard candy. Oils, butter, etc., can be used to provide added tenderness and flavor while reducing stickiness, but they also lead to dullness and opacity of the coating. The patent of Rebane (1971) describes the composition of two such coatings: (1) 52 lb of granulated sugar, 16.6 lb corn syrup, and 2.5 lb salt, and (2) 45.4 lb sugar, 25.7 lb corn syrup, and 8.6 lb coconut oil. In both cases the syrups are cooked to 290°F.

Hard candy syrup adheres well to popcorn, but the layer of fat on roasted nuts leads to an uneven coating that causes many of the nuts to separate in the package.

According to a patent of Doan and Lepley (1965), adherence of the candy to nuts can be greatly improved by pre-coating the nuts with a small amount (e.g., 0.2%) of a mutual binder. Suitable binders are certain plant-derived gums such as acacia and tragacanth, and polyhydroxy esters of fatty acids (mono- and diglycerides, lecithin, etc.). The syrup mix described by the inventors has the following composition.

	%
Sucrose	67.93
Corn syrup (43°Be)	10.73
Honey	0.25
Sodium acetate	0.09
Glycine	0.04
Water	20.96

The syrup is heated to 315°F, which should reduce the moisture content to less than 1%.

Fruit-flavored coatings for popcorn can be made from hard candy recipes. The increased hygroscopicity of these coatings, resulting from their content of small amounts of citric acid and invert sugar, cause unacceptable lumping and stickiness under moderately severe storage conditions. Rebane (1971) patented a method of distributing a layer of powdered edible organic acid over an oil-based flavor solution which is applied to popcorn by spraying. The candy glaze is then applied on top of the acid layer. These products have the tart flavor expected of fruit flavored confections without undergoing substantial inversion of the sugar.

Large quantities of cheese-coated popcorn are centrally prepared and sold in supermarkets, bars, amusement stands, etc. Many flavor variations have been developed but they never have achieved the widespread popularity of the cheddar cheese type. Composition of these coatings is described in the ingredient chapter on Cheese and Other Dairy Products. They generally consist of dried cheese, whey solids, salt, artificial flavor, and artificial color.

Candy coatings at 2 to 5% moisture, as commonly applied, should contain some invert sugar and must be put on at high temperatures in order to get good coverage. In humid atmospheres, these coatings will absorb moisture and become sticky. It has also been proposed to coat breakfast cereals with sugar coatings containing 35% water. If these coatings contain 1 to 8% of nonsucrose sugars, such as dextrose or invert sugar, the development of crystals upon subsequent drying is still avoided and a clear glossy appearance can be retained. It is said (Vollink 1959) these coatings do not become sticky under humid conditions. Of course, the high moisture content would cause some shrinkage of popcorn when initially applied and the volume will not be recovered upon drying. The quoted patent suggests the use of thickeners such as carboxymethylcellulose in the syrup to reduce its penetration into the cereal body.

An automated caramel corn production system will include, as a minimum, a continuous dry popper, a sifter, a caramel mixing kettle with automatic controls and metering pump, a film-type caramel concentrator, a caramel corn coater, a cooling separator tunnel, and a continuous belt-type cooling conveyor to deliver the finished product to the packaging equipment.

Typically the caramel coating will be made from liquid sugar, corn syrup, color, and flavor continuously mixed and heated to between 150 and 180°F in a steam-jacketed kettle. This dilute syrup will be metered continuously at a controlled rate into a film concentrator which, in a matter of seconds, delivers candy at a suitably low moisture content.

Salt and soda can be metered into the molten candy as it is delivered to the continuous coating reel. The outside jacket and mixer shaft of the coating device will be steam heated to maintain the coating in a state sufficiently plastic to facilitate uniform and complete coating of the popped corn. Lecithin or oil can be sprayed onto the product about midway through the coater to reduce the formation of clusters, if this is desired.

The hot caramel corn is passed through a rotary perforated stainless steel cooling reel where room temperature air cools the kernels as they tumble down the cylinder. The partially cooled and substantially nonsticky product is finally dropped on to a continuous belt conveyor where it undergoes some additional temperature decrease as it is carried to the packaging equipment hoppers (de Muesy and Stinson 1971).

BIBLIOGRAPHY

BROWN, G. K. 1966. Signal: popping expansion. Smoke Signals 66-3, Wyandot Co., Marion, Ohio.

BROWN, G. K. 1968. Popcorn storage problems. Smoke Signals *7-68*, Wyandot Co., Marion, Ohio.

BROWN, G. K. 1973. Personal communication. March 13. Marion, Ohio.

DAME, D., JR., STINSON, W. S., JR., and CAPOSSELA, A. C. 1972. Snack food product and process. U.S. Pat. 3,647,474. Mar. 7.

DE MUESY, E., and STINSON, W. S. 1971. One man makes 550 lb caramel snacks per hour. Food Process. *32*, No. 1, 21–22.

DOAN, C. A., and LEPLEY, W. D. 1965. Method for producing a candy coated mixture of nut meats and puffed cereal particles. U.S. Pat 3,184,316. May 18.

DODGE, M. N. 1939. Confection. U.S. Pat. 2,181,109. Nov. 21.

HERZKA, A. 1947. Notes on maize. J. Soc. Chem. Ind. *66*, 396–397.

KRAUCER, P. 1972. Popcorn product. U.S. Pat. 3,704,133. Nov. 28.

LYALL, A. A., and LUNDY, C. N. 1974. Cereal coating composition and process. U.S. Pat. 3,792,183. Feb. 12.

LYERLY, P. J. 1940. Some factors affecting the quality of popcorn. Ph.D. Thesis. Iowa State College, Ames.

MCALISTER, R. E. 1972. Microwave puffing of cereal grain and products made therefrom. U.S. Pat. 3,682,651. Aug. 8.

MIDDLETON, J. C. 1972. Personal communication. July 13. Kansas City, Mo.

RASMUSSON, B. E. 1971. Method of preparing oil-milk-sugar clad cereal particles and the resulting product. U.S. Pat. 3,582,336. June 1.

REBANE, A. 1971. Puffed food product and method of producing. U.S. Pat. 3,617,309. Nov. 2.

REEVE, R. M., and WALKER, H. G., JR. 1969. The microscopic structure of popped cereals. Cereal Chem. *46*, 227–241.

SCALES, H. 1972. Popcorn—the irrepressible snack. Snack Food *61*, No. 9, 31–35.

VOLLINK, W. L. 1959. Process of producing a candy coated cereal. U.S. Pat. 2,868,647. Jan. 13.

CHAPTER 13

Puffed Snacks

INTRODUCTION

Following the lead provided by popcorn, there were efforts to puff other whole kernels. Consistent expansion to low densities could not be achieved with other cereal grains when the simple processing methods used for popcorn were applied. Good results were obtained, however, when moisturized grain was superheated (i.e., above the boiling point of water at atmospheric pressure) and then suddenly transferred to a region of substantially lower pressure. These methods became commercially successful, and breakfast cereals "shot from guns" resulted.

It was soon recognized that variations in shape, size, and composition were severely limited as long as whole kernels had to be used as the raw material. The desire for greater freedom in product design led to the development of methods for puffing pellets made from corn meal.

The breakfast cereal industry was the first to make use of methods for puffing corn meal products. The puffing equipment was the same as had been used for puffing kernels of wheat, rice, etc. Pellets were prepared by adding water to a farinaceous material containing a substantial proportion of dried corn particles (i.e., corn meal, corn flour, or corn cones), drying the pellets to a moisture content suitable for puffing, expanding the dried pellets in a puffing gun, and further drying the puffed pellets.

Similar techniques have been used in the snack industry. Half-products are pieces cut from doughs which have been cooked and extruded, but not puffed, in continuous mixers. These doughs are sometimes shaped by a second extruder. The half-products can be expanded by baking or frying without using complex high-pressure extrusion apparatus.

The logical next step was puffing extruded material directly. Enough energy and heat must be applied to the cereal-based feed material to thoroughly gelatinize or cook the ingredients. All extruders perform this function. Extrusion puffers are designed to compress the dough and heat it to more than 212°F so that the water in the extrudate suddenly bursts into steam as the extrudate emerges from the die. The dough then expands, becomes cellular and porous, and finally sets up due to cooling and drying. The shape and size of the product are determined by the die design and the speed of the knife which cuts the strands. Temperature, extrusion rate, pressure, moisture content, and composition of the dough also affect the appearance of the piece.

The preferred feed stock is number two grade dry-milled corn meal with hull and germ removed. The fat content will be less than 1%. Particle size is also important. Combinations of rice and corn may be used. Rice provides a crisper and blander product. Milled second head rice was found to be suitable. Certain pregelatinized starches can also be extruded into puffed forms.

Moisture content of the meal is critical in determining extrusion temperature, pressure, and product texture. As feed moisture is increased, extrusion temperature drops and less expansion occurs in the extrudate. The pores in the product become larger and have thicker walls. After it is baked, the product is crisper or crunchier in texture. High moisture results in a dense and hard product due to incomplete gelatinization of the starch. Such products are suitable for frying, under some conditions.

As feed moisture is reduced, extrusion temperature rises, the extrudate expands more, and the pores get smaller with thinner walls. After baking (drying) the collet is softer and has less crunch. The collets will start to darken and scorch as the moisture approaches a low level. Production rate is also affected by the moisture content of the meal. The throughput can be increased by decreasing the moisture content, but this generally has an adverse effect on the product quality.

Moisture should be evenly distributed throughout the meal. Gross nonuniformity can cause stratified areas in the collets, scorched particles, and other product defects. Ideally, any moisture added as water or aqueous solutions should be allowed to equilibrate throughout the bulk material before it is extruded, even though acceptable results can sometimes be obtained by dripping moisture into the extrusion chamber, particularly if only very small amounts are involved.

Moisture contents of 13 to 14% are generally recommended. The product collected from the extruder will normally reach an overall moisture content of about 8% and this is further reduced to 4% or less in hot air ovens or deep fat fryers.

Further developments in the last 10 yr have included the introduction of modified starches which permit the formation of half-products with relatively simple equipment. This has further blurred the dividing line between extrusion puffed snacks and baked snacks, since the mixing and forming methods closely resemble those used in cookie and cracker baking. In this book, baked snacks will be defined as those products leavened (i.e., expanded) primarily by carbon dioxide from yeast or sodium bicarbonate and heat processed at ambient pressures. They are discussed in Chapter 15 while puffed snacks formed by high-pressure extrusion and by frying or baking half products are discussed in this chapter.

FORMULATIONS AND PROCEDURES

Puffable Materials

The following flours and meals have been used for puffed snack products.

Rice Flour.—Expands readily into a low-density, white, and bland tasting product of crisp texture.

Corn Meal or Flour.—Expands very well into crisp pieces with the typical corn flavor.

Oat Flour.—Due to its relatively high fat content, this cereal requires high moisture and high temperature for adequate expansion. The puffs have a fairly soft texture.

Wheat Flour.—Relatively high moisture and high temperature are needed to obtain satisfactory puffing performance. Does not expand as well as corn, rice, etc.

Potato Flour.—Requires high moisture and high temperature, but extrudes well under these conditions, and forms collets of excellent texture.

Tapioca Flour.—Gives bland tasting puffs when treated at high temperature and moderate moisture content.

Soy Flour.—Can be used as auxiliary ingredient with generally adverse effects on color, flavor, and texture.

Unmodified Cereal Starches.—Can be puffed at medium to high temperatures using either steam or water as moisturizers and longer dwell time in the pressurized chamber. They can serve as bland tasting bases for formulated products including nutritional snacks.

Granular Starch Systems.—Uncooked granular starches are suitable bases for half-products. A simple formula would include fat and monoglycerides, color, flavor, and 16% water. The ingredients would be preblended and processed through a continuous cooker extruder at 250 to 350°F. Conditions must be sufficiently rigorous to rupture the starch granules. After cooking, the dough is formed into the desired shape and size by any convenient method. As with most half-products, the dough pieces can either be baked immediately or dried to the specific moisture and deep fat fried. Depending upon the conditions of cooking and expansion, and the amount and type of ingredients, structures will vary from a low-density large cell foam to a brittle dense network.

Pregelled Starches.—Certain pregelatinized starches have been offered specifically as bases for half-products. Many of these can be satisfactorily formed by conventional laboratory, pilot plant, and commercial sheeting and cutting equipment after mixing in conventional vertical blenders. The starch ingredient is first mixed with the other raw materials such as flour, color, flavor, and spices. About 20% (15 to 25%)

water is then added and the mixture is fed into an extruder, from which the desired shapes are obtained. The formed piece can be baked in microwave ovens, or in standard ovens at 375 to 425°F. When dried to 10 to 15% moisture, the dough pieces can be deep fat fried. This method is very suitable for addition of flavors in the dough because of the blandness of the starch base and the relatively low processing temperature.

Puffing Behavior of Various Starches

The use of cereal flours and meals as the predominant ingredients in puffed snacks has been a natural result of the low cost and excellent expansion of such materials as corn meal. A demand for blander base materials which could also give structures and textures not obtainable with cereal or root flours has led to extensive experimentation with purified or modified starches.

Available cornstarches range from high (50 to 70%) amylose through regular dent (25 to 27% amylose) to the waxy maize varieties that are virtually 100% amylopectin. Sorghum starches with amylopectin contents from 100% to about 17% are also offered, but they are blander in flavor than cornstarches.

The extent of puff and the texture of the finished snack are influenced by the amylose:amylopectin ratio. High amylopectin content starches tend to give fragile products of low density. Some amylose is needed to give adequate resistance to breakage and textures that are acceptable. On the other hand, products containing only corn, red milo, and tapioca starches will be hard in texture and too high in density. The texture can be softened somewhat by the addition of plasticizers such as sucrose, dextrose, or sorbitol, but normally 50% or more amylopectin is needed for a good quality product. A starch system containing 5 to 20% amylose was suggested as most suitable (Feldberg 1969).

In baked-type puffed snacks, pre-extrusion moisture contents in the range 20 to 35% and a starch having about 80 to 100% amylopectin content are necessary to yield acceptable products.

Some modified and derivatized starches have found limited application in puffed snacks. Chemical modification processes include crossbonding, pregelatinizing and forming phosphate, acetate, and hydroxypropyl derivatives.

Some writers (manufacturers) describe semi-products or half-products as third-generation snacks. They are produced by extrusion cooking of certain somewhat modified wet milled starches to form cooked extrudates which are then passed directly into a sizing extruder to produce shaped but unpuffed pieces such as onion rings, twists, scoops, etc. These pieces are puffed by exposure to heat, almost always by french frying procedures.

The two main types of extrusion cooking systems have been described as high temperature/short time extrusion cookers and pressure cooking extruders.

Extrusion Methods

Puffed snack products may be prepared commercially by at least two methods.

(1) An intermediate piece of material composed in large part of gelatinized starch is formed without significant expansion and maintained at a moisture level of, e.g., 12% until it is puffed by frying, baking, or other application of high temperatures. It should be noted that some direct expansion snacks are also cooked in hot fat to give the typical french-fried flavor, but no further puffing is expected in these cases.

(2) Direct expansion puffing, in which the desired volume increase occurs as the material containing gelatinized starch emerges from a pressurized chamber into the atmosphere. The puffed piece still contains excess water and must be dried by frying or baking. Further expansion is not obtained in these drying steps.

Half-products have been defined as special food formulations which, upon immersion in hot frying oil, rapidly expand into a low density product. These half-products, also sometimes called intermediates, are often produced by gelatinization of a starchy dough which is then shaped into chip form and dried to a horny consistency.

A representative process has been revealed in the patent of Campfield (1964). A raw starchy material such as corn, potato, tapioca, wheat, rice, sorghum, or oatmeal undergoes a simultaneous cooking and kneading operation so that the starch granules are swollen. A typical dough would consist of 81.5% corn flour, 16.5% water, and 2.0% salt. The dough is directly transferred to an extrusion chamber under pressure sufficient to cause rupture of some of the starch granules. The pressures are of a magnitude sufficiently great to cause the extrudate to expand upon emergence to normal atmospheric conditions, but expansion is avoided by cooling the dough. The extrudate having a moisture content of about 15 to 35% issues in a compact, densified condition. It is usually cut or shaped into desired shapes before further processing.

The strand of dense, gelatinized dough is dried to a moisture content of less than 12% (e.g., 9%). Equilibration or case-hardening may be introduced at this point. If conditions are satisfactory, the horny chips will puff upon immersion in hot (e.g., 400°F) frying oil for 5 to 10 sec. The resultant snack base has a light vesicular crisp texture and a pleasant flavor.

Further details of cold forming and low-pressure cooking procedures follow.

Cold Forming.—Doughs of relatively low apparent viscosity can be extruded at low pressures. If the orifices are of reasonable size or the extrusion rates are low, pressure build-up is avoided, shear is nominal, and temperatures can be held close to ambient. Bakery doughs and masa for corn chips are typically handled in equipment of the cold-forming type. Further discussion of these methods will be deferred until the next chapter.

Low-pressure Cooking and Forming.—In these processes, the dry ingredients are mixed with water and fed to a cooking extruder. High-temperature fluid is circulated through the jacket and, in some designs through the screw, while additional heat is generated by the work performed on the dough. Temperature and time are controlled to get the necessary degree of starch gelatinization in the product. The dough is cooled, usually by a refrigerated die, before it is extruded into the atmosphere, so that the water it contains does not flash into steam. As a result, the extruded dough is compressed and substantially free of bubbles instead of being expanded and foam-like.

High-pressure Cooking and Forming.—These procedures require raising the starchy dough to temperatures substantially above 212°F. Energy input is from the heated jacket and internal friction. Compression of the plastic mass within the chamber by the tapered screw prevents vaporization of the water content (Schaeder *et al.* 1969).

Pressure represents a resistance to flow of the extrudate leaving the extruder. The meal or dough is held in the extruder for a longer time so that more mechanical energy is absorbed and the temperature rises, i.e., higher pressure results in higher product temperatures. As a result, the extrudate expands more and has smaller pores and a softer texture. Lower back pressure leads to lower extrusion temperatures, other conditions being equal, and there is less expansion with larger vesicles, coarser walls, and a harder texture.

There are a number of ways to adjust the pressure in an extruder. As the number of die openings (or their size) is increased, the pressure falls. Increasing the rpm of the extruder will increase pressure. A secondary backup die through which the product must flow before it reaches the extrusion orifices will increase the temperature.

The strand of dough passing through the orifice contains water at a temperature far above its boiling point at atmospheric pressure, but the water is held in liquid form by the high pressures existing in the chamber. As soon as the dough emerges from the orifice, the excess pressure is released and the free liquid vaporizes almost instantaneously. The dough is elastic and very viscous at this point and the large amounts of steam are temporarily trapped within its structure and force the dough to expand. At some point, the elastic limit of the structure is reached, vesicles burst, and the hot gas escapes. The bubbles tend to collapse and the dough piece to shrink as the internal pressure drops precipitously and the walls of the vesicles contract. This process normally stops far short of a complete return to the original density, however, because the entire structure "sets up" or becomes relatively firm and rigid as a result of loss of water and reduction in temperature. It is noteworthy that the structure is normally still under stress at the time it sets up, and that renewing the mobility of the intramolecular bonds, as by absorption of moisture, will permit additional shrinkage to occur. Retention of maximum specific volume of pieces is facilitated by factors which accelerate the time of set-up, probably including, among others the following.

(1)   Minimum amount of moisture consistent with attainment of desired expansion.
(2)   Minimum exit temperature consistent with desired expansion.
(3)   Small piece size (or, rather, at least one dimension being held minimal).

Drying

Extrusion puffed pieces normally reach a moisture content of about 8% as they set up, representing a loss of just a few percentages of water. To gain the necessary crispness, it is necessary to dry them to about 4% moisture content in a hot air oven, or some equivalent type of heater. Lower moisture contents are not considered necessary and may even lead to the accelerated development of oxidative rancidity. In some cases, lower moisture contents may also cause a powdery mouth feel as the structure becomes excessively friable. The exact level to which the product must be dried depends to some extent upon its composition and its surface area, but for most expanded snacks composed mostly of starch, 4% is a reasonable target. The 4% is, of course, calculated on the basis of the puff, not the finished product with added oil, salt, and seasonings.

Controls in Extrusion Puffing

According to Smith (1971), steps which can be taken to affect texture, density, mouth feel, solubility, and form of extrusion puffed snacks include the following.

(1)   The method of feeding and preconditioning of ingredients and mixtures.
(2)   The method and point of moisture application.

(3) Control of temperature and moisture contents of product entering the extruder.
(4) Control of temperatures within each extruder section.
(5) Control of the point within the extruder where maximum dough viscosity is attained.
(6) Control of extrusion speeds.
(7) Control of time and temperature relationships within each section of the extruder.
(8) Control of the time during which product temperatures are elevated to maximum extrusion temperatures.
(9) Control of final extrusion temperatures.
(10) Selection of the shaping and sizing devices.
(11) Selection of the type, dwell time, drying temperatures, and the velocities within the drier and cooler, and the desired final product moisture.
(12) Point and method of flavor application.

Large differences in the physical character of the product can be achieved by modifications in the conditions of the process. As an example, there are two main types of corn curls, the fried type which is denser, crunchier, and more irregular in shape, and the baked type which is more uniform, more highly puffed, and softer in texture. The original corn curl was of the fried type, but it was largely replaced in the late 40's and early 50's by the baked collet. The feed stock is the same for the two products, with the usual exception of a slight variation in moisture content, and the very significant difference in organoleptic quality is achieved solely by adjusting the extrusion conditions.

Examples and Modifications.—*Normal Extrusion Puffing.*—The following text, adapted from a patent of Corbin *et al.* (1972), describes the main steps in an extrusion puffing operation.

The moisture content of a given lot of corn meal in the particle size range of 40 to 60 mesh was adjusted to 12.5% moisture by addition of the calculated amount of water. The moistened meal was stored in a closed container for 3 to 24 hr to allow equalization of the moisture content before feeding to an extruder at the rate of about 400 lb per hour with the metering in of additional water to increase the total approximately 25%.

The extruder was preheated by means of a high-pressure steam jacket. After operation started the head of the extruder was cooled and the flow of steam adjusted to maintain a temperature of approximately 315°F. Product from the extruder was transferred to a hot air dryer operating at 375°F and brought to a moisture level of less than 2%. The hot dry collets were transferred to a flavoring reel where a mixture of oil, salt, and cheese flavoring was added by means of a meter pump and spray nozzles so that the final product contained 15% fat, 9% cheese flavor, and 3.5% salt. These flavored collets were then cooled and packaged.

Puffing Half-products in a Hot Air Stream.—Clausi and Vollink (1969) described a method for puffing cereal pellets from 1.5 to 3 times their original size by heating in a high-velocity hot air steam. Although the patent indicates the application is for breakfast cereals, it is obviously suitable for certain kinds of snacks. A cereal dough containing from 22 to 27% moisture is prepared by mixing corn, wheat, barley, or flour with flavoring syrup. The dough is cooked, as in a closed pressure vessel, at approximately 20 psig for 6 to 30 min.

The cooked mixture is formed into pellets which are case hardened by blowing ambient air through a bed of them until the critical moisture content of 16 to 21% is reached. Drier pellets will be hard and will not puff, while pellets having a moisture content above about 21% will puff into very fragile hollow spheres.

The case-hardened pellets are reduced in thickness by passing them through flaking rolls adjusted in separation distance to give pellets thicker than conventional corn flakes since thinner pellets cannot be puffed. The bumped pellets are then heated at temperatures of from about 350°F to about 750°F for 8 to 35 sec while being conveyed by air having a velocity of at least 500 ft per minute.

The moisture content of the cereal immediately after puffing is usually in the range 3 to 7%. The moisture may be further reduced and the product toasted at 225 to 500°F for about 2 to 10 min.

Simplification of the Forming Process for Half-products.—Popel (1974) patented a method for producing a puffed snack food item from starch-containing foodstuffs without the use of expensive high-pressure extrusion equipment and in a continuous process without the need for a holding or refrigeration step prior to forming the product and drying it. He also claimed improved flavor resulted from the use of this process.

A relatively dilute mixture of starchy material, water, and flavor is prepared and then heated sufficiently to gelatinize a substantial part of the starch. After a thin sheet of the dough has been formed, the surface is partly dried and it is subdivided into pieces of convenient size before the completion of drying. The pieces are deep-fat fried to puff them.

Workable ingredient ranges described in the patent are 60 to 90% water and 10 to 40% total starch. One unique feature is the inclusion in the starting mixture of a relatively high proportion of starch with a high setback or congealing power. The presence of 20% or more of amylose in the starch is recommended. Specifically, in selecting the added starch for the mixture, corn, wheat, or sago starches are found to give the desired characteristics, when mixed with the foodstuff in the amount of about 25% solids. For operating in the higher-solids range where the viscosity of the extruded gel may tend to be too high, acid-modified or "thin-boiling" starches from any source may be substituted.

The inventor claims particulate substances such as grains, potato pieces, etc., can be added to the gel to add interesting flavor and texture contrast to the finished puff.

Modification of Dough Rheology.—When pregelatinized potato flour is used to prepare a dough, the developed dough is hard to handle and will set to form a hard plastic mass which may not be capable of extrusion with available equipment. Singer and Beltran (1970) patented a method for improving the handling characteristics of doughs containing pregelatinized potato flour and for providing snacks having a low fat and low calorie content but with the texture of snacks containing a relatively high fat content. The essential steps are (1) forming a dough from pregelatinized starch, a saturated monoglyceride, and water, (2) extruding the dough, (3) forming the dough into pieces of the desired shape, (4) drying the dough to a suitable moisture content to form a half-product, (5) allowing the half-product to cool or age, and (6) puffing the half-product by exposing it to a uniform field of high intensity infrared radiation. Chips containing less than 5% fat simulate the texture and much of the eating quality of snacks containing up to 50% fat.

Recommended conditions are (1) 32 to 45% moisture content in the dough, (2) dough temperature of 110°F and die temperature of 180°F, (3) stock pressure of 1,000 psig, (4) dough sheet 0.4 mm thick, (5) half-product moisture content of 5 to 8%, and (6) puffing temperature of 230°F, reached in 5 to 10 sec. Hot air ovens and microwave radiation do not provide satisfactory results.

A snack product having the shape and other desirable characteristics of a potato chip was prepared by the above method and the following formula (solids basis):

	%
Potato flour	83.74
Carrageenan	0.3
Cottonseed oil	0.78
Myverol 18-07	1.05
Phosphated monoglyceride	0.3
BHT antioxidant	0.03
Food color	0.02
Monosodium glutamate	2.6
Sucrose	0.73
Flavor, smoked meat type	4.8
Salt	2.0
Powdered hydrogenated cottonseed oil	3.2

Simulated Popcorn.—Extrusion puffing can be used to produce a very reasonable facsimile of popped corn from corn meal. The product is denser than most popped corn and is missing the hulls, of course, but the appearance is entirely satisfactory in some cases. The patent of Wood *et al.* (1969) describes a suitable method. Cornmeal having a moisture content between 10 and 12.3% is forced at high velocity, e.g., 950 to 1800 in. per minute, by means of a high-pressure feeder through orifices in one or more extrusion plates. The size of the collecting pocket in the proximal die face and the extrusion rate must be closely controlled. The orifice size and cut-off length are obviously critical. The preferred particle size is no less than 20% on a No. 12 screen and 5% through a No. 40 screen. Temperature of the mass before it emerges from the die will be in the range 250 to 400°F. The puffs formed under these conditions have smaller cells and are expanded more than corn curls. The small cells make the product whiter and provide eating qualities similar to popcorn.

Snacks with Reduced Fat Content.—The low-calorie puffed snack described in the patent of Corbin *et al.* (1972) contains only about 5 to 18% (typically 10%) oil as compared with the 25 to 45% oil normally found in a snack of the corn curl type. Less than normal amounts of oil usually result in poor adhesion of the salt but the patented method circumvents this difficulty by spraying an aqueous salt solution on to the puffed piece before it is dried.

Improvements in Gun-puffing of Dough Pellets. Sticking or clustering of the dough pellets is often encountered in gun puffing. Chilling of the pellets prior to puffing somewhat reduces these tendencies. Coating the pellets with an edible oil is also effective if the oil can be retained on the pellets during the heating process. Using a small amount (e.g., 1.5%) of a fatty acid monoglyceride as an ingredient will reduce or eliminate the sticking problem but apparently causes unevenness of color in the finished product. It is not entirely clear whether or not this is due to poor distribution.

ADDITION OF FLAVORS AND COLORS

Added Flavors

Flavoring materials added to extruder feed stock undergo significant changes during puffing and most of the changes are undesirable. Volatile flavor components flash off. Interactions and decomposition occur as a result of the high temperatures. There are some indications that improved results are obtained when encapsulated flavors are used (Kinnison and Chapman 1972). In some cases, the flavoring materials interfere with texture development, this being particularly true if they intro-

duce fatty substances. For these reasons, the usual practice is to add flavors to the puffed and dried collet. A popular combination is oil, salt,
and cheese powder, and the ingredients are often mixed together in
stainless steel kettles before being dribbled or sprayed on to the collets
in a tumbling-type coater. The vegetable oil is generally coconut oil or a
combination of coconut with soybean or cottonseed oil.

Half-products made with integral flavoring ingredients may retain
them reasonably well through the oven puffing or frying processes. The
chemical stability and volatility of the flavor components are the chief
determinants of success in this approach. Natural flavoring materials
such as cheese powder will have noticeable effects on the texture and
amount of expansion.

Certain waxy cornstarches designated as cold-water swelling can be
used to prepare workable doughs containing flavors from which pieces
can be cut for later baking or frying into snacks having about 1.5 to 10
times the specific volumes of the original. Some suggested formulas
(Belshaw 1972) are:

Cheese Flavored Snacks

	%
Starch	53.0
Spray-dried cheddar cheese	18.0
Salt	2.0
Water	27.0

Peanut Butter Flavored Snacks

	%
Starch	55.0
Peanut butter	20.0
Water	25.0

Potato and Onion Flavored Snacks

	%
Starch	29.6
Potato granules	27.8
Salt	2.3
Onion powder	0.2
Water	34.6
Soy concentrate	5.5

The dry ingredients are blended in a planetary mixer, then worked up
into a homogeneous dough. The dough is formed into desired shapes by
sheeting and cutting, extruding, or other methods, and either baked at
400 to 450°F for a short period of time, e.g., 4 min, or fried at 375°F.

Many kinds of natural flavoring ingredients can be added to half-
products. Walter (1972) patented a method for making chip-type foods
containing flavoring materials derived from seafoods, meats, fowl, vegetables, and fruits. A typical dough formula is shown in Table 13.1.

About 1 part of shucked clams to 3.5 parts (by weight) of water are
heated at about 212°F for 5 to 20 min in a closed vessel and then al-

TABLE 13.1

DOUGH FORMULA FOR HALF-PRODUCTS

	%	
	Minimum	Maximum
Clams, fresh shucked	25	51
Wheat starch, powder form	44	70
Salt, powder form	2	5
Baking powder	0.05	2
Monosodium glutamate, pure powder form	0.15	3
Soy sauce	0.085	0.17
Lemon, pure juice	0.025	0.068
Water	250	650

lowed to cool before being mixed with the other ingredients, including the remaining water. The clams are chopped into finely divided form. The entire mixture is then cooked to a specific viscosity as determined by Brookfield viscometer testing, and the slurry is poured on to a thin and relatively smooth supporting surface so as to form a layer of about 0.1875 in. thickness. The slurry is baked on the supporting surface until both the upper and lower surfaces of the slurry harden to a skin-like formation and the moisture content is reduced to between 7 and 9.75%. The product of the heat treatment may be further reduced in shape and size. The thickness of the dehydrated pieces should be relatively thin as compared with their surface areas, however. The intermediate product will, if properly packaged, keep almost indefinitely. When it is desired to finish off the snack, the pieces are fried, preferably in peanut oil held between 425 and 450°F. Under optimum conditions, a 4- to 5-fold expansion is obtained.

Added Colors

Extrusion puffed snacks can be satisfactorily colored with food dyes in some cases. Between 30 and 600 ppm of FD&C pigments may be required (Kinnison 1971) to achieve the desired results. Higher levels often lead to products that are gaudy and unnatural looking.

Colors can be added by a dry blending process prior to extrusion. Dye solutions are sometimes metered at the feed screw ahead of the pressure chamber. If added moisture cannot be tolerated, it may be possible to use other polar solvents such as propylene glycol or alcohol. Titanium dioxide, FD&C lakes, and iron oxides being insoluble cannot be metered in as solutions.

Fading of color in extruded and expanded snacks is a common complaint and can be related to four key factors: (1) excessive heat, (2) reaction with various proteins, (3) reaction with reducing ions such as iron and aluminum, and (4) reactions with reducing sugars.

FD&C Red No. 2 is sensitive to moist heat and is not recommended for this type of processing. FD&C Red No. 2 and FD&C Yellow No. 5 are particularly susceptible to protein effects. FD&C Red No. 2 is also the most sensitive to reducing sugar interaction.

There is also a physical cause of fading. The foam structure causes a refraction of light which whitens or lightens the basic color of the material. The smaller the bubbles (or cells) the lighter the color.

BIBLIOGRAPHY

ANON. 1969. The rise of onion flavored rings. Snack Food *58,* No. 7, 39–40.

BAKER, R. J. *et al.* 1972. Expanded food product. Brit. Pat. 1,288,193. Sept. 6.

BATESON, R. N., and HARPER, J. M. 1973. Apparatus and process for puffing food products. U.S. Pat. 3,746,546. July 17.

BELSHAW, F. 1972. Special starch simplifies nutritional snack preparation. Food Process. *33,* No. 11, 40.

BENSON, J. O. 1969. Process for producing a tubular puffed product. U.S. Pat. 3,462,276. Aug. 19.

BENSON, J. O. 1970. Process for producing onion flavored ring snack. U.S. Pat. 3,540,890. Nov. 17.

BENSON, J. O., and PEDEN, M. F. 1970. Process of making snack products. U.S. Pat. 3,539,356. Nov. 10.

BRETCH, E. E. 1972. Snack food and method of producing same. U.S. Pat. 3,703,378. Nov. 21.

CAMPFIELD, W. W. 1964. Simultaneous gelatinizing and kneading. U.S. Pat. 3,150,978. Sept. 29.

CLAUSI, A. S. 1969. Preparation of ready-to-eat puffed cereal. U.S. Pat. 3,453,115. July 1.

CONWAY, H. F., and ANDERSON, R. A. 1973. Protein-fortified extruded food products. Cereal Sci. Today *18,* 94–97.

CORBIN, D. D., MARQUARDT, R. F., and GABBY, J. L. 1972. Process for preparing a low calorie snack. U.S. Pat. 3,682,652. Aug. 8.

D'ARNAUD GERKENS, D. R. 1973. Method of making crisp snack food product. U.S. Pat. 3,753,735. Aug. 21.

FAST, R. B., and MORCK, R. A. 1969. Process for making a puffable chip-type snack food product. U.S. Pat. 3,451,822. June 24.

FARRELL, R. J. 1972. Available extrusion equipment: types and applications. Presented at the 57th Annual Meeting of the American Association of Cereal Chemists, Miami Beach, Fla., Oct. 29 to Nov. 2.

FELDBERG, C. 1969. Extruded starch-based snacks. Cereal Sci. Today *14,* 211–212, 214.

FRITZBERG, E. L. 1972. Vacuum puffed foods. U.S. Pat. 3,650,769. Mar. 21.

GLICKSMAN, M., KLOSE, R. E., and KIRKEBY, R. D. 1972. Method of making an extruded expanded protein product, U.S. Pat. 3,684,521. Aug. 15.

HARMS, V. D., JENSEN, E. R., and LANGAN, R. E. 1973. Method for preparing food snack compositions. U.S. Pat. 3,753,729. Aug. 21.

HEILAND, W. K., and MERCALDO, R. G. 1969. Automatic explosive puffing apparatus. U.S. Pat. 3,456,567. July 22.

HERZKA, A. 1947. Notes on maize. J. Soc. Chem. Ind. *66,* 396–397.

HESS, J. 1973. Puffed—the magic snacks. Snack Food *62,* No. 4, 35, 38–39, 42.

HOLTZ, W. E., JR., and REINHART, R. R. 1972. Method for producing an improved puffed cereal. U.S. Pat. 3,660,110. May 2.

HRESCHAK, B. O. 1973. Food product and method of making same. Brit. Pat. 1,321,889. July 4.

KELLEY, E. F., and THOMAS, F. J. 1969. Method for producing an expanded food product from cereal grain. U.S. Pat. 3,458,322. July 29.

KINNISON, J. W., and CHAPMAN, R. S. 1972. Extrusion effects on colors and flavors. Snack Food 61, No. 10, 40–42.

LA WARRE, R. W., Sr. 1973. Snack food production. U.S. Pat. 3,711,296. Jan 16.

MAXWELL, D. L. 1969. Method of preparing a ready-to-eat puffed product and apparatus. U.S. Pat. 3,454,403. July 8.

NADISON, G. 1969. Seasoning blends for expanded snack product. Cereal Sci. Today 14, 215–216.

PEDEN, M. F., JR. 1965. Process for preparing a snack product with a rippled surface. U.S. Pat. 3,190,755. June 22.

REEVE, R. M., and WALKER, H. G., JR. 1969. The microscopic structure of popped cereal. Cereal Chem. 46, 227–241.

ROSSEN, J. L., and MILLER, R. C. 1973. Food extrusion. Food Technol. 27, 46, 48–53.

SANDERUDE, K. G. 1973. Should you be using E.S.P.? Snack Food 62, No. 4, 44–45.

SCHAEDER, W. E., FAST, R. B., CRIMMINS, J. P., and DESROSIER, N. W. 1969. Evolving snack technology. Cereal Sci. Today 14, 203–204.

SHATILA, M. A. 1973. Fabricated onion ring with dehydrated potato shell. U.S. Pat. 3,761,282. Sept. 25.

SINGER, N. S., and BELTRAN, E. G. 1970. Process of making a snack product. U.S. Pat. 3,502,479. Mar. 24.

SMITH, O. B. 1971. Why use extrusion. Presented Feb. 12, 1971. Symposium on Extrusion Cooking. American Association of Cereal Chemists, St. Louis, Mo.

STRAUGHN, R. O., ELOFSON, G. L., and REINHART, R. D. 1973. Method and apparatus to make potato stick snack. Can. Pat. 929,406. July 3.

STROMMER, P. K., VALENTAS, K. J., and DUNNING, H. N. 1972. Process and apparatus for controlling the expansion of puffable materials. U.S. Pat. 3,656,965. Apr. 18.

TSUCHIYA, T., and PERTTULA, H. V. 1969. Process of preparing puffed cereal product. U.S. Pat. 3,464,827. Sept. 2.

TOEI, R., AONUMA, T., WATANABE, H., and YUASA, T. 1973. Method for producing expanded food stuffs by gaseous conveying heating. U.S. Pat. 3,754,930. Aug. 28.

WALTER, H. P. 1972. Process for producing a chip-type food product. U.S. Pat. 3,684,527. Aug. 15.

WILLARD, M. 1973. Fabricated potato snacks. Snack Food 62, No. 4, 52–54.

WISDOM, L., FOWLER, D. P., and ZINN, R. E. 1971. Method for making center-filled puffed food product. U.S. Pat. 3,615,675. Oct. 26.

WOOD, D. H., JR., GIBNEY, G. O., and SMITH, R. 1969. Process for preparing expanded cornmeal extrusions. U.S. Pat. 3,476,567. Nov. 4.

WRIGHT, E. S., ANGSTADT, J. W., and GARROW, G. L. 1973. Apparatus for deep-fat cooking. U.S. Pat. 3,754,468. Aug. 28.

ZIEMBA, J. V. 1972. Strides in forming. Food Engineering 44, No. 11, 66–70.

CHAPTER 14

Corn Chips and Simulated Potato Chips

INTRODUCTION

The products discussed in this chapter are similar in that they are prepared by low-pressure extrusion (or sheeting by rollers) and fried under conditions leading to relatively minor puffing. They differ from the products discussed in the preceding chapter because high-pressure extrusion is no part of the process and a high degree of expansion is not achieved. They differ from the baked snacks described in the following chapter in that baked products, as the term is used in this book, refers to nongelatinized nonextruded doughs usually leavened by chemical systems or yeast fermentation.

Corn chips of the Frito® type are among the oldest kind of fried cereal snacks. They are still very important commercially. Simulated potato chips are, by contrast, among the newest snack products but they have assumed great commercial importance in just a few years.

CORN CHIPS

Corn chips have a close relationship to tortillas. They are made from the same alkali-treated corn dough, and they have similar flavor and texture. It is instructive to compare a modern tortilla production method with corn chip processing. A description of the preparation of tortillas in a mechanized plant was published by Havighorst (1971). The cooking cycle begins by charging a kettle with 2,000 lb white corn and a measured amount of water. The kettle contains an upper perforated ring from which water is sprayed over the corn. A steam injection ring in the bottom part of the kettle maintains the contents within the desired temperature range and agitates the corn.

The corn is first hydrated with water maintained at about 120°F. Then the temperature is raised to 165°F and the corn is gelatinized after which the heat is shut off. During the cooking cycle, hot water is sprayed over the corn and continuously recirculated by a pump connected to the discharge valve. When the temperature drops below 140°F, the recirculation is discontinued and the corn is allowed to steep undisturbed overnight. The steep water contains 12 oz hydrated lime for each 100 lb corn.

After the steeping period is completed, the corn is flushed to a draining conveyor with fresh water. The conveyor discharges to a screw-type elevator in which the corn is again washed to remove free starch before milling.

The mill consists of a stationary lower stone disc and a rotating upper stone of 16-in. diameter and 4 in. thick. This buhr mill of conventional design is powered by a 30 hp motor and grinds about 3,000 lb of corn per hour.

The ground alkalized corn dough, or masa, is conveyed to the hopper of a tortilla cutting head by a screw-type extruder. A thick sheet of masa is extruded between sizing rolls and the resulting ribbon is cut into discs of appropriate size.

Gas-fired ovens operating at 600°F bake the tortilla blanks in 30 to 32 sec. Tortillas discharged at 175°F are cooled to 85° to 90°F in an atmospheric multi-tiered conveyor.

The conventional process for making corn chips begins with a mixture of white and yellow corn of the dent type which is added to a vat of heated tap water containing a proportionate amount of lime. The ratio of white and yellow kernels is varied depending on the color desired in the final product. The mixture is heated to the boiling point, then the heat is cut off and the contents of the vat are allowed to stand undisturbed for 10 to 20 hr. During the heating and steeping steps, the hulls are hydrated and partially hydrolyzed. They become jelly-like in consistency and are easily removed by agitation and the water jets used later in the process. The starch is gelatinized and the kernels absorb moisture so that about 50% by weight is water at the end of the steeping period. Small amounts of sulfite may be added to the steep water to slow down bacteriological activity.

Hulls are removed in a washer, where jets of water wash off the hulls and remove any remaining lime. The washed kernels are then transferred to a stone mill where they are ground into dough. Hand forming of the masa into large cylindrical loaves precedes filling of hydraulically powered extrusion presses. The cylindrical chamber of the press contains a closely fitting piston which forces the dough through a die plate having series of slot-like ports about ½ in. wide. A cutting device severs the extruded strands into pieces of the desired length, usually 1.5 in. Alternatively, the dough can be rolled out into a thin sheet from which fancy shapes can be cut.

Dough pieces fall directly into cooking oil held at about 375°F. After the moisture content has been reduced to a few percent, the chips are salted, cooled, and packaged.

According to Murdock (1970), the conventional process suffers from the following disadvantages. Irregularities can be caused due to a variation in the size of the initial corn kernels. The smaller the kernel, the more complete the penetration of water and alkali. Also, the starch in a kernel having a cracked or broken hull becomes completely gelatinized, since there is a clear path for the water into the starchy layers. If there are too many cracked kernels, the moisture content of the dough may increase to the point where inferior quality chips result through puffing and excessive oil buildup. Also, corn usually contains extraneous material, and even so-called "clean" corn may require further cleaning. In the cooking vat, extraneous material rises to the surface and has to be skimmed off.

Time and scheduling can be a problem. The conventional operation outlined above is one which requires 24 hr or more notice to produce product. Many hours elapse between cooking the corn and getting finished product.

A considerable amount of labor is required in the conventional process, for cooking the corn kernels and for the grinding and frying steps. The conventional process also usually requires quality control personnel to periodically test the moisture content of the steeped corn, and to release it when ready for further processing. There is an optimum moisture pickup range for the steeped corn kernels, above which the finished product is puffy and oily, and below which the chips are hard and slate-like.

Water consumption is considerable in the conventional process, due to the steps of cooking and washing. Sometimes, considerable losses of material occur in the washing step, due to broken pieces of kernel being discarded with the hydrolyzed skins. This is particularly true when the method of conveying the steeped corn to the washer results in breakage.

A substantial capital investment is required for the large amount of basic equipment necessary: cooking vats, steeping vats, washers, grinders, and fryers. All of this equipment is of considerable size, and as a result takes up appreciable space. Power costs add further to the total production costs.

The common corn chip offered in the marketplace today probably conforms rather closely in composition to the traditional formula described above. Several modifications have been suggested to improve flavor, speed up the process, or achieve other advantages. Weiss (1967) patented a composition of corn grits, water, sucrose, and an alkali such as sodium bicarbonate, potassium bicarbonate, sodium carbonate, potassium carbonate, or sodium phosphate. Following a steeping period of ½ hr or more, sucrose is added, the ingredients are cooked, and the dough is formed into pieces which are partially dehydrated, tempered, and fried.

Cereal grains other than corn can be formed into chips and fried. The alkalizing process is omitted in these cases. Shredded wheat (pressure cooked whole wheat extruded in thin strands) formed into multiple layers, salted, and baked form the basis of the popular Triscuit®. Ball and Demeny (1972) describe a method whereby individual kernels of wheat or rye are sliced, combined with water and flavoring ingredients, and pressure cooked before forming into thin dough pieces, drying to 8 to 14% moisture, and frying to a final moisture content of 0.2 to 3.0%. A substantial portion of rice flour may be used in the dough to modify the texture and appearance of the finished chip. The dough piece undergoes a minor amount of puffing under these conditions. Salt is dusted on to provide 0.5 to 3.0% in the chip.

Anon. (1972) describes a method for preparing edible tubular corn snack products by extruding masa through a tubular forming die, cutting the extruded tubes into predetermined lengths and frying the pieces so as to form products having case-hardened inner and outer surfaces but a flaky open interstitial structure.

Extrusion Presses

The presses used for corn chips (e.g., Fritos) use hydraulically or pneumatically activated pistons to force the corn meal dough through the shaping orifices. They are batch presses; i.e., the piston must be withdrawn and a charge of dough inserted into the chamber at the start of each cycle. There are a number of similarities to the old-style macaroni product extruders and to sausage stuffers. Pressures reached in the chamber are much less than the pressures generated in the continuous extrusion chambers used for puffed snack products such as corn curls. Shear is low and the temperature does not rise significantly. As a result, little or no steam evolution occurs within the dough and it does not expand appreciably as it exits from the die. Chip size is controlled by varying the speed of the cut-off knife. The variety of shapes which can be generated in this manner are rather limited since the three-dimensional effect obtainable in puffed snacks is lacking.

SIMULATED POTATO CHIPS

According to Willard (1973), all of the fabricated potato snacks can be classed generally in four groups: (1) dry collet processes, (2) single screw extrusion of dry potatoes, (3) forming and frying a moist dough, and (4) forming a high solid content dough into a thin sheet which is then cut into pieces and fried. It is the latter process which is being used for making the simulated potato chips enjoying the greatest commercial success today.

What Willard calls dry collet processes are similar in principle to the half-product methods described in the preceding chapter. Potato flour or ground dehydrated potatoes is mixed with other starchy material such as tapioca starch or potato starch and then gelatinized by some cooking technique. Under certain circumstances, ground blanched fresh potatoes can be used as the predominant ingredient. Flavoring materials such as ground shrimp can be added to the mixture. After cooling, the solidified gel is sliced and dried. When the discs are deep-fat fried, they puff to about six times their original thickness. In another process, a mixture of dehydrated potatoes and water is gelatinized and kneaded in a single screw press, extruded, and cut into pellets. These pellets are flattened between bumping rolls, and partially dried before finishing.

Direct extrusion puffing of dry potatoes and auxiliary ingredients leads to products having the internal structure of corn curls and similarly prepared snacks. The commercial versions have been in the shape of french fries, i.e., in elongated strands of square cross section, but there is no obvious reason why disc-shaped pieces could not be made.

The third method described by Willard—forming and frying potato snacks directly from a wet dough—has apparently achieved only limited success in delivering a product closely resembling potato chips. At least, there is no known commercial product being made by extrusion and slicing of gelatinized dough masses.

Simulated potato chips may be formed by methods resembling those used in the cookie and cracker industries. It is necessary to formulate a dough having physical properties making it suitable for rolling out into a thin sheet. This is generally achieved by the inclusion of substantial amounts of other cereal flours, or other rheologically active additives. Another approach is to use ungelatinized starch in the dough, then steam treat the pieces after all forming operations have been completed.

After sheeting, the dough is cut into pieces of the desired shape and size, and then fried or baked. In the case of the Pringles® chip of Procter and Gamble, each piece is transferred to a curved mold which controls its shape as the piece is being fried (Liepa 1971). In this way, uniform contours are obtained in the finished product, making it possible to stack the chips in an efficient manner.

Ingredient statement on the currently distributed Pringles declares dehydrated potatoes, vegetable shortening, salt, mono- and diglycerides, dextrose, ascorbic acid, sodium phosphates, sodium bisulfite, and BHA. The mono- and diglycerides are undoubtedly used as starch complexing agents, to simplify handling of the gelatinized dough. Ascorbic acid and sodium bisulfite may be used to reduce darkening of the dehydrated potatoes or the final mixture.

According to the patent previously cited, the dough is formed into thin, flat wafers by a reciprocating cutting device. These pieces are transferred by air jets to a curved, perforated mold surface which is then covered by another mold having the same general configuration. The dough piece is in this way inclosed in a space 0.03 to 0.08 in. thick, with perforated metal sheets on both sides, in which it passes through the frying oil.

The Procter and Gamble potato "chips" made from reconstituted dehydrated potatoes and other ingredients have an apparent bulk density of 0.16 gm per cubic centimeter when stacked in the cylindrical container, while conventional potato chips have an apparent bulk density of 0.056 gm per cubic centimeter. The true density of potato chips is approximately 1.165 gm per cubic centimeter.

BIBLIOGRAPHY

ANON. 1972. A tubular snack product. Brit. Pat. 1,259,758. Jan. 12.

BALL, M. E., and DEMENY, L. M. 1972. Food chip and process for making it. U.S. Pat. 3,656,966. Apr. 18.

BENSON, J. O., and PEDEN, M. F. 1970. Process of making snack products. U.S. Pat. 3,539,356. Nov. 10.

HAVIGHORST, C. R. 1971. Mechanizes age-old process. Food Engineering 43, No. 6, 60–61.

KORTSCHOT, C., and ADAMS, P. F. 1972. Method of preparing textured snack food products. U.S. Pat. 3,698,914. Oct. 17.

LIEPA, A. L. 1971. Preparation of chip-type products. U.S. Pat. 3,576,647. Apr. 27.

MURDOCH, G. B. 1971. Process for making flour from corn, and processes utilizing the same. Can. Pat. 878,424. Aug. 17.

SUCCO, J. A., and YOUNGQUIST, R. W. Potato chip product and process. U.S. Pat. 3,519,432. July 7.

WEISS, V. E. 1967. Snack product and process. U.S. Pat. 3,348,950. Oct. 24.

WILLARD, M. 1973. Fabricated potato snacks. Snack Food 62, No. 4, 52–54.

CHAPTER 15

Baked Snacks

Although most kinds of baked foods are adaptable to use as snacks, certain kinds are consumed almost exclusively outside normal meal periods. Pretzels, many kinds of crackers, and most cookies fit into the snack pattern. Because of space limitations, the discussions in this chapter are restricted to rather general treatments of formulation theory and processing methods with a few examples of formulas for some of the more important items. Detailed discussions of all important baked snack products can be found in *Cookie and Cracker Technology* (Matz 1968).

SALTY-SAVORY BAKED SNACKS

Soda Crackers

Soda crackers and variants such as saltines, oyster crackers, etc., are themselves used as snack foods and also serve as the basis of combination snack foods either made in the home (e.g., crackers and cheese) or by manufacturers (cheese crackers, peanut butter and cracker sandwiches, etc.).

The soda cracker is made from a "lean" fermented dough. It does not contain much shortening, sugar, or milk. Table 15.1 summarizes the range and average composition of several published formulas.

Since flour may be present to the extent of 80% or more of the finished product, its qualities are the principal controlling factors in machining quality of the dough. It is also an important texture determinant. However, due to the bland flavor, it does not supply, except in unusual circumstances, the dominant flavor note, even in unsalted crackers. Appearance, insofar as it can be separated from machining response, is also considerably affected by ingredients other than flour.

The specifications of flours suitable for cracker production are narrower than those for flours intended for cookies. The sponge flour should be relatively strong, unbleached, with an ash of 0.39 to 0.42%, a protein content of 8.5 to 10.0%, and an acid viscosity value somewhere in the range of 60 to 90°M, the exact value depending upon the product and the conditions. The dough flour should be weaker, with an ash of about 0.40%, a protein content of 8.0 to 9.0%, and an acid viscosity reading of 55 to 60°M.

Cookie bake tests are suitable for evaluating some of the properties of cracker flours, but they do not give enough weight to the gluten strength

TABLE 15.1

SODA CRACKER FORMULAS

	Average	Range	
		High	Low
Sponge Ingredients, Lb			
Flour	70	80	60
Yeast	0.23	0.5	0.06
Water	30	34	28
Shortening	4	8	0
Diastatic malt	0.02	0.1	0
Sponge Time, Hr	18	20	16
Dough Ingredients, Lb			
Flour	30	40	20
Shortening	5.8	10	0
Salt	1.4	1.6	1.25
Sodium bicarbonate	0.63	0.7	0.52
Malt syrup	0.92	1.5	0
Water	0.8	2	0
Totals			
Flour	100	100	100
Yeast	0.23	0.5	0.06
Water	31	34	29.5
Shortening	9.5	10.5	8
Malt	0.02	0.1	0
Malt syrup	0.92	1.5	0
Salt	1.4	1.6	1.25
Sodium bicarbonate	0.63	0.7	0.52
Fermentation			
Time, hr	4	5	3
Temperature, F	82	84	80

factor since the conditions of the test allow little opportunity for gluten development. Mixograph and pup loaf tests may be valuable, especially if a long series of results from supplies by the same miller are available for comparison.

Strong flours tend to increase oven spring but the crackers are often tougher. Weak flours lead to lesser amount of spring, and to a tender, more friable cracker. The effect of fermentation is to mellow the gluten. Weak flours and lengthy fermentation combine to yield flat, tender crackers.

Flour for thick saltines (120 count) should be stronger than that for thin crackers (160 to 170). The thicker crackers need a sponge flour of about 9.0 to 10.5% protein, 0.41 to 0.45% ash, and a viscosity of about 95 to 125°M. Thin saltines require a weaker flour in the doughs—a protein content of 8.0 to 8.5%, 0.43% ash, and 55 to 60°M viscosity. Alternatively, a certain percentage (determined by trial) of the strong flour is replaced by cookie flour.

Lard and oleo (the liquid fraction of beef fat) are widely used shortenings. The flavor of crackers containing lard is probably superior to those made with oleo. Plastic shortenings are not essential, so liquified fats handled by bulk transfer and measuring systems are common. Hydrogenated shortenings have the advantage of improving spring while lard contributes tenderness and frequently detracts from the oven expansion. Emulsifiers are commonly added.

The amount of topping salt is not shown in the quoted formulas. Based on dough weight, about 2.5% is a good average figure. Salt suppliers sell a size specifically intended for this purpose. Different brands are probably distinguished mainly by the percentage of fines, which should be at the minimum it is feasible to obtain. Dough salt should be of a finer granulation, though this is not as important in crackers as in cookies. It has been said that flake salt, used in the dough, causes crackers to have slightly more spring in the oven.

All crackers of the type discussed in this chapter are made from laminated doughs. Formerly, this step was performed on reversible dough brakes. At present, automatic laminators are used throughout the industry. The number of layers formed is somewhat variable, but must be at least 6 or 7 to secure any benefit from this operation.

Laminating of cracker doughs is usually done without the benefit of an interleaving ingredient such as is used in puff pastry. In cream crackers, a mixture of flour and shortening is added between the dough sheets.

Shortening is shown as being added to the sponges. One of the advantages of this alternative is that the shortening is certain to be adequately distributed. If any crust forms on the sponge, it is made softer by the shortening. Some authors also say that resistance to rancidity is increased as a result of including the shortening in the sponge. The fermentation rate is probably not affected appreciably by the fat.

Fermentation

There are special considerations involved in cracker sponge fermentations which need to be examined more fully.

Micka (1955) showed that the yeast added to cracker sponge and the bacteria from ingredient flour or from deposits retained in the trough from previous doughs will grow for 10 to 15 hr. After this period, both the yeast and the bacteria are retarded, but the bacteria are inhibited more than the yeast. Acidity increases are largely due to bacteria and are favored by low percentages of yeast. The converse is also true. Sterile troughs retard bacteria and yeast fermentation as well as development of acidity. When the yeast addition is greater than 0.50% or the trough is sterile, acidity is retarded to such an extent that the finished cracker is of high pH and has an undesirable flavor.

The rapidity of gas and acid development is obviously related to the temperature of the sponge, and for a given formula is a function of the temperature at which the sponge is set and the temperature of the fermentation room. It is also related to dough composition, as follows.

(1) Absorption—the greater the percentage of ingredient water, the faster the fermentation.
(2) Salt—this ingredient inhibits fermentation.
(3) Amylolytic enzymes—the yeast first uses up the monosaccharides in the dough. After these are exhausted, a more or less quiescent adaptation period ensues, and then maltose split off from starch by amylolytic enzymes can be metabolized. Amylases are found in flour, but are present in much larger quantities in malt and fungal supplements. Most bacteria can utilize maltose but some strains cannot.
(4) Added sugars—sugar, whether in the form of corn syrup, sucrose, or invert syrup is consumed rapidly by both the yeast and bacteria. Bakers' yeast does not utilize lactose. Many bacteria do so, however.

The soda percentages shown in the formulas are only estimates, and are not constant. The correct amount of sodium bicarbonate to be added to the trough contents at the doughing-up stage is the quantity necessary to assure the obtaining of a predetermined pH in the baked cracker. The addition will therefore be related to the amount of acid produced during fermentation. As this indicates, the amount of acid produced cannot always be predicted.

One problem which has still not been solved to everyone's satisfaction is that of insuring uniform fermentation in all the troughs of sponges or, if it is not possible to achieve this goal, compensating for the different levels of acid produced in different troughs. The difference in rate of fermentation between troughs is evidently due to varying levels of inoculation left by preceding batches. The difference is particularly noticeable when some troughs have been left idle for a time while others have been in constant use. Furthermore, some troughs seem to retain a heavy inoculation better than others. Some technologists have attempted to overcome the uncertainties of predicting proper fermentation time and proper soda addition by numbering both the troughs and their usual location in the fermentation room. Assuming the temperature and time are kept constant and the troughs are in continual use, the development of a known amount of acidity in a given trough can be expected. The soda addition should then be predictable from day to day.

The acidity which must be compensated for by addition of soda can be measured. It is related in a general way to pH, and the pH readings on the dough can be used as a rough guide for adjusting the soda supplementation. A better and more direct indicator is the total titratable acidity. This figure is more difficult to determine accurately and the analysis is more time-consuming than is the pH test and it is not much used in practice. The pH determination, though it is only indirectly correlated with the amount of acid which has been developed, is probably the most useful index for actual fermentation room practice. Sturdy, accurate pH meters, requiring only daily standardization by quality control personnel, can be placed near the mixer used to dough up the sponges. Measurements with temperature compensated electrodes can be taken directly on the sponges immediately prior to remix, and the amount of soda determined by reading the appropriate figure from a chart.

Other procedures which have been suggested are pH tests on lumps of dough or sponge sent through the oven ahead of the rest of the trough contents. This technique can be expected to compensate for some of the unknown responses of fermentation-derived chemicals in the oven which might not be accurately predicted on the basis of raw dough pH.

As the dough ferments, it rises in temperature. The rate and extent of heat production are undoubtedly related to the same processes by which acids are elaborated. However, quantitation of this relationship is uncertain at best. Basing soda addition on degrees of rise in sponge temperature is likely to lead to some undesirable variations in pH of the cracker.

Material is lost through fermentation. Ethanol and other volatile materials are produced from starch and sugars. Carbon dioxide is also lost, both from the trough and in the oven. The losses may amount to between 2 and 3% in a normal operation. In a low profit item such as soda crackers, it is important to keep these losses at a minimum. However, reduction in loss without reducing flavor or other desirable changes is very difficult. One thing that can be done is to keep the fermentation time at the shortest possible length consistent with a quality cracker. Over-fermenting for convenience of scheduling should be avoided, if possible. Adding ripe sponges or a fermented broth for flavor purposes are possible approaches to minimizing fermentation losses. No doubt we will ultimately see the widespread use of pure bacterial cultures and chemical dough modifiers employed in conjunction with a short (perhaps continuous) dough fermentation to yield flavorful crackers with minor fermentation losses in a perfectly controlled system. Such systems are already widely used in bread manufacture, and only the limited market for equipment is delaying their modification to cracker production.

Sprayed Crackers

This rich cracker, often in a round shape, is usually made from a chemically leavened dough. Actually, many representatives of this class could be considered a cross between a cracker and cookie, since they are not only chemically-leavened but also sweeter than saltines. In a few plants, a small amount (about 0.25% FWB) of old sponge, made up of flour, water, and a fractional percentage of yeast, is fermented for 36 hr or more and then added to the dough for flavor. Perhaps the original sprayed cracker was prepared by a sponge and dough process very similar to that used for soda crackers. In any case, the leavening system must be adjusted to bring the pH of the finished product below neutrality, a pH of 6.5 being regarded as desirable by many authorities.

The sponge-and-dough formula shown in Table 15.2 is to be used in a procedure similar to the soda cracker method, i.e., with a long sponge

TABLE 15.2

FORMULAS FOR SPRAYED CRACKERS

	Sponge and Dough	Chemically Leavened
Sponge Ingredients		
Flour	50	—
Yeast	0.25	—
Water, variable	24	—
Dough Ingredients		
Flour	50	100
Sugar	2.5	5
Malt	2.5	1.5
Shortening	8	7.5
Salt	1.5	1
Sodium bicarbonate	0.52	0.9
Nonfat dry milk or dried buttermilk	—	2.5
Invert syrup	—	2.5
Monocalcium phosphate	—	0.75
Hot water, variable	4	28

fermentation. A short method uses about 2% yeast and more water, the sponge is allowed to stand in the trough for about 30 min, and then the rest of the ingredients are mixed in. The dough is fermented about 5½ hr before machining.

Sprayed crackers are coated with 20 to 25% of a bland shortening. Coconut oil (76F) is preferred, but peanut oil or even hydrogenated vegetable shortenings have been used. The oil is applied either by spraying at 150 to 160°F, or by a device similar to an enrober in which the crackers pass through a flowing curtain of oil. In an old laborious technique, the crackers were dumped while hot into wire baskets and dipped into a tank of melted fat. The oil must be applied while the crackers are hot, but to avoid checking and other problems, the crackers must be annealed or equilibrated for a few minutes before spraying. The crackers may be salted very lightly on the cutting machine.

Cheese Crackers

Cheese crackers are usually made from fermented doughs. The flavor elaborated during fermentation is complementary to the cheese flavor and provides an inexpensive way of getting the desired strength and character. These crackers must be on the acid side of neutrality to yield a typical cheese flavor. Formulation is based on soda cracker formulation, except that the fat and moisture added in the fresh cheese must be compensated. Fermentation conditions are very similar to those in cracker processing with a sponge stage of about 18 hr and a dough fermentation of perhaps 4 to 6 hr depending on the temperature.

Representative formulas are given in Table 15.3. They are based on sponge and dough processes. Straight doughs with slightly increased

TABLE 15.3

CHEESE CRACKER FORMULAS

	Blue-cheese Type	Cheddar Type		
		Average	High	Low
Sponge Ingredients, Lb				
Flour, cracker sponge	80	75	80	60
Yeast	0.2	0.3	1.5	0.2
Cheese (blue or cheddar)	17.5	10	20	0
Salt	1	—	—	—
Shortening	—	7.5	12	4
Water	21	23.5	25.0	22.5
Dough Ingredients, Lb				
Flour, cracker dough	20	25	40	20
Shortening	20	—	—	—
Sodium bicarbonate	0.75	0.45	0.58	0.31
Malt syrup	—	1.0	1.25	0.8
Water	—	0.8	4.0	0
Cheese	—	4.0	20	0
Salt	—	1.0	1.25	0.8

yeast percentages and fermentation conditions of 5 to 6 hr at 90 to 94°F have also been suggested.

Cheese itself will not add enough color to make the cracker distinctive in appearance. Paprika—about 0.25%, flour weight basis (FWB)—may be used in the cheddar cracker to intensify the color and to add a flavor which is complementary to the cheese. A very small amount of cayenne is included in some formulas. Caraway or other spices and poppy seeds add to the blue-cheese cracker a character thought by many to be desirable. One author advocates a small amount of sage in this product.

If natural cheese is being used as the flavoring agent, the rind and any large areas of mold should be removed, and it should be ground as fine as possible immediately before putting it into the mixer. Ground cheese which has been allowed to dry out may never incorporate properly into the dough. The formulas given here show the addition of cheese to the sponge. This allows full hydration of the cheese and aids dispersion during mixing of the dough. Some manufacturers make a premix of a ground cheese and shortening, using the spindle mixer, and keep it in the fermentation room for 24 hr. This is said to develop more flavor and lead to better dispersion of cheese in the dough. This premix can be added at the doughing stage. Force-cured cheeses can be bought to specification. These products are cured under higher than usual temperatures.

Although volume producers of cheese crackers may wish to buy aged cheddar as the flavoring component, the quality of the natural unmodified material fluctuates so much that it is usually better, and often less expensive in the long run, to buy a compounded cheese powder from a reputable supplier. Storage and incorporation of cheese flavor into doughs create fewer problems when a powder is used. Most manufacturers offering powders have invaluable experience in selecting and blending cheese of standardized quality so that the uniformity of the product can be relied upon. When using powders in cracker formulas, ask for the supplier's recommendation as to the percentage required. Modifications of the recommended amount may be necessary depending upon the results of consumer surveys or other reliable guidance.

Cheese powders usually contain artificial color. There are several very powerful, and fairly typical, artificial cheddar and blue cheese flavors on the market. These should be used wherever possible.

Pretzels[1]

Most of the equipment used for producing pretzels in the United States has been made by Reading Pretzel Machinery Co., American Machine and Foundry, and Hinkle Machinery Co.

Pretzel doughs are made very stiff so that they will withstand the punishment of machining without becoming sticky or misshapen. The sponge is fermented for a shorter time than cracker doughs, about 10 hr, and might consist of 20 lb of flour, 10 lb of water, and 1 to 2 oz of compressed yeast. At the dough stage, 80 lb of strong flour, perhaps 25 lb of water, 1.2 lb of salt, and up to 3 lb of shortening are added. Doughs may receive a short proof stage, but frequently are made up without additional fermentation. The machining steps, including formation of the

---

[1] This section based in part on a discussion by Raymond J. Sell.

pretzel, are handled automatically in all but a very few small plants. The characteristic gloss of the pretzel is the result of a lye dip. The dip solution contains about 0.5% sodium hydroxide or 2% sodium carbonate and is maintained at about 210 to 212°F. Immersion time is about 10 sec. The solution may also be applied by spraying.

In Reading Pretzel Machinery equipment the dough is placed in a hopper from which a helix forces it through a slot in the face plate of the extruder. The dough is cut into small strips as it is extruded. The dough drops on a canvas belt which carries it under a second belt. Between the two belts the dough is rolled to the desired thickness. At the end of the rolling process the string of dough has the ends clipped so that the length is uniform. The dough strip then enters the twister. As the shaped pretzel dough leaves the twister it passes under a roller which exerts a slight pressure that sets the knots.

The raw pretzels are placed across a proofing belt approximately 40 ft long, by means of a reciprocating conveyor. From the proofing belt the dough passes through a caustic bath.

The caustic section consists of two tanks. There is a smaller tank through which the pretzels travel and a larger make-up tank, usually at a lower level. The caustic solution is pumped from the make-up tank to the upper or immersion tank. The level in the upper tank is maintained by adjusting the overflow pipe. This system keeps the volume in the upper tank constant. The caustic solution of 1.25 ± 0.25% sodium hydroxide is maintained at 186 to 195°F. If the caustic concentration becomes too high there is not a complete conversion to sodium bicarbonate in the baking and drying cycles and the pretzels will be hot to the taste due to the residual sodium hydroxide. There appears to be no FDA regulation on the amount of sodium hydroxide in the caustic solution.

Immediately after the pretzels leave the caustic solution they are salted. The salter consists of a supply hopper from which the salt is dispensed by means of a grooved roller. Salt slides down a chute until approximately 2 in. above the pretzels and then drops the rest of the way. The general aim is 2% salt on the finished product, but it is necessary that the initial application be at the rate of 8 to 10% due to losses in processing.

The pretzels then enter the oven, which in the case of the Reading Pretzel Machine is usually a 50-ft oven. The bake section is the top portion of the oven and has burners over and under the band that carries the pretzels. The temperature of the bake section is quite variable; it might be 475°F as the pretzels enter and 425°F at the exit of the bake section. The time in the bake section is controlled by a variable speed drive and is between 4 and 5 min. The moisture at the end of the bake period should be about 15%.

The pretzels then go down a slide to the drying section which is underneath and separated from the bake section by heavy insulation. The pretzels that cling to the baking belt are removed by means of a doctor blade. The belt in the drying section travels in the opposite direction of the belt in the bake section. The speed of the drying belt is also variable but is run much slower than the belt in the bake section. The pretzels form a bed several inches deep and remain in the drying section from 25 min up to 90 min. The temperature is held in the range of 225 to 250°F. There is much debate over the drying time and its total effect. Other than reducing the moisture to the desired 2.0 to 2.5%, many claim that the long drying time is needed to temper the pretzel so that it will not break too easily during packaging.

From the drying oven the pretzels are conveyed to the packaging machine. The sooner the pretzels can be packaged, the less breakage there is likely to be. Most companies strive to keep the breakage less than 15% at the time of packaging. Covering the drying belts to protect the pretzels from cold drafts and to facilitate equilibration of moisture vapor is a useful procedure for reducing checking. The relatively large amount of checking in twisted pretzels is due to the moisture gradients set up by the slower bake-out of the thicker knotted parts.

Stick pretzels are extruded using a group of 5 extruding heads containing 10 to 12 holes per extruding head. The dough is forced through the extruding head by means of a helix, and falls on to the proofing belt. As the dough nears the end of the proofing belt it is cut into the desired length by a group of reciprocating knife blades. These blades are circular and travel across the belt cutting the dough. When the knives reach the edge of the belt they rise and return to their starting point. The stick pretzels pass through a caustic and salting operation similar to that for twist pretzels. The temperature is usually kept at a constant 420°F and the time in the bake oven is between 4 and 5 min. The drying section is run at 225 to 250°F with the sticks exposed for approximately 55 min.

Logs and nugget type pretzels are made similarly to the sticks except that they are cut off at the extruder head.

It should be noted that Reading Pretzel Machines also use a 25 ft oven for stick pretzels. All of their ovens are composed of modules approximately 5 ft long, so that ovens can be shortened or lengthened to the customer's specifications and available space. As supplied, the ovens require manual lighting and adjusting of each individual burner.

Water is adjusted to suit varying flour and climatic conditions. The stick pretzel can be made using almost any flour. However, the flour used in twisted pretzels is very critical.

On the American Machine and Foundry equipment the twisters work differently than the Reading Pretzel twisters. American Machine and Foundry have ovens similar to the Reading Pretzel ovens, but they also make or distribute a single pass oven that is approximately 90 ft long. In this oven the pretzels enter a bake section that is about 30 ft long and run at 450°F. The pretzels then enter a drying section that covers the remaining 60 ft and has a temperature range of 225 to 250°F.

TABLE 15.4

TYPICAL PRETZEL FORMULA

	Twist Lb	Stick Lb	Stick %
Flour	160	160	69.18
Shortening	2	4	1.73
Malt (nondiastatic)	2	4	1.73
Yeast	2/5	2/5	0.17
Ammonium bicarbonate	1 oz	4 oz	0.11
Sodium bicarbonate	3 oz	—	—
Water	8 gal.	7.5 gal.	27.08
Yeast food	As required		
			100.00

It must be stated that variations in any and all phases of pretzel production can be found in any operation that is visited. This is true, not only between companies, but also between plants within any single company. At the present time the production of pretzels is more of an art than science, and therefore the pretzel manufacturer is very reluctant to discuss his operations with outside technologists.

Methods for making filled pretzel sticks or nuggets have been patented (Nelson and Nelson 1972, 1973). These usually require the drilling of a hole in a completely baked pretzel stick, and then extruding a paste-like filling into the hole. Fillings can be based on mixtures of peanut butter or cheese with oil and sugar or some nonsweet powder such as lactose or dextrins.

SWEET BAKED SNACKS

Plain Cookies

Plain cookies, as the term is used here, means cookies that are made in one operation, i.e., it does not include filled, coated, sandwiched, and other multiple component cookies. In this section, the principles of formulation and examples of recipes are given for base cakes and plain cookies in the categories of rotary molded, wire-cut, deposited, rotary cut, and stamped goods. There is, of course, some overlapping, as in most classification schemes. It is also true that there are special types of cookies not accurately defined by these broad groups.

It should be understood that the formulas given as examples are intended to serve as points of departure in developing recipes suitable for a given plant, and not as rigid guidelines which can be followed explicitly in every case. The interaction of personnel, equipment, and environment, not to mention the unpredictable fluctuations in ingredient characteristics (especially of the flour), make such universal formulas impractical. Therefore, variations in absorption, sugar, leaveners, or shortening may be necessary in order to get a workable dough based on the sample formulas.

Function of the Ingredients.—The continuous structure of the cookie arises from the flour. The basic framework is tenderized by sugar, invert sugar, egg yolk, ammonia, soda (or baking powder), and shortening. It is firmed or toughened by water, cocoa, egg white, whole egg, milk solids, and the leavening acids. Flavors and spices are usually not present in sufficient quantity to affect texture. Salt is usually considered to be a toughener.

Sugars and especially syrups in large amounts tend to make the dough sticky and hinder release from dies and wires. Shortening is one of the principal agents for increasing tenderness, at least so far as the rich, sweet biscuits are concerned. Too much shortening may lead to a greasy, smeary cookie which is susceptible to rancidity because the free fat soaks into the package, although these effects can be largely overcome by using plastic trays and cello overwrap. Too much sugar leads to hardness and excessive sweetness in the finished cookie.

A wide variety of flours are being used, from a soft cookie flour to a rather strong sponge flour. The stronger the flour, the more shortening and sugar must be used to obtain an acceptable texture. High protein contents lead to hardness of texture and coarseness of internal grain and surface appearance. Chlorine bleached flours are not recommended for soft-type cookies where relatively large amounts of tenderizing and moisture-retaining ingredients such as sugar, shortening, and egg yolk are used (Flick 1964). If the flour is decreased too much, as when large amounts of enriching ingredients are added, the cookie will lack body and may become too fragile.

Whole frozen eggs contribute a better structure and a more delicate texture than do dried eggs, especially in semi-batter wire-type cookies. Frozen eggs seem to give greater volume and a more open grain (Velzen 1963). Whole eggs either fresh or frozen cream readily and seem to provide better structure in drop cookies. Use of egg yolk as a part or complete replacement for whole (using a smaller percentage than whole) will produce a tender cookie with excellent eating quality, but the grain or internal structure of the cookie may not be as good as with the whole egg (Flick 1964).

Velzen (1963) described the ideal relationship of shortening and sugar in three types of doughs as follows: cutting machine, 15% shortening and sugar variable; rotary, 30% of each of these ingredients; and wire-cut 50% sugar and 50% shortening, all on a flour weight basis. Problems in machining are encountered when the shortening content is raised much above that indicated for rotary and cutting machine doughs, but this is not necessarily so for wire-cut doughs. Modern techniques have altered some beliefs; for example, it is possible to run certain shortening-free doughs on cutting machines.

Lecithin seems to increase the shortening effect of fats. It also promotes a tendency for the fat to cover or spread among slightly moist particles of sugar, flour, etc., which would otherwise repel the fat.

Granulated, fine granulated, and powdered sugar can be used alone or in combination to adjust spread and machining properties. For better uniformity of spread a specific granulation should be established. No two suppliers of sugar have the same screen analysis even though the nomenclature is the same. Finer sugars require less mixing than do the coarser varieties and they may reduce sticking to the band. A fine granulated sugar creams better than powdered. Dextrose can often be substituted for up to 20% of the sucrose. Invert syrup must be used with care. According to Wittenberg (1965), it makes wire-cuts, especially wafers, soft, light, and spongy, with an open texture. The crust color is often brighter and develops earlier in the baking.

Milk blends or ameliorates harsh flavors without contributing much flavor of its own. About 5% of the flour weight as skim milk powder is a good average figure for securing the full benefits of the ingredient. Crust color and gloss are generally improved. The protein components bind water and make the dough stiffer and somewhat stickier. The toughening effect on the finished product is minor, in most cases.

Whey has similar results, except that the stiffening, water-binding, and toughening effects are negligible.

Proportions.—The range of ingredient additions in three kinds of cookie doughs are shown in Table 15.5. These data were accumulated by

TABLE 15.5

CONTENTS OF MAJOR INGREDIENTS IN THREE KINDS OF COOKIE DOUGHS

	Water %[1]	Shortening %[1]	Sugars %[1]	Syrups %[1]
Rotary moulded	5–15	10–40	20–45	0–20
Cutting machine	10–25	5–20	15–50	0–20
Wire cut	10–40	10–50	30–85	0–20
Range	0–61	9–98	9–98	0–85
Average	22.2	33.7	43.7	16.9

[1]Based on flour as 100%.

Wittenberg (1965) in a survey of most of the cookie producers in the United States.

Deposit Cookies.—This variety is the machine-made counterpart of the hand-bagged cookie and many of the latter formulas can be successfully adapted to automatic production. Deposit cookies will have about 35 to 40% sugar, 65 to 75% shortening, and 15 to 25% liquid whole eggs. The flour should be from soft wheat, unbleached, with 8 to 8.5% protein and 0.35 to 0.40% ash. It should have a viscosity of 40°M or more and a spread factor of 79 to 80.

The flour must be able to carry the sugar and shortening without too much spread so that the top design is preserved through baking. At the same time, the flour or other ingredients must contribute enough adhesive properties to the dough so that it will adhere to the band and pull away from the main tube of the dough in the deposit stage.

Table 15.6 gives typical basic formulas for three types of deposit cookies. Modifications by adding small amounts of flavors and colors

TABLE 15.6

FORMULAS FOR DEPOSIT COOKIES

	Peanut Butter	Spritz	Star
	Lb	Lb	Lb
Moderately strong flour	50	50	100
Bleached cake flour	50	50	—
Sugar	80	50	45
Shortening	32	40	46
Whole eggs, frozen	12	8	3
Salt	1.5	1.25	1.5
Sodium bicarbonate	0.75	0.4	0.15
Ammonia	0.25	0.125	—
Invert syrup	5	1.25	—
Butter or margarine	—	5	—
Dried sweet whey	—	1	4
Baking powder	—	—	0.5
Vanilla extract	—	0.5	0.25
Water or ice	35	6	16
Peanut butter	85	—	—

may generally be made without other adjustments. For example, orange oil, lemon oil, or butter flavors can be used to advantage in Spritz and star cookies. The amount of sugar given in the formula should be divided between powdered and granulated with the proportions being chosen to give the proper dough consistency and spread. A recipe for basic almond macaroon is 5 lb almond paste, 5 lb powdered sugar, 1.75 lb egg whites, and 0.75 lb white corn meal. Macaroons can be made by the cold process, hot syrup process, or cooked process (McGee 1955). They should not be made by the cold process, however, unless they are to be consumed within three days. The possibility of soapiness developing due to lipase activity is doubtless remote, but it has happened often enough to be a danger.

Wire-cut Cookies.—The dough composition can be varied over a wider range for wire-cut cookies than for any other type. In these doughs, it is necessary to have the material sufficiently cohesive to hold together as it is extruded through an orifice and yet be nonsticky and short enough so that it separates cleanly as it is cut by the wire. Formulas may contain up to several 100% sugar based on the flour, and shortening up to 100% or more, based on the flour content. Doughs may be almost as soft as cake batters or too stiff to be easily molded by hand. The very soft doughs overlap deposit doughs in consistency, while the other extreme is close to rotary moulded type doughs. Advantages over rotary moulded cakes are a more open texture, and over deposit goods, a more uniformly shaped cookie. Disadvantages over the rotary moulded piece are the lack of design, and somewhat less uniformity.

Wittenberg (1965) classified wire-cut cookies as shown in Table 15.7. McGee (1955) further subdivides soft cookies into (1) drop type, as those to be used in sandwich cookies filled with a marshmallow or imitation cream and usually having an amount of sugar equal to the flour, (2)

TABLE 15.7

FORMULATION OF WIRE-CUT COOKIES

	Sugar %	Flour Type[1]	Shorten-ing %	Liquid Whole Eggs %	Final Moisture %
Low-cost promotional cookies	40–50	A	20–25	Little if any	4–5
Standard market shelf cookies	50	A	50	10	4–5
Soft cookies	up to 75	A & B	60	up to 20	12–15
Specialty high quality cookies	35–40	A & C	65–75	15–25	—

[1]Flour Types: A = soft wheat unbleached with 8 to 8.5% protein, 0.35 to 0.40% ash; 40+ viscosity; spread factor 79 to 80; B = cake flour; C = bread flour.

sugar cookies, molasses cookies, coconut, raisin, date, and honey varieties, (3) shortbreads, in which the shortening is usually $\frac{1}{2}$ to $\frac{3}{4}$ as much as the flour, and (4) macaroons, with little or no flour and large proportions of sugar. Such varieties as lady fingers, based on pound cake recipes, would not be included in the above classifications.

Flick (1964) prefers to categorize soft cookies as (1) filled, (2) old fashioned sugar cookies, (3) drops, and (4) bars. For soft cookies he recommends a medium strong soft red winter wheat, lightly bleached with chlorine, with an ash content of about 0.39% and a protein content of about 9.5%, or a strong soft red winter wheat, unbleached having an ash of about 0.41%, and a protein of about 9.5%. The shortening should be a good quality creamable fat. The sugar should be predominantly granulated with perhaps 18 to 24% invert (FWB). Eggs should be present at the 14% level, or more (FWB). Emulsifiers may be necessary.

The vanilla wafer formula is perhaps the simplest version of the wire-cut cookie. A representative formula is:

	Lb	Oz
Flour	100	
Granulated sugar	72	
Invert syrup	5	
Shortening	25	
Nonfat milk solids	1	8
Salt	1	8
Ammonium bicarbonate		4
Egg yolk, dried	1	
Soda		10
Water, variable, about	55	

This formula will yield a finished cookie having a pH of about 7.5 to 8.0. A richer cookie could be made by using larger quantities of egg, especially frozen whole eggs, and by increasing the shortening or using part butter. If the above formula is changed by reducing the granulated sugar by 12 lb, doubling the nonfat milk solids, replacing the egg yolk with 4 lb of whole eggs, and adding enough monocalcium phosphate to bring the pH of the finished cookie within the 6.5 to 7.0 range, a typical sugar cookie will result. Some authorities recommend a pH of 7.0 to 7.6 for sugar cookies, to yield better baked color and texture. Of course, the water must also be adjusted, and in this case might have to be reduced to as low as 22 lb.

To the sugar cookie can be added molasses, oatmeal, raisins, various kinds of nut pieces, coconut, cocoa, etc. To the shortbread can be added slivered almonds, cocoa, etc. Oatmeal cookies require approximately equal parts of flour and oats; against the total of flour and oats should be added at least 65% sugar and 55% shortening. A peanut butter cookie can be formulated on the sugar cookie base using 100% to 105% sugar (part brown), 60% shortening, 75% peanut butter, and 12% eggs. The quality would be improved by also adding 3 to 5% nonfat dry milk.

Table 15.8 summarizes some of the preceding comments and includes additional material enabling a comparison of the formulas for a brown-edge wafer with vanilla wafer and sugar cookie formulas. Flavorings have been omitted. A wire-cut cookie at pH 7.0 will spread and color well if amply fortified with enriching ingredients, but the leaner it is,

TABLE 15.8

COMPARISON OF THREE WIRE-CUT COOKIE FORMULAS

	Sugar Cookie		Vanilla Wafer		Brown-edge Wafer	
	Lb	Oz	Lb	Oz	Lb	Oz
Flour	100		100		100	
Granulated sugar	60		75		85[1]	
Invert syrup	6		5		—	
Shortening	25		25		55	
Nonfat dry milk	3		2		5	
Salt	1		1		1	
Ammonium bicarbonate	—			4		2
Monocalcium phosphate		10	—			4
Frozen whole eggs	4		1		15	
Sodium bicarbonate		12		10	—	
Water or ice, variable	22		55		46	

[1]Powdered sugar.

the higher the pH have to be in order to induce spread. If the pH is too high, any overbaking adversely affects the color and flavor. The difference between brown-edge wafers and vanilla wafers is principally the result of a lower pH of the former resulting from adding monocalcium phosphate and omitting the ammonia. The pH values are about 5.0 to 6.0, and 7.5 to 8.0, respectively. The sugar and shortening have also been raised in the brown-edged wafer in order to secure adequate spread. The sugar cookie is near neutrality in pH, and it actually has a little less sugar. The batter is stiffer, giving a thicker cookie (Bohn 1956).

It sometimes proves difficult to get an attractively cracked top surface on sugar cookies and similar varieties. The type of flour, the granulation of sugar, and the balance of the leavening agents affect this feature. In some cases, the desired appearance can be obtained by holding part of the granulated sugar out of the cream-up stage, and adding it to the dough just before it is taken from the mixer. Others find that use of steam in the oven, or adjustment of the temperature and damper pattern will secure the preferred cracking.

Cutting Machine Goods.—The principal distinguishing feature of this method of machining doughs is that the pieces are cut from a continuous web of dough by a cylindrical die or a reciprocating cutter. A rather large variety of dough types can be handled. It is necessary that the dough be sufficiently cohesive to form the continuous sheet from which the blanks will be cut and to hold the scrap (if any) together, as it is lifted from between the blanks. Reciprocating cutting machines, or stampers, are often discussed as though there were many points of similarity between them and rotary cutters. In fact, there are so many differences between the two types of equipment that it is not particularly in-

TABLE 15.9

CUTTING MACHINE DOUGHS

	Sugar Cookie Lb	Peanut Snap Lb	Milk Cracker Lb	Ginger Snap Lb	Chocolate Lb
Soft flour	100	100	100	100	100
Sugar	30	32	5	12	45
Invert sugar syrup	6	—	—	38[1]	5
Corn syrup	12	—	—	—	—
Shortening	13	10	11	15	18
Salt	0.75	1	0.5	1	1
Malt	1.5	6	1	5	—
Ammonia	0.75	0.25	0.3	—	0.4
Sodium bicarbonate	0.5	0.75	0.75	2	0.625
Acid cream	0.625	—	0.75	—	—
Nonfat milk solids	1	2	1	—	1
Water	12	19	30	8	25
Peanut butter	—	30	—	—	—
Whole egg, frozen	5	8	—	—	—
Flavor	—	1[2]	—	2[3]	15[4]

[1] Molasses.
[2] Vanilla extract.
[3] Usually about 3 parts ginger to 1 part cinnamon.
[4] 5 lb cocoa and 10 lb chocolate liquor.

formative to consider them together. It is true, however, that many formulas can be run on either type with similar results. Stamping machines are mostly used for soda crackers, graham crackers, and the like. They seem to require a somewhat more elastic and cohesive dough than is needed for the rotary cutters. The latter is generally better suited to making small pieces and elaborate designs.

Some examples of typical formulas suitable for running on the rotary cutter are given in Table 15.9. Many modifications will suggest themselves to the reader. The basic sugar cookie formula should be flavored with vanilla, orange, lemon, or mace. To the sugar cookie formula, 50 lb of currants may be added to make currant cake. Coconut might be added, or nut pieces.

The general procedure for mixing these doughs is to cream the salt, sugar, and fatty materials, then mix in the syrups, eggs, and milk. Water with ammonium bicarbonate is added next, and, finally, the flour, soda, and acid creams are mixed to the proper stage.

Rotary Molded Cookies.—In formulating a dough for rotary molded cookies, the consistency must be such that it will feed uniformly and readily fill all of the crevices of the die cavity under the pressures existing in the feeding hopper. The dough blank must be capable of being extracted from the cavity without undergoing distortion or forming tails of considerable size, but it must adhere to the die roll enough to prevent

TABLE 15.10

COMPARATIVE FORMULAS FOR PLAIN ROTARY MOLDED COOKIES

| | Sandwich Base Cookies | | | Butter Cookies Lb |
	Butterscotch Lb	Vanilla Lb	Chocolate Lb	
Flour	100	100	100	100
Powdered sugar	20	32	33	23
Shortening	25	27	24	22
Acid cream	0.5	—	—	0.3
Invert syrup	—	—	5	—
Sodium bicarbonate	0.5	0.5	0.8	0.3
Salt	1.5	1.5	1.5	1.5
Ammonium bicarbonate	—	—	0.25	0.35
Frozen whole eggs	3.5	3	1	3
Sweetened cond. skim milk	6	6	3	6
Butter	1.2	—	—	4
Cultured butter flavor	0.15–0.30	—	—	0.5–1.0
Vanillin	—	0.05	0.05	0.05
Lecithin	0.3	0.25	0.35	0.4
Malt	1.4	—	0.4	1.2
Water variable	10	8	13	8
Dutched cocoa	—	—	5	—
Chocolate liquor, natural	—	—	8	—

the dough from falling out before it reaches the extraction roller. The blank must have sufficient cohesion to hold together and not break up at any of the transfer points before or after baking. The dough must

TABLE 15.11

FORMULAS FOR SPECIALTY ROTARY COOKIES

| | Variety | | | |
	Molasses Lb	Almond Short Lb	Coconut Lb	Sugar Lb
Flour	100	100	100	100
Shortening	35	20	15	35
Sugar	30	28	13	48
Invert syrup	4	2	3	7.5
Frozen whole eggs	—	4	—	2
Sweetened cond. skim milk	10	3	6	6
Brown sugar	—	—	20	—
Molasses	10	—	—	—
Acid cream	—	0.2	—	—
Sodium bicarbonate	1	0.5	0.62	0.35
Ammonium bicarbonate	—	0.25	—	0.5
Salt	1.5	1	0.75	1.5
Butter	—	15	20	—
Water	5	5	6	7
Almond slices	—	6	—	—
Extra fine coconut	—	—	5	—

flow very slightly or smooth out during forming and baking so that woodiness or undersirable irregularities· in the surface are not apparent in the finished cookie. Usually, the spread and rise should be minimized so as not to blur or distort the design.

Doughs formulated to these requirements are usually fairly high in sugar and shortening, and low in moisture. The development of gluten is definitely to be avoided. Table 15.10 gives comparative formulas for plain rotary cookies. To these may be added artificial butter flavors, spices (e.g., mace or nutmeg), pure vanilla, or traces of orange or lemon oil to modify acceptability. The desirable volatiles of pure vanilla are largely lost in a typical rotary cookie bake. Yellow color will be required in butter and vanilla varieties, and carbon black in the chocolate. About 2 oz of dissolved gelatin added per 100 lb of flour improves the gloss of chocolate rotary doughs. Gelatins of low Bloom strength are cheaper and just as effective as the stronger types. They are also easier to dissolve and disperse. Table 15.11 summarizes the formulas for a few specialty items.

Most manufacturers use flour of about 8.1 to 8.2% protein for rotary base cakes, although a range of 7.1 to 9.2% has been reported. Ash should be about 0.415, with a known range of 0.33 to 0.47% being used satisfactorily. Oleo shortening added in the liquid condition is suitable for most of these doughs, but vegetable shortening can also be used. Powdered sugar and sugar syrups are the preferred sweetening ingredients. Nonfat milk solids are often added, but it is thought that condensed milk is preferable since the liquid ingredient removes any possibility of lumps appearing in the finished cookie. Lecithin at about the 0.4% level will improve machineability.

One stage mixing is often perfectly satisfactory, but a creaming operation with most of the minor ingredients added before the flour goes in gives added assurance that lumps of undistributed ingredients will not appear in the cookie. Dough temperatures from 72 to 90°F are being used for rotary sandwich bases. Generally, rotary and cutting machine doughs need higher temperatures than do vanilla wafers and brown-edge wafers, which should be machined below room temperatures.

The following faults can sometimes be corrected by increasing absorption: (1) dough too sticky—won't come out of die cavities; (2) dough too stiff—won't fill die cavities; and (3) surface irregular.

BIBLIOGRAPHY

BOHN, R. 1956. Production of wire cut cookies—basic types—basic formulas. Proc. Am. Soc. Bakery Engrs. *1956*, 306–309.

FLICK, H. 1964. Fundamentals of cookie production, including soft type cookies. Proc. Am. Soc. Bakery Engrs. *1964*, 286–293.

FREEDMAN, J. 1972. Quinlan's non-stop pretzel production. Candy Snack Ind. *137*, No. 8, 36–37, 40.

HESS, J. 1973. Out of the horse and buggy era. Snack Food *62*, No. 11, 25–28.

MCGEE, O. L. 1955. Soft cookies. Proc. Am. Soc. Bakery Engrs. *1955*, 251–260.

MATZ, S. A. 1968. Cookie and Cracker Technology. Avi Publishing Co., Westport, Conn.

MICKA, J. 1955. Bacterial aspects of soda cracker fermentation. Cereal Chem. *32*, 125–131.

NELSON, R. L., and NELSON, W. P. 1972. Apparatus for making food item. U.S. Pat. 3,666,485. May 30.

NELSON, R. L., and NELSON, W. P. 1973. Apparatus and method for making filled food item. U.S. Pat. 3,763,765. Oct. 9.

VELZEN, B. H. 1963. Production of wire cut cookies. Proc. Am. Soc. Bakery Engrs. *1963*, 243–250.

WITTENBERG, H. L. 1965. Wire cut cookies. Biscuit Bakers' Inst. Training Conf. *1965*.

CHAPTER 16

Nut-Based Snacks

Nuts are very frequently used to upgrade popcorn-based snacks and are themselves sold for snacks, either as the individual variety or as mixed nuts. Peanut butter is used in relatively small quantities as an ingredient in commercially prepared snacks, e.g., cracker and peanut butter sandwiches.

According to Woodroof (1967), the salted nut trade uses about 9% of the shelled domestic tree nut production. The proportion of shelled peanuts sold in this way is nearer 16%.

PEANUTS

Sorting

Since peanuts are an agricultural commodity, they can arrive at the receiving platform contaminated with a wide variety of foreign substances. In addition, shrunken, moldy, and discolored kernels must be removed if a quality product is to be made. Separation of impurities can be made automatically on the basis of size, weight, and visual appearance.

A typical air separator classifies products into two or three groups and works on the principle that desirable kernels are either lighter or heavier than the remaining particles. Uncleaned products are floated through a rising stream of air.

Electronic sorting machines measure the reflectance of a light beam from individual nuts and eject those kernels which are significantly darker or lighter than standard.

Blanching

For peanuts, and other nuts, the term "blanching" means the process of removing the red skin or testa. Nuts are often roasted immediately after blanching, but in some cases are held in storage at a separate blanching plant or at the roasting plant for some time before further processing. Conditions of the heat treatment which is part of the blanching procedure have a definite effect on the flavor profile of the roasted nuts (Pattee and Singleton 1971). Storage of blanched peanuts results in pronounced flavor changes, most of them being undesirable.

In the literature can be found descriptions of blanching methods involving rubbing between two rubber belts, slicing with knives, soaking in water, and treating with alkaline solutions. Heating the peanuts and then rubbing them between brushes or ribbed rubber belts is probably the most common procedure. The brushes and belts need frequent cleaning and occasional replacement in order to maintain maximum efficiency. It has also been proposed to water blanch peanuts by cutting the skins of the individual kernels longitudinally as they roll between sharp stationary blades and then running the kernels through a hot water spray to further loosen the loosened skins. Finally, the skins are removed on a knobbed conveyor as the kernels pass under an oscillating canvas-covered pad. Spin blanching is an adaption of the water blanching method wherein the hot water spray is bypassed and the kernels are conveyed directly to the driers where they are quickly dehydrated at a lower than roasting temperature to loosen the skins. Barnes and Holaday (1974) describe a method particularly applicable to small lots of peanuts and depending for its effectiveness entirely upon the action of air currents. A downwardly pointing conical screen has a nozzle at its lower apex. Air at 50 psig is directed into the peanuts at this point. The loosened skins are carried off by a vacuum hood separated from the lower cone by a few inches. A rubber seal joins the open ends of the upper and lower funnels to provide an air tight connection.

Roasting

In roasting, the kernels become dehydrated and a browning reaction occurs throughout the kernel. The texture, appearance, and flavor change markedly. Dry roasting of peanuts for butter or salted nut production may be continuous or in batches. In one version of the batch method, peanuts are subjected to 400°F air in a revolving oven. The peanuts are held at 320°F for 40 to 60 min. Continuous roasters are usually rotating inclined cylinders with louvers directing the nuts in a gradual flow to the exit, but fluidized beds and bin-type roasters have been used.

Oil roasting may also be of either the continuous or the batch type. Batch-type roasters are of the same general type used for french frying potatoes, in which wire baskets are dipped into vats of hot oil. The continuous roasters are elongated tanks with constant circulation of oil, usually externally heated and filtered, and frequently supplemented by additions of fresh oil.

Various research workers have indicated that the unique nutty flavor of roasted peanuts results mostly from browning reactions of the Maillard type. Glucose and fructose originating from the hydrolysis of sucrose are apparently the major carbohydrate reactants. The type of roasting seems to have an effect on the course of the browning reaction, probably because of the variation in rates of heat penetration and therefore in duration of the exposure to high temperatures in different parts of the nut. Young (1973) indicated that microwave roasting might be preferable to more conventional methods, on the basis of the flavor developed. Dry roasted peanuts scored lower than oil roasted for flavor (Young *et al.* 1974).

Salting

It is often difficult to obtain adequate adherence of salt on roasted nuts. The problem is greatest when the oil is of low melting point. When nuts are fried in hydrogenated shortening or 76°M coconut oil salt is applied before they have cooled to room temperature, and while the oil on the surface is still molten. As the fat cools and sets up, the salt granules are held firmly.

Where nuts are dry roasted, the slight binding action observed is caused by the small amount of nut oil which comes to the surface. Adhesion is much improved by adding about 2% of a coconut oil as a dressing to the warm nuts, and immediately thereafter applying the salt.

The type of salt used is very important in obtaining maximum adherence. Small particles of irregular configuration are desirable. Most salt manufacturers can supply salt made specifically for this purpose.

Product "Shrinkage" During Processing

A factor of considerable economic importance in nut processing is the amount of product which is lost, i.e., the difference between the weight of material filled into retail packages and the weight of nuts and other ingredients entering the line. These losses consists of: (1) moisture and other volatile substances released as a result of heat treatment; (2) defective product intentionally removed and discarded; (3) dust and other very fine particles lost to the atmosphere or screened out; (4) shells, skins, and other nut components intentionally or unintentionally removed; (5) nut oil which leaks out and coats equipment; (6) scrap or masses of unmarketable material which have been ruined by erroneous processing methods; (7) samples removed for quality control checks; (8) pilferage; and (9) overfill of packages. Gains in weight occur as the result of addition of salt and other seasonings, and from adherence of oil in the roasting process.

Representative weight changes observed in a large nut processing plant are shown in Tables 16.1 and 16.2.

Low Calorie Peanuts

There have been several attempts to lower the calorie content of peanuts so as to make them more conformable to the needs of people who

TABLE 16.1

PRODUCT WEIGHT CHANGES IN BATCH OIL ROASTING OF NUTS

Product	Oil Consumed (%)	Salt Consumed (%)	Product Yield[1] (%)	Water Lost[2] (%)
No. 1 Spanish peanuts	4.75	1.75	101.70	4.80
Jumbo Spanish peanuts	5.31	1.90	101.83	5.38
Medium Virginia redskins	5.14	1.95	101.73	5.36
Extra large Virginia redskins	5.87	2.19	101.23	6.83
Blanched medium Virginias	3.66	2.30	103.15	2.81
Blanched extra large Virginias	4.91	2.06	102.80	4.91
Nonpareil almonds 23/25	5.22	1.93	102.49	4.66
Jordanolo blanched almonds	3.72	2.17	102.54	3.35
Chipped Brazil nuts	4.04	1.83	103.75	2.12
Scorched whole cashews	5.28	1.75	103.16	3.87
240 Cashews	3.92	2.12	103.12	2.92
Cashew pieces	7.12	2.44	105.12	4.44
Unblanched extra large Filberts	8.76	1.79	105.47	5.08
Pecan halves	15.51	3.36	112.25	6.62

[1]Based on raw net weight of nuts as 100%.
[2]Calculated by subtracting finished weight from total input.

are weight conscious. The most obvious approach is to extract some of the oil with a fat solvent such as hexane. For example, Pominski *et al.* (1964) removed about 81% of the oil from roasted peanuts by this method. The usual observation is that both the flavor and texture of solvent defatted peanuts are nontypical and consumer acceptability is substantially reduced.

A novel method was developed at the Southern Regional Research Laboratory of the U.S. Department of Agriculture in the early 1960's.

TABLE 16.2

NUT SHRINKAGE IN THE PACKAGING OPERATIONS

Product	Package Size (Oz)	Screenings, Sweepings, and Salvage (%)	Shrinkage[1] (%)	Total Lost (%)
Salted in shell peanuts	3	0.68	2.25	2.93
Salted in shell peanuts	6	0.24	2.40	2.64
Walnuts	10	0.93	2.21	3.14
Pecans	10	0.25	2.21	2.46
Blanched Virginia peanuts	1.5	1.08	0.91	0.91
Pistachios in the shell	0.75	0.27	4.00	4.27
Cashews	0.75	0.70	5.29	5.29
Mixed nuts	7	0.55	2.26	2.81

[1]Mostly due to overfill.

Whole raw kernels are compressed until about 80% of the oil is forced out. When they are immersed in hot water, the nuts assume approximately their original shape. They are then roasted and salted. The flavor is superior to that of solvent extracted nuts. Apparently the FDA has objected to calling this product "low calorie peanuts" since the caloric content has not been reduced 50% and the product is not, strictly speaking, a peanut after the treatment.

Salting Peanuts in the Shell

Peanuts are salted in the shell, before roasting, by applying a salt solution and then drying. To speed penetration, a small amount of wetting agent may be added to the water and pressure or vacuum is applied intermittently to force the solution through the shells. A procedure which has been found to be satisfactory is described below.

About 240 lb of unshelled peanuts are dumped into a basket which fits inside a 150-gal. pressure kettle. A retort of the type used for sterilizing canned goods is suitable. The pressure vessel is filled about ⅔ full of a concentrated salt brine (about 27 lb salt in 9 gal. of water) so that the peanuts are completely immersed. The basket must have a cover to prevent the nuts from floating.

The vessel is closed and about 120 psi hydraulic pressure is applied for 4 to 8 min. The pressure is released to allow air to escape from the shells, and then applied again. This pressurization may be repeated several times.

After the peanuts have been brined, the salt solution is drained into a storage tank and the peanuts are flushed with one clear water wash but not allowed to soak. The wash water is drained off and discarded, and the peanuts are removed from the pressure vessel.

Usually, the wet nuts are centrifuged to remove the mechanically held water before they are transferred to the dryer. Drying can be accomplished by heating in a gas-fired rotary dryer 45 to 60 min to adjust the moisture to about 6%. Roasting can be done in gas-fired rotary roaster, for example at about 325°F for 75 to 90 min.

Intermittent application of vacuum would probably be even better than the pressure method for forcing the brine solution through the shells. Perhaps a vacuum pan such as is used for concentrating hard candy mixes would be satisfactory.

Woodroof (1966) suggested the use of either sodium tetraphosphate or Calgon as the wetting agent. The nuts are soaked in a 1% solution for 5 min at 70°F. The detergent treatment is followed by a quick rinse in tap water to prevent an excess of the chemical being drawn into the shell.

OTHER NUTS

Pecans

About 15 million pounds of pecans are consumed annually as toasted and salted nuts, i.e., directly as snacks. This includes pecans used in mixed nuts. Although the total amount does not approach the usage of peanuts in snack form, it is a substantial economic factor, amounting to well over $30 million at retail. Much larger quantities are used in candy, baked goods, and other formulated foods.

Pecans differ from most other nut kernels in that blanching is never required. Pecan meats can be either dry toasted or cooked in oil. Much less heat treatment is required than for peanuts both because the initial moisture content of pecans is lower and the development of a true roasted flavor is seldom desired. Frying is said to give a more uniform roast and cause less mechanical damage. There is also an uptake of oil which offsets the loss of about 4% moisture. Use of antioxidants in the frying medium will help stability.

Optimum roasting conditions will vary with the kernel dimension, but are said to average 12 to 15 min at 375°F (Woodroof 1967). Shorter heat treatments develop insufficient aroma and flavor, while more stringent conditions lead to objectionable dark colors and bitter flavors.

After roasting, the kernels are quickly cooled in tunnels or perforated bins with forced air circulation to remove any undesirable odors that have accumulated. The cooled product is cleaned to remove any debris, foreign matter, and scorched nuts.

Roasted kernels are generally coated with about 1% of finely ground salt. A relatively small amount of toasted pecans are sold in flavored form, that is, coated with cinnamon sugar, barbecue flavored powder, etc. Formulation and application of the powders are conventional in nature.

Storage stability of toasted pecans is very limited and vacuum packaging in metal or glass containers is highly recommended.

Almonds

Almonds are a common constituent of mixed nut packs, and are fairly popular as a separate variety when roasted and salted. Their greatest use is as an ingredient in candy, bakery goods, and other confectionery, but such applications are outside the scope of this volume.

For salted almonds, the kernels are generally not blanched. If blanching is required, the nuts are soaked in hot water until the skin loosens sufficiently to be removed by rubbing. Woodroof (1967) described another method, originated by Leffingwell and Lesser, which consists of passing kernels through a heated solution of 1 oz glycerol and 6 oz sodium carbonate per gallon of water. The loosened skins are removed by a stream of water and any remaining alkali is neutralized by dipping the blanched nuts in a weak citric acid solution. The wet nuts are usually dried before roasting.

Almonds can be dry roasted or fried. Oil roasting is probably the more common method. Hot air flowing through a rotating perforated cylinder or applied to a fluidized bed can be used to roast nuts. In the process the internal moisture decreases to about 1% and the temperature rises to as high as 300°F. Oil roasting can be continuous or batch type. The nuts are immersed in oil (coconut oil being preferred) heated to between 250 and 350°F. When removed from the fat they are cooled, salted, and dressed with oil. Dielectric and microwave heating have been used experimentally for roasting almonds and other nuts.

A considerable quantity of almonds is processed into "gourmet snacks" by coating with smoke flavored salt, cinnamon sugar, and similar flavored powders.

SUGARED AND SPICED NUTS

Woodroof (1967) gives the following formula and method for preparing sugared and spiced pecans.

Sugared and Spiced Nuts (Pecans)

Nuts, whole halves or large pieces	12 lb
Sugar	12 lb
Water	1½ cups
Cinnamon	2 tbsp
Nutmeg	1 tbsp
Cloves	1 tbsp
Antioxidant (Tenox II)	0.5 oz

Heat sugar and water to 240°F in revolving pan, and add nuts and spices, and revolve until nuts are coated. Dry for 1 hr at 120°F, or 2 hr at room temperature, in circulated air. Pack in hermetically sealed tin cans, glass jars or plastic containers, preferably under vacuum. Hold at room temperature if they are to be eaten within a month, otherwise store under refrigeration.

For small batches where revolving pans are not available, the sugar, water, and spices may be heated, added to the nutmeats, and stirred or tumbled until the meats are uniformly coated. The product may then be dried by spreading on a table in front of a fan for 2 hr.

If the nuts are not dried to 4.0% moisture before processing the coating will become damp and sticky upon storage.

The general method is applicable to almonds, macadamia nuts, walnuts, etc. Variations in spicing will readily suggest themselves to the reader.

For adequate shelf-life, the nuts should be packed in vacuum or refrigerated.

BIBLIOGRAPHY

BARNES, P. C., JR., and HOLADAY, C. E. 1974. Pneumatic apparatus for blanching heated or roasted peanuts. U.S. Pat. 3,808,964. May 7.

COX, D. C. 1947. Electric sorting peanuts. Peanut J. Nut World 26, No. 4, 27–28.

HARRIS, N. E., WESTCOTT, D. E., and HENICK, A. S. 1972. Rancidity in almonds: shelf-life studies. J. Food Sci. 37, 824–827.

LAWLER, F. K. 1965. Presses calories from nuts. Food Eng. 37, No. 9, 138–139.

McWATTERS, K., HEATON, E. K., and CECIL, S. R. 1971. Storage stability of peanut pie mixes. Food Prod. Develop. 5, No. 1, 69, 71, 73, 74.

PATTEE, H. E., and SINGLETON, J. A. 1971. Observations concerning blanching and storage effects on the volatile profile and flavor of peanuts. J. Am. Peanut Res. Educ. Assoc. 3, No. 1, 189–194.

POMINSKI, J., PATTON, E. L., and SPADARO, J. J. 1964. Pilot plant preparation of defatted peanuts. J. Am. Oil Chem. Soc. 41, 66–68.

WOODROOF, J. G. 1966. Peanuts: Production, Processing, Products, Vol. 1 and 2. Avi Publishing Co., Westport, Conn.

WOODROOF, J. G. 1967. Tree Nuts: Production, Processing, Products. Avi Publishing Co., Westport, Conn.

WOODROOF, J. G. 1973. Peanuts: Production, Processing, Products, 2nd Edition. Avi Publishing Co., Westport, Conn.

YOUNG, C. T. 1973. Influence of drying temperature at harvest on major volatiles released during roasting of peanuts. J. Food Sci. 38, 123–125.

YOUNG, C. T., YOUNG, T. G., and PERRY, J. P. 1974. The effect of roasting methods on the flavor and composition of peanut butter. Proc. Am. Peanut Res. Educ. Assoc. 6, No. 1, 8–16.

Equipment

CHAPTER 17

Extruding Equipment

The definition of Rossen and Miller (1973) can be considered satisfactory for food extrusion in general: "Food extrusion is a process in which a food material is forced to flow, under one or more of a variety of conditions of mixing, heating, and shear, through a die which is designed to form and/or puff-dry the ingredients." Many kinds of foods other than snacks are processed by such methods.

EXTRUDER DESIGN AND FUNCTIONS

Many kinds of extruders are used in the food industry, but in this chapter only those having special application to the manufacture of snack foods will be discussed. The extruders used in processing snack foods have four functions: mixing, cooking, shaping, and puffing. Various combinations of these functions can be performed simultaneously by a single piece of equipment.

Mixing will occur in most extruders except those designed to receive a pre-mixed plug or cylinder of dough, such as those used in shaping corn chips. To achieve good mixing in these extremely viscous materials special agitator designs are required, however. If cooking is to be performed, jackets are provided for circulating steam to supplement the Btu's originating from the work performed on the doughs. Shaping or forming is a result of the configuration of the orifice through which the material is extruded and the relative speeds of the cut-off knife and the dough strand. If the dough temperature is substantially in excess of 212°F when it leaves the pressurized chamber, the extruder can also function as a puffer, since the dough will expand under the influence of the pressure of steam. In many extruder designs, two or more of these functions are performed simultaneously.

Macaroni presses and meat grinders have been used to extrude snack type foods (Ziemba 1972), but specially designed extruders are generally used because of the high temperature required for efficient processing of modern snacks. Onion flavored rings and similar tube-shaped products

Courtesy of Autoprod, Inc.

FIG. 17.1. AUTOPROD EXTRUDED SHAPE DEPOSITOR WITH NINE HEADS

are being commercially manufactured from seasoned cereal mixes by modified Bock (Haskon) fillers. These rings are generally breaded before frying. Another typical low-pressure extruder is the Autoprod (see Fig. 17.1). Golden Dipt Division of DCA Food Industries extrudes a dough containing diced onions from specially designed low-pressure equipment. A free-flowing liquid coating containing sodium alginate is first applied to the rings and then set by applying a calcium chloride solution. Many different kinds of extruders are being used for doughs and for cookie batters. Details of such extruders may be found in *Cookie and Cracker Technology* (Matz 1968).

Some snacks are made by low-pressure (i.e., atmospheric) cooking and forming. Dry ingredients are mixed with water and heated in a cooking extruder. Heat is applied to the dough from the fluid circulating through the jacket and from the mechanical work exerted by the screw. The cooked dough is then processed through a cooking (less than 212°F) section and forced through a die. Expansion or puffing does not occur. Final shaping and cutting may be accomplished away from the die face. The dough may be processed by frying or toasting, or be puffed by extrusion from another extruder maintained at higher temperature.

The first step in producing conventional corn chips is the processing of dried raw corn kernels (often a mixture of white and yellow corn) with lime water to impart the characteristic flavor and put the kernels in a condition such that they may be further reduced in size. Usually, the corn is cooked for a period of time in the lime solution and then permitted to steep for an extended period of time. Brown and Anderson (1966) showed the extended steep period could be avoided and made continuous by cooking the corn kernels under pressure with sufficient lime or other alkalies. The minimum amount of lime required in their process is 1.5% by weight of the dry corn. Because masa is thoroughly cooked and the starch gelatinized before the dough is placed in the extruder, further heat treatment is not required, and puffing does not occur when the dough strip emerges from the extruder. Some puffing or blistering does occur after the raw piece is placed in the frying fat.

The presses used for corn chips (e.g., Fritos) use hydraulically or pneumatically activated pistons to force the corn meal dough through the shaping orifice (see Fig. 17.2). They are generally batch presses, i.e., the piston must be withdrawn and a charge of dough inserted into the chamber at the start of each cycle. There are a number of similarities to the old-style macaroni product extruders, and to sausage stuffers. Pressures reached in the chamber are much less than the pressures generated in the continuous extrusion chambers used for puffed snack products such as corn curls. As a result, little or no steam evolution occurs within the dough and it does not expand as it exits from the die. Chip size or length is controlled by varying the speed of the cut-off knife. The variety of shapes which can be generated in this manner is rather limited since the three-dimensional effect obtainable in puffed snacks is lacking.

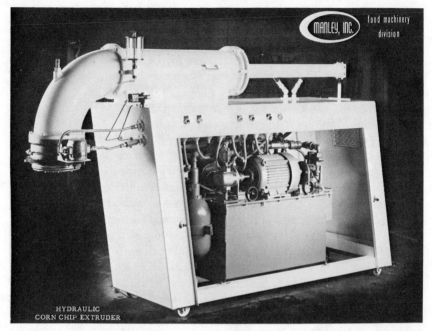

FIG. 17.2. HYDRAULIC CORN CHIP EXTRUDER

Half-products or intermediates are gelatinized starchy doughs which have been formed into chips or other shapes and then dried to a horny consistency. When immersed in frying oil, these shapes rapidly expand into a low-density product.

The half-products are made by heating and pressurizing a cereal pulverulent material until it reaches a plastic gel condition, extruding the plastic material, chopping it to form rings, slices, etc., and subsequently baking or frying the pieces until the moisture content has been reduced to about 1 or 2% by weight and the snack has taken on its characteristic crisp texture and cooked flavor.

Obviously, not all gelatinized doughs puff appreciably when fried. The amount of expansion is determined by the type and percentage of starch, the moisture content, the shape and dimension of the pieces, the temperature of the frying oil, and other factors not well understood.

Some general features of the manufacture of half-products are described in the patent literature, e.g., in the prior art discussion by Marotta *et al.* (1972). Raw, starch-containing material such as flour of corn, potato, tapioca, wheat, rice, sorghum, or oats, is subjected to a simultaneous cooking and kneading operation wherein the starch composition is cooked to the extent of swelling the starch granules, and then directly transferred to an extrusion chamber under high pressures which rupture some of the starch granules. The extrusion pressures are such that the extrudate would normally expand on release to normal atmospheric conditions; but this expansion is avoided by cooling the plastic mass so that the extruded material issues from the orifice in a compacted densified condition with a moisture content of 15 to 35%. The preferred moisture content depends upon the working characteristics of the mass, e.g., its apparent viscosity.

After extrusion, the pieces are dried to a moisture content of less than 12%.

A representative mix for use in this process is 81.5% ungelatinized corn flour (containing about 12% moisture), 16.5% water, and 2.0% salt.

Rossen and Miller (1973) classified extruders both on the basis of their thermodynamic properties and according to the way in which pressure is generated. The thermodynamic scheme is as follows:

(1) Autogenous (nearly adiabatic) extruders. The work exerted on the contained mass generates the necessary heat. The temperature is not adjusted by circulating heating or cooling fluids through the jacket, screw, or die head. Some puffed snack extruders approach this condition, while others require added heat.

(2) Isothermal extruders. The temperature developed in the food mass as a result of the conversion of mechanical energy is maintained at a more or less constant level by circulating a cooling fluid through the jacket or other parts of the equipment.

(3) Polytropic extruders. These devices are intermediate between the autogenous and isothermal types. In a strict sense, all practical food extruders would fall in this category although most are capable of being classified as primarily either autogenous or isothermal.

When classified as pumps, or on the basis of the means of generating pressure, the following categories are applicable.

(1) Direct (positive displacement)
  (a) Hydraulic or pneumatic ram. Has very low shear. Used for corn chip forming.
  (b) Intermeshing twin screw. Generally operates in the low shear range. Seldom if ever used in the manufacture of snack foods, but is used in the chewing gum and chocolate industries.

(2) Indirect (viscous drag)
  (a) Rollers. Low shear. Used for cookies and confectionery.
  (b) Single screw. High shear capabilities.

(3) Hybrids of direct and indirect. Nonintermeshing twin screw mechanisms which can create high shear conditions. Sometimes used for extruding sheets of dough.

The snack foods industry utilizes equipment falling into classes 1a, 2a, and 2b, for the most part.

The ram or piston type extruder is used primarily for cold forming corn chips before they are fried. This type of equipment exerts little or no shear on the dough before it passes through the die. The masa plug is placed in a cylinder which is positioned in front of a piston. Hydraulic or pneumatic pressure on the piston forces the dough through a die containing the properly shaped slots.

Intermeshing twin screw extruders are also of the positive displacement type, but the plastic mass is subjected to some shear due to viscous drag between the intermeshing flights and to a high degree of mixing. Heat is transferred through the jacketed barrel. Long retention times can be achieved through slow speed drives.

The nonintermeshing twin screw extruder is intermediate between the direct or positive displacement type and the indirect or viscous drag flow type represented by the single screw extruder. Since there is some back flow, both extensive (flight to flight) and intensive (layer to layer) mixing can be achieved with nearly positive displacement pumping action. Equipment of this type can be run with a reverse pitch section of flight to perform a highly effective mixing action (Farrell 1972).

According to Smith (1971), the following factors of an extrusion cooking and puffing process may be altered in order to modify texture, density, mouthfeel, solubility, and form of the product in a desired direction.

(1) The method of feeding and preconditioning of ingredients and/ or mixtures.
(2) The type and point of moisture application.
(3) Temperature and moisture content of product entering the extruder.
(4) Control of temperatures within each extruder zone.
(5) Control of point within the extruder where maximum dough viscosity is obtained.
(6) Control of extrusion speeds.
(7) Control of time/temperature relationships within each section of the extruder.
(8) Control of period during which product temperatures are elevated to maximum extrusion temperatures.
(9) Control of final extrusion temperatures.
(10) Selection of the shaping and sizing components of the system.

(11) Selection of the type, dwell time, drying temperatures, and the velocities within the drier and cooler and the desired final product moistures.

(12) Point and method of flavor application.

According to Farrell (1972), the cooling of a screw will give the effect of a screw with a shallower flight depth while heating or cooling by means of a jacketing arrangement (conductive or convective) has no great effect on mixing or conveying action except where screw velocities are low enough to permit a reasonable time for heat transfer to take place. Jacket heating can be used on start-up to bring the extruder to equilibrium sooner but grain type extruders usually employ jacket heating only as an insulating medium and jacket cooling to help prevent burning of the material.

Puffing Extruders

The use of pressurized extrusion systems for making puffed or expanded snack products is a fairly recent development. The original snack was, of course, popcorn. Attempts to obtain similar textures and novel appearances from other starchy bases led to more complex processing methods and equipment. Puffed wheat, puffed rice, and puffed corn obtained their first success as breakfast cereals. Much of the reservoir of knowledge which was tapped for puffed snack technology resulted from engineering for macaroni products where high-pressure extrusion of stiff doughs has been used for many decades.

Puffing of whole grains by heating them in high-pressure chambers, usually with the addition of water, before releasing them to the lower pressure atmosphere, was the next step. Some expansion can be achieved by treating certain gelatinized kernels such as rice with high temperature air.

It was then discovered, more or less by accident, that expanded pieces of desirable texture could be made by extruding high-temperature plasticized and gelatinized cornmeal from a pressurized chamber into the atmosphere. The first corn curl processing equipment was essentially the same device used for extruding cattle food pellets. From this adventitious beginning, specialized equipment has been developed for puffed snack food production.

Puffing extruders are generally designed to convert number two grade dry-milled cornmeal (hull and germ removed) into various shapes and sizes of expanded snack pieces. Small additional amounts of water can supplement the natural moisture of the meal to achieve the desired expansion. This ingredient water can be added by tumbling the meal in a simple mixing device while the liquid is sprayed or dripped into the chamber, or by injecting water or steam into the feed part of the extruder. Often, an amount of water calculated to be slightly below the optimum is added to the meal, and then the amount added to the feed is adjusted until the desired results are observed. Food dyes can be added to the water.

The conventional extruder used for puffing or expanding cereal mixes consists of a spiral, tapered feed element rotating inside a strong, closely fitted chamber. The space for the feed material becomes increasingly restricted as it moves toward the outlet. This leads to a great increase in pressure. Temperature also increases by absorption of kinetic energy from shear forces and may be further heightened by heat applied to the jacket or die. The particles of the feed material form a plastic mass under these conditions. As the compressed mass is released from the extruder orifices, the lower pressure allows the water to flash into steam, expanding the mass into a low-density cellular structure. After the puffed pieces are dried in ovens or fried in hot oil baths, flavoring materials such as salt, cheese, and oil are applied.

COMMERCIALLY AVAILABLE EXTRUSION EQUIPMENT

Several firms offer extrusion equipment which can be used by the snack food manufacturer. These will be briefly surveyed in this section of the chapter.

Manley, Inc., of Kansas City, Missouri, makes three models of corn chip extruders. The pneumatic corn press, model 179, with a capacity of 480 lb per hour has a 6-in. cylinder and is manually loaded. Their hydraulic corn chip extruders have integral hydraulic units which automatically cycle in a sequence of fast approach, slow extrusion, and fast return. A variable speed drive for the cut-off knife controls chip size. Cylinders with 8- or 10-in. diameters give outputs of 840 to 1,200 lb per hour. The automatic corn chip extruder model 141, pneumatically activated and automatically loaded and packed, has a production capacity of about 1,800 lb per hour.

The Baker Perkins Ko-Kneader has been suggested for use in preparing and forming masa pieces for corn chips. This piece of equipment is a kneading extruder having a rotor or screw with interrupted flights inside a closely fitting cylindrical chamber. The casing has three rows of spaced teeth projecting into the lumen. By a specially designed reduction gear drive unit the screw is both rotated and reciprocated so that the stationery teeth affixed to the jacket pass through the openings in the helix. As a result, some of the corn dough is temporarily held back by each of the fixed teeth, increasing the kneading, working, and mixing action. Ultimately, however, all of the mass will be carried forward and delivered to the orifice under great pressure.

The design causes individual particles in the plastic mass to trace out loops, giving a maximum opportunity to undergo shear while at the same time absorbing heat from the steam circulated in the jacket. The screw is also internally heated by pressurized steam. Cold tap water is circulated through the tempering head which is provided at the extrusion end of the Ko-Kneader in order to bring the dough below 212°F and prevent its expansion as it emerges into the atmosphere at a moisture content of, e.g., 23%. A cut-off knife traverses the orifice at a rate which can be adjusted to produce pieces of the desired length.

The "Model R-150" puff type extruder manufactured by Richeim Co. (South Bend, Ind.) is shown in Fig. 17.3. It is powered by a 25-hp motor and can produce 200 to 300 lb per hour of puffed corn collets. Rice, corn, and soya mixtures can be processed. Some of the shapes which have been produced are corn curls, flats, "hulless popcorn," scoopers, balls, and onion rings. This company also manufactures an extruder for fry-type corn collets.

The Sprout-Waldron (Muncy, Penna.) continuous pressure cooker is designed to cook or gelatinize cereal grains and other starchy materials and to create an expanded product by forcing the cooked material through a relatively thin die. The temperatures necessary for gelatinization of the starch are obtained by introducing steam into a continuous cooker rather than by converting mechanical energy into heat. This method is said to substantially reduce energy requirements of the extruder.

According to the manufacturer, there are seven essential steps in the operation. (1) Material is introduced from a bin into the feeder inlet; (2) the feeder screw delivers the material into a pressure sealing unit which passes it into the cooking section without allowing the high pressure steam to escape; (3) steam (and water, if necessary) are added to the pressure cooker through a bustle; (4) during a retention time of 3 to 5 min the material is activated by a slow speed paddle agitator controlled by a variable speed drive; (5) at the cooker exit a scraper on the agitator positioned over the outlet provides a positive discharge; (6) the material drops from the cooker into a high speed extruder which forces it toward a single stationary die; and (7) the material is extruded through holes in the die.

The number of die openings and the area of each opening is related to the through-put and cooking pressures. If through-put remains constant and cooking pressure is increased, the open area of the holes must be decreased to maintain the "plug" in the extruder. This can be accomplished by sealing some of the holes in order to decrease the total open die area. Increasing the feed rate without sealing any of the die holes will accomplish the same effect.

Courtesy of Richheim, Inc.

FIG. 17.3. MODEL R-150 PUFF-TYPE EXTRUDER FOR CORN MEAL

A variable speed cut-off knife assembly is located at the die discharge. When the die openings have been drilled on a horizontal line, the drive shaft of the cut-off knife assembly is set perpendicular to the extruder shaft. When die openings have been drilled in a radial arrangement, the cut-off drive is rotated 90° and its shaft becomes parallel with the extruder shaft. A pivot arrangement makes it possible to switch from one cut-off method to the other.

To clean out the chamber, the cooker cover bolts are removed so that the hinged cover can be swung back out of the way. The extruder flights are welded on to a stainless steel tube which slips on to the keyed extruder drive shaft.

The 8-in. extruder offered by the Bonnot Co. (Kent, Ohio), is designed to take cooked doughs containing 25 to 30% moisture and extrude them at pressures up to 3,000 psig. Production rates can approach 1,200 lb per hour. Extrusion temperatures and pressures can be closely controlled by circulating cooling or heating fluids through the jacketed barrel and the hollow screw. Temperatures in the range of 450 to 500°F can be obtained. When used in conjunction with a continuous cooker, the extruder can perform a cooling function.

In all cases cooking is performed as a separate operation. Extrusion rates will vary widely depending upon the type of material being processed, the speed setting of the variable speed drive, and size of the die opening, but production rates of 40 to 120 lb per hour can be anticipated for the 2-in. machine.

In order to clean the machine, first the die is removed by unbolting six retaining bolts. The torpedo head of the extrusion worm is then unscrewed and an eye bolt put in its place. The extrusion screw can then be pulled out, opening the inside of the machine. The unit can be filled with water and allowed to soak for a short time, after which a long brush is used to clean out the bore.

The two models of Wenger cooker extruders are the X-150, capable of processing 6,000 lb per hour or more with a 150 hp motor, and the X-200 which can extrude about 10,000 to 14,000 lb per hour using a 200-hp motor. The Wenger process utilizes atmospheric steaming and water addition with a 1 to 5 min retention time in the feed section followed by high-speed shearing for perhaps 10 to 12 sec just prior to the extruder inlet. The feed material will enter the extruder at 200 to 210°F and will be slowly heated by screw compression (decreasing pitch and decreasing flight depth) and by a series of steam locks which prevent blow-back from higher pressure regions of the extruder (see Fig. 17.4). Temperature builds up quickly in the final cone shaped section. The Wenger plant is located in Sabetha, Kan.

Another company which makes nearly all of the equipment (except packaging equipment) needed for a corn puff line is Manley, Inc., of Kansas City, Mo. (see Fig. 17.5). Their model 140-A collet machine will accept about 300 lb per hour input of cornmeal from a stainless steel hopper with vibrator. It has a rectangular tube construction with stainless steel face plate, hopper, and guard. An automatic refrigeration unit

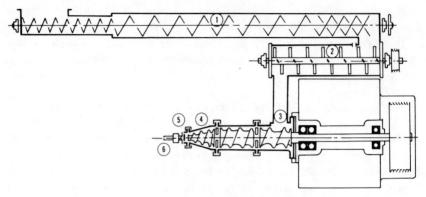

FIG. 17.4. MODEL 150 CONTINUOUS COOKER EXTRUDER

1—Section for steam preconditioning of feed material at atmospheric pressure. 2—Moisture addition and mixing zone. 3—The first section of the extruder assembly works the wet mixture into a dough. 4—The second section of the extruder assembly elevates the dough temperature to 240 to 350°F within 10 to 20 sec. 5—Die with one or more orifices. 6—Cut-off knife mechanism.

of 0.5-hp capacity supplies cooling water to the head. There is a water feed adjustment with a blender-agitator to condition the meal during the feed cycle for maximum expansion. The cut-off knife has a variable speed drive. A 25-hp motor with variable speed drive powers the extruder. The Manley Co. offers a variety of forming dies for the manufacture of cylindrical pieces, balls, chips, hollow tubes, etc.

Manley also manufactures a collet extruder designed for research and development work. It is more fully instrumented and controlled than the standard equipment.

Dorsey-McComb, Inc., manufactures equipment for complete puffed snack lines. Output of 300 lb per hour, and more, is claimed for their corn puff extruder. Features are somewhat similar to those of the equipment previously discussed.

The Krispy Kist Corn Machine Co. of Chicago offers a puffing extruder for cornmeal snacks. It appears to be of conventional design, but further details are not available (see Fig. 17.6).

COMPLETE PLANTS

As a minimum, a line for producing expanded corn meal snacks would include a conveyor feeder for the collet machine, an extruder, conveyor from the collet machine to the oven, an oven or dryer, a coating device with tumbling conveyor and equipment for applying powder and oil, and whatever packaging equipment is selected. A line suitable for pro-

Courtesy of Manley, Inc.

FIG. 17.5. COLLET MACHINE MODEL 140

ducing 300 lb per hour of finished cheese curls might have the following equipment (descriptions taken from commercial literature).

(1) Auger feeder for conveying corn meal to the collet machine, with a 400-lb hopper and 0.75-hp motor.

(2) Collet machine with a 200-lb per hour input, equipped with a vibrated stainless steel cornmeal hopper. Die head will be water cooled with a temperature controller. Cut-off knife will have means for adjusting speed. One or more die plates must be available.

(3) A flighted transfer conveyor to feed the oven. A stainless steel hopper receives the collets from the extruder and deposits them

Courtesy of Krispy Kist Korn Machine Company

FIG. 17.6. EXTRUDE FOR CORN MEAL SNACKS

on a mesh belt with stainless steel side guides. It will have a distribution plate to insure uniform feeding to the oven intake belt.

(4) A 300-lb per hour collet oven consisting of a 4 ft × 10 ft oven box and a 36 in. × 12 ft high carbon mesh belt with roller chain edge and stainless steel product guides. A 400,000 Btu recirculating air flow system incorporating a 1.5-hp fan provides the necessary heat. Capacity is based on 9% moisture infeed material, a bulk density of 3.5 lb cu ft, collets piled 2 in. high on the belt, and a dwell time of 3 min in the oven.

(5) A flighted transfer belt to take the dried collets from the oven to the coating tumbler.

(6) A coating tumbler 30 in. in diameter, 10 ft long, and rotating at a rate of 10 rpm to 94.5 rpm. The variable speed drive operates from a 0.5 hp motor. Three 1,000-watt calrod heaters with reflectors can be used to provide heat control within the tumbler.

(7) Two 80-gal. stainless steel jacketed kettles provided with high-speed mixers. These are used to blend the oil and other components of the coating mixture.

(8) A pumping system for the oil and cheese mixture. Generally includes a gear-type pump with about 0.25 hp variable speed drive, a spray tube assembly with dual nozzles, and a piping network from kettles to pump and pump to coater.

(9) A salter with stainless steel hopper, vibrator feed, and warm air blower.

(10) A flighted transfer conveyor to feed the packaging machine.

At the time this is being written, such a line would cost in excess of $60,000 exclusive of ancillary devices and installation. An example of such a line is shown in Fig. 17.7.

SPECIAL PROCESSES AND RECENT IMPROVEMENTS

Greater uniformity in the physical characteristics (especially size and density) of collets can be obtained by advancing the cereal grain material into an extrusion head by a screw flight rotating within a confined tubular sleeve which is internally threaded or grooved oppositely to the screw flight. The groove in the internal surface of the sleeve is transversely curved and defined by three different radii of curvature. The net effect of this design seems to be that the material is held back or circulated by the groove design, creating additional grinding and frictional heating so as to insure uniform plasticizing and gelatinizing of all the cereal particles. An improved extrusion head coupled with the sleeve acts to compress the material to an extent sufficient to assure the production of collets having uniform dimensions.

Onion flavored snack rings are made from a mixture of waxy maize starch, fat, and emulsifiers that is brought to about 30 to 31% moisture, cooked or gelatinized in one kind of extruder and then formed into a half-product by another extruder. The primary or gelatinizing extruder is typified by the Bonnot machine. It has a 25/1, L/D, 8/1 compression ratio with constant pitch and increasing root diameter. The barrel has five jacketed segments, the first and last being water cooled while the center three are heated by 100 psig steam. The screw is hollow and can be heated or cooled as required by the process. A 2.25-in. diameter extruder operating at 10 to 15 hp will generate 400-lb pressure at the die head.

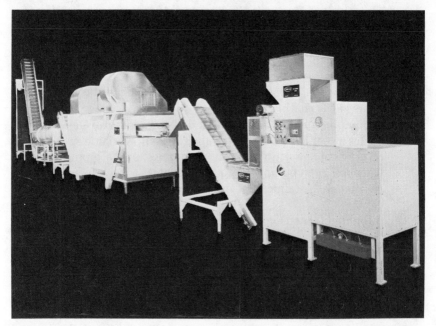

Courtesy of Manley, Inc.

FIG. 17.7. COMPLETE COLLET PRODUCTION SYSTEM—FROM RIGHT TO LEFT: PUFF-ING EXTRUDER, OVEN FEED CONVEYOR, DRYING OVEN, TUMBLER FEED CONVEYOR, TUMBLER FOR APPLYING SEASONINGS, AND CONVEYOR TO PACKAGING MACHINE HOPPER

The forming extruder has only two jacket sections but otherwise is similar in construction to the cooking extruder. It may be necessary to use two or more forming extruders to accept the output of one primary extruder.

Sticks of square cross section similar in appearance, flavor, and texture to French fried potatoes can be made from a mixture of 75 parts granulated potato solids and 25 parts corn grits with enough water to raise the moisture content to about 19% (Straughn *et al.* 1971). After the water is added, the mixture is tempered for a time sufficient to in-sure uniform distribution of the moisture, then it is transferred to a col-let extruder where it is worked at a controlled temperature up to 340°F. The plastic mass is adjusted to a temperature in the range of 255 to 340°F and a pressure between 800 to 1,500 psig before it is extruded. The compression screw length to diameter ratio falls between 1:1 and

3:1. The expanded sticks are toasted-dried and coated with an edible oil and salt. If the die orifice is shaped like a cross ("intersecting slits"), the sticks will be square in transverse section. By allowing the extruded strands to hang straight downward before cutting and until they set or solidify, curling is prevented. The finished product will have a typical composition of 53.25% potato solids, 17.25% corn solids, 25% coconut oil, 2.5% salt, and 2% moisture by weight.

A method and apparatus for cooking the meal in the shaping process, obviating a separate cooking process and giving rings or other forms of improved conformation, was patented by La Warre (1973). The meal mix must contain at least 5% of water, on a bone-dry basis. Additional amounts up to 25% can be added for increased puffing. The equipment must be capable of grinding the meal to a powder and subjecting it to elevated temperatures and pressures sufficient to cause it to go into a liquid flow state. A sequence of zones for feeding, compressing, and pumping must be provided. In the first compression zone a temperature of at least 200°F and a pressure between 4,000 and 15,000 psig acts to plasticize the ingredient mix, while a second compression zone is maintained at 200°F or above and at a pressure of 5,000 to 80,000 psig. Moisture is expelled in these compression zones until the content in the mass is no greater than 3%. Under the recommended conditions, complete cooking will take place with cornmeal in the liquid state at a temperature of between 375 and 400°F in less than 10 sec. The die is maintained at a temperature of 300 to 400°F. The cutting knife must have a clearance of about 30 mils from the face of the die plate.

A continuous system for making corn chips from alkali treated whole corn, without the hulls removed has been described by Cunningham *et al.* (1962).

Treated corn containing hulls and some whole kernels, is delivered by conveyor to the hopper of an extruder. The extruder comprises a pair of delivery rolls situated at the bottom of a hopper. The delivery rolls guide the mass to a gear pump which kneads it and forces it through a tapering extrusion passage terminating in a die head with one or (usually) more orifices. Each orifice is provided with a mashing and cutting element mounted for sliding movement in a vertical direction over the front face of the extruder head.

Masa extruded from the die orifices emerges in the form of thin ribbons, each of which is cut into sections of appropriate length by the sliding knives. The bevelled lower surfaces of the cutting elements are wide enough to exert a mashing action on the emerging dough. Because of this, any lumps or whole kernels which may occur in the dough are forced into the space below the orifice and are mashed each time the knife is actuated.

Cunningham *et al.* (1966) in a later patent described a slightly different method of forming sheets of masa. A hopper receives the corn dough and delivers it to a pump which forces it through a short tapered chamber ending in a die having an orifice of about 16 in. width and $\frac{1}{16}$ in. height. As the sheet of masa emerges from the orifice, it falls on to the surface of an endless conveyor belt. A complex system of circular cutters severs discs of dough from the continuous sheet and holds them until they are ejected into the frying fat. In this procedure it will be important to have the masa consist of uniformly fine particles, since any large lumps or pieces of hull tend to hang up in the die and create slits in the emerging dough sheet.

BIBLIOGRAPHY

CONWAY, H. F. 1971. Extrusion cooking of cereals and soybeans. Part I. Food Prod. Develop. *5,* No. 4, 27, 29, 31.

CUNNINGHAM, F. A., KLATT, J. O., BROWN, R. F., and SLOVACEK, W. W. 1962. Preservation without chemicals. U.S. Pat. 3,020,162. Feb. 6.

CUNNINGHAM, F. A., KLATT, J. O., PINSON, A. G., and BROWN, W. B. 1966. Apparatus delivering masa under high pressure. U.S. Pat. 3,294,545. Dec. 27.

FARRELL, R. J. 1972. Available extrusion equipment—types and applications. Presented at the 57th Annual Meeting of the American Association of Cereal Chemists, Miami Beach, Fla. Oct. 29–Nov. 2, 1972."

LA WARRE, R. W., Sr. 1973. Snack food production. U.S. Pat. 3,711,296. Jan. 16.

MAROTTA, N. G., ZWIERCAN, G. A., and BOETTGER, R. M. 1972. Manufacture of starch-containing food products. U.S. Pat. 3,652,294. Mar. 28.

MATZ, S. A. 1968. Cookie and Cracker Technology. Avi Publishing Co., Westport, Conn.

ROSSEN, J. L., and MILLER, R. C. 1973. Food extrusion. Food Technol. *27,* No. 8, 46, 48–53.

SCHENKEL, G. 1966. Plastics Extrusion Technology and Theory. American Elsevier Publishing Co., New York.

SMITH, O. 1971. Why use extrusion. Presented at February 12, 1971 Symposium on Extrusion Cooking. American Assoc. Cereal Chemists, St. Louis, Missouri.

STRAUGHN, R. O., ELOFSON, G. L., and REINHART, R. D.1973. Method and apparatus to make potato stick snacks. Can. Pat. 929,406. July 3.

ZIEMBA, J. V. 1972. Strides in forming. Food Eng. *44,* No. 11, 66–70.

CHAPTER 18

Equipment for Frying, Baking, and Drying

INTRODUCTION

This chapter contains discussions of the equipment designed specifically for applying heat to the product at some stage in its manufacture. Other chapters provide descriptions of equipment and processes which heat the ingredients, but usually incidentally to the performance of another function such as mixing, forming, or puffing. In any case, the distinction is not pretended to be a sharp one, however.

Specialized equipment for popping corn or roasting nuts is also described in other chapters.

HEAT TRANSFER MECHANISMS

Heat may be transferred from one location to another by convection, conduction, or radiation. The relative effectiveness of each of these means of transfer varies with oven design as well as with the distribution of product size and conformation. The response of the product pieces, and thus many of the qualities of the finished product, is affected by all of these factors.

Radiation

Transfer of energy by electromagnetic radiation is a significant factor in most ovens or dryers. These radiations are not in themselves heat, but are converted into heat through absorption by and interaction with absorbing molecules. Light, which is an electromagnetic radiation, is an extremely poor means of transferring heat. Invisible infrared radiation is quite effective. The radio frequency waves generated by electronic ovens are also efficient transmitters of energy. The quantity of heat transmitted between two bodies by radiation is directly proportional to the differences of the fourth powers of the absolute temperatures and inversely proportional to the square of the distance between the two bodies.

Radiation has two characteristics different from the other means of heat transfer which affect its action on the heated particles. (1) It is subject to shadowing, and (2) it is very responsive to changes in absorptive capacity of the material, e.g., to changes in coloration in the case of infrared radiation or to changes in water content in the case of radio waves of certain frequencies.

Changes in color affect thermal profiles principally through increases in the absorption of infrared rays. The darkening accompanying baking indicates an increase in the absorption of visible wavelengths. An increase in absorptive capacity for infrared rays, though not apparent visually, is an almost invariable concomitant of the visible changes. As a result of the increase in heat absorption in the darkening areas, there is a tendency for the color changes to accelerate after the first browning appears. This means that ovens relying on radiant energy for a relatively large proportion of the heat transfer will tend to accentuate color differences. Such effects may be either good or bad, depending upon the characteristics desired in the finished product.

The shadowing effect is most apparent with radiation near the visible range (i.e., infrared) and is much less with radio frequency waves. The infrared radiation comes from all angles in the oven, although its intensity will vary depending on the temperature of the source. The radiation originates from burner flames and all the metal parts in the oven sides, top, band, etc. Radiation from below the band is completely intercepted by the band and partly reradiated as infrared energy and partly converted into conducted heat. If the product is in a container, such as a baking pan, the sides will shadow the piece and greatly reduce the effectiveness of radiant heating.

If a food particle is shaped approximately like a segment of a sphere, with a relatively smooth surface, the reception of radiation during its trip through the oven will be approximately equal over all parts of its surface. Absorption will also be approximately equal as long as the color (related to absorptive capacity in the infrared) remains the same over the entire surface. This also applies to flat-surfaced pieces covering most of the band. Products with irregular surfaces, such as wire-cut cookies with holes in the middle, corn curls, popcorn, pretzels, etc., will have some parts of their surfaces in a better position than other parts to intercept radiant energy. If these surfaces are sufficiently absorptive, heating will occur at an accelerated rate in the prominences. The shadowed areas will receive less energy and will tend to heat slower and dry or bake slower. It must be understood that all of these effects of radiant energy can be offset by complementary inputs of conducted or convected heat.

The specific effects of·microwave radiation will be discussed in this chapter.

Convection

Convection is the transfer of heat from one part to another within a gas or liquid by the gross physical mixing of one part of the fluid with another. In the oven, molecules of air, water vapor, or combustion gases heated by whatever means, circulate throughout the oven, constantly mixing with the other gases and transferring heat by conduction when they contact solid surfaces. Within the food piece, convection occurs as the result of the movement of water vapor and other gases. Furthermore, some translocation of liquid water, melted fat, and other liquids cause a transfer of heat from one region of the dough to another.

If the overall results of convection can be generalized, it would be as a smoothing or evening effect of heat distribution. The gases within the oven mix readily as long as there are no mechanical barriers and tend to make more uniform the temperature throughout the chamber. When different temperatures are required in different zones of the oven, it is necessary to isolate these zones in some manner to retard heat transfer between them by convection.

Within the food material, one of the principal effects of convection is a smoothing or blurring of temperature differentials. The hotter areas of the particle give off more water vapor than the cool ones, and the loss of this vapor in some areas and its gain in others, tends to minimize the differences.

Convection affects exposed areas first, of course. The bottoms of the pieces, protected as they are by the conveying band, do not participate directly in this type of heat exchange. It is doubtful that much convective transfer occurs at the bottoms even when mesh bands are used. All exposed parts probably participate about equally in convective heat exchange. No doubt some protective effect of protuberances could be shown, but the high degree of turbulence of the gases known to exist above the band and the relatively low relief of most dough pieces suggest that variations in reception of convective heat by the different parts cannot be very great. On the other hand, products at the edge of the band may receive substantially more heat because of more rapid flow of hot gases and higher temperatures in the area.

Convection is effective in speeding up baking because it strips away the layers of relatively cold air which surround the product and serves to "even out" the temperature throughout the oven space. Since the hottest air is at the burners and the coldest air next to the product, this "evening out" constantly brings hotter air into contact with the dough.

In addition to distributing heat more uniformly, convection also conveys moisture vapor. In the absence of injected steam, the highest concentration of water vapor will be found next to the piece.

Conduction

Conduction is the transmittal of heat from one part to another part of the same body, or from one body to another in physical contact with it, there being no appreciably displacement of the particles of the bodies. The fundamental differential equation for heat transfer by conduction is called Fourier's law:

$$q = \frac{dQ}{d\theta} = -kA\frac{dt}{dx}$$

where q is the rate of heat flow in Btu per hr, Q is the quantity of heat transferred in Btu, θ is the time (hr), A is the area normal to the direction in which the heat flows (sq ft), and dt/dx is the rate of change of temperature (°F) with the distance (ft) in the direction of heat flow, that is, the temperature gradient. The factor k is called the thermal conductivity and is dependent upon the material through which the heat is flowing and upon temperature. The negative sign indicates that heat flows in the direction of lower temperatures. The thermal conductivity of water at 32°F is 0.343 and at 100°F is 0.363 Btu/sq ft/°F/ft.

In baking in a band oven, conduction of heat to the dough occurs only through the band. The band receives its store of energy from radiation, from convection, and from heat conducted through the supports on which it rides. Because of the localized nature of conductive transfer, steep gradients of temperature are set up within the dough piece, the hottest areas being the ones in contact with the pan and particularly where the pan is in contact with the band. Unwanted differences in the rates of heat-catalyzed reactions can easily occur unless these gradients are carefully controlled. Conduction from one part of the dough to another is a force tending to reduce temperature differentials.

To summarize the effects of the three types of heat transfer during baking or drying, it can be said that conduction and radiation tend to cause localized temperature differentials, conduction acting to raise the temperature of the bottoms and radiation acting to raise the temperature of exposed surfaces (and especially darkened areas or protuberances), while convection tends to even out temperature gradients within the oven. In frying systems, radiation plays a much smaller role and most of the heat transfer occurs by convection within the frying medium and by conduction from fat to product.

Further details of the relative contributions of the three modes of heat transfer to baking effects in a multi-zone band oven can be found in the article by Kaiser (1974).

OVENS

Ovens are used for baking cookies, crackers, pretzels and snack foods of similar types. Some half-products can be puffed in conventional ovens although most of them are finished in fryers. Band ovens are the common types found in all large cookie and cracker plants today, and other designs are of only historical interest or perhaps are used in some small specialty shops.

Band ovens are usually classified in accordance with the type of heating system. They may, for instance, be classified as being heated by gas, electricity, oil, or other means. Because the fuel or energy source does not tell us much about interaction of the oven with doughs, it would seem to be more meaningful to categorize them on some other basis. The following classification scheme is preferred by the author.

(1) Ovens in which the primary heat source is outside the baking chamber.
   (a) Combustion gases circulate in closed ducts around the baking chamber.
   (b) Steam or other gases flow through numerous radiator tubes inside the baking chamber.
(2) Ovens in which the primary heat source is inside the baking chamber.
   (a) Ribbon burners (strong convection component).
   (b) Radiant heaters, electric or gas.
(3) Ovens in which heat is generated inside the dough piece (electronic ovens).

Flexibility, accuracy of temperature control, economy of operation, safety, and direct effects on product characteristics are all influenced to some degree by the method of heating. The differences can best be understood in terms of the relative contribution of the three different types of heat transfer described earlier in this chapter.

The modern band oven is an outstanding triumph of engineering and technology. With very little maintenance, an oven 300 ft long and 39 in. (1,000 mm) wide will routinely bake 2,000 to 3,000 lb of cookie dough per hour continuously for long periods. On certain doughs, much higher outputs can be obtained. Assuming proper adjustment, the millions of units put out in 24 hr of operation will be uniform enough to satisfy the most discriminating of consumers. The feeding and removal of dough pieces is routinely automated so as to require minimal attention by the operators.

Baking Chamber Construction

The oven chamber consists of a frame supporting the necessary rollers, guides, burners and the like, together with insulation on top, sides, and bottom. Vents with dampers, clean-out doors, and observation ports are included as necessary.

Fiberglass wool is a common insulating material. When this is used, the oven is cased in enameled metal such as 10 gauge steel with a white vitreous enamel. A representative arrangement would be 10 in. of 3 lb density fiberglass on top, 8 in. on the sides and 6 in. on the bottom.

Sheet material composed essentially of asbestos fibers and binders can be used both as insulation and structural component. In this case, an outer covering is not necessary.

The baking chamber is usually provided in modular units, for example, 10 ft lengths. Ovens up to 400 ft are in operation, while ovens as short as 60 ft are used for special items.

Bands

The baking surfaces can be either solid steel bands, perforated steel bands, or wire-mesh bands. The earlier mesh bands had a fairly open weave, but the more recent versions have had a finely-textured closely-woven structure. Perforated bands and wire mesh allow steam to escape from the bottoms of dough pieces and help prevent gas pockets and distortions. They also tend to retard spread.

According to a survey reported by Suarez (1963), solid steel bands now being used range in thickness from 0.032 in. to 0.092 in. but most are 0.062 in. Thinner bands allow use of smaller drive drums and assure rapid transmittance of heat. Wire mesh bands reported in this survey had an average thickness of $\frac{7}{32}$ in.

Most steel bands are low carbon alloy steels which have been cold rolled according to given specifications. The cold working process generally increases the yield strength and tensile strength, and reduces ductility. The hardened steel is tempered to restore the ductility (Lugar 1962). The necessity for a straight, flat band is obvious. When properly made and maintained, a split band provides satisfactory service. Patched bands are highly undesirable. Uniform thickness is also essential. The expansion and contraction of the band caused by temperature changes is compensated for by pneumatic cylinder adjustments or by heavy counterweights on one of the drums.

On occasion, cookies may stick to the band. Presence of particles of sugar or milk solids in contact with the band will inevitably lead to sticking. Certain doughs are much more troublesome than others. Lean doughs which are relatively high in water are more prone to sticking. Sticking can also be accentuated or alleviated by band characteristics.

Dirty bands with built-up carbon deposits and bands with numerous pits and scratches are conducive to sticking. The band should be slightly oily from previous runs of rich cookies. If it is not, it may be necessary to grease it lightly or dust it with flour. Use of liquid shortening in the dough helps to eliminate sticking, though it may cause other difficulties.

New bands are usually conditioned by applying sufficient shortening to thoroughly saturate the surface. Sometimes beeswax or other special materials are used. The shortening is usually rubbed on at the inlet and the excess rubbed off at the exit end of a warm oven (300 to 350°F).

Temperature is then increased in 100°F steps until the highest operating temperature is reached. Lard, lard plus lecithin, corn oil, and coconut oil are some of the band conditioning materials which have been used.

The band is powered by a variable speed motor. Relatively low horsepower units are required. The time of baking is adjusted by varying the speed of the band.

Band Guides

In long band ovens, even a slight amount of sideways movement will soon accumulate to the point where the band will be damaged and the oven jammed unless corrective measures are taken. This sideways movement is prevented by band guides which may be of several kinds. The simplest, and least satisfactory, are angle irons welded at strategic places along the band length. They will force a wandering band back into place, but they can also cause serious damage to the band in the process. Friction quickly wears out these simple guides.

Spring-mounted vertical spools fixed at numerous places along the band are reasonably satisfactory. They do cause edge wear. Canting the rollers supporting the band will guide it without edge wear, but this system requires close attention by the operator. Overcompensation, or failure to correct the faulty condition soon enough, can ruin the band and damage the oven. A combined system utilizing spools as a sensing device to warn the operator when to adjust the support rollers is probably the best compromise solution. A recently patented system in which the spools themselves cant the rollers when the spools are driven off center by a wandering band may be even better.

Heating Means

The two main types of conventional ovens are those with ribbon burners in the baking chamber itself and those with combustion chambers located outside the oven proper. Ribbon burners are placed approximately 2 ft apart while the recirculating ovens have a heating unit about every 50 ft. There has been a great deal of controversy about the relative desirability of these two systems, but it is well known that both types are being used satisfactorily. Open flame or surface combustion (ceramic) elements have been used, but the former type is more common.

Several burner systems have been devised to obtain maximum safety and control of the heat sources within the oven. The open inspirator system uses gas at 1 to 5 lb pressure and room air drawn into the inspirator by the Venturi effect. Good combustion is obtained at or near the maximum output, but at lower settings adequate air for complete combustion may not be entrained. The question arises as to the fate of solid particles drawn in with the room air. Smaller particles will undoubtedly pass through the jets and be burned without causing any problem. Large particles which might clog the burner can be entrapped by filters fitted over the intakes to the inspirator. Even so, some accumulation of deposits inside the burner occurs, and these systems are designed for easy and fool-proof cleaning of the inspirator and other critical parts.

The zone proportional mixer system uses air pressurized to approximately 1 psig to draw gas at substantially atmospheric pressure into a proportional mixing unit. This gas is depressurized to zero gauge by a special valving arrangement. The combustible mixture is distributed to several burners, usually to a complete zone. Control is effected by changing the volume of air going into the proportional mixer. Any change made affects all of the burners in the zone, making this system rather inflexible. The air is filtered before being pressurized.

A modification of the zone proportional system mixes the air and gas in proper ratio at each burner. It is necessary to lead a zero pressure gas line and low pressure air line to each burner and to have a separate proportional mixer and gas valve for each burner. This system provides excellent flexibility although it may be questioned as to how often the high level of control achievable by it is actually necessary.

Gas and air for all the burners in an oven are premixed in a special device in the premixed gas system. The disadvantage of this method is the presence of an explosive mixture in relatively large quantities throughout the distribution system. Each burner must be fitted with a fire check to prevent backfiring. The system is flexible because each burner can be adjusted separately, but because of the potential dangers is not much used. Some ribbon burners can be adjusted so as to give different flame height in three sections across the band.

In the exterior-combustion chamber type of oven, several arrangements are possible. Generally, there is one combustion chamber per zone, which restricts the number of adjustments which can be made. On the other hand, relatively close control of the temperature of the circulating gases can be achieved. The type of burner is more-or-less immaterial so far as the effect on the dough piece is concerned.

Zone Control

Division of the oven into zones which can be independently controlled in temperature, at least to a considerable extent, provides the cookie and cracker technologist with the means for adjusting heat treatment to the particular requirements of a given cookie type. The burners can be regulated separately for the top and bottom of each zone; stationary or movable baffles separate the zones; and the exhaust control dampers are separate for each zone. The number of zones in an oven will vary according to requirements, but separation into more than eight zones is rarely done. It is often constructive to consider the oven as divided into three sections even though more than this number of zones are present.

The recommended temperature pattern for most wire-cut cookies is a fairly low temperature in zone 1, where the spread and rise of the dough takes place; a considerable increase in temperature in the intermediate zone(s) to set and bake the cookie; and a slightly lower temperature in the final zones to give the desired color and dehydration. Relatively low bottom heat, as compared to the top heat, is used in the first zone to promote good spread and a finer, more even cell structure.

In baking saltines, the first zone (or section) of the oven is adjusted for high bottom heat and relatively low top heat. This arrangement leads to proper bottom development and starts the gas evolution necessary for adequate spring while at the same time maintaining a soft uncrusted top which will allow moisture and gases to escape. Shelliness and premature development of blisters result from excess top heat in the first zone. In the middle section, the top heat is brought up and the bottom heat is usually reduced somewhat. The top of the cracker is set, and the final shape of the biscuit develops, although moisture is still being driven off. In the third section, the moisture is reduced to the point where dehydration can be satisfactorily completed during cooling, and the coloration is begun.

Rotary goods and some rich wire-cut doughs require low heat in the first zone, slightly increased heat in the middle section, and the correct intensity of heat for bake-out and coloration in the final zone. This provides the gentle, slow development necessary for these doughs, and prevents excessive or premature rise. Higher bottom heat in the early zones tends to set the bottom early and may facilitate the removal of sticky doughs from the band.

Production Rates

Band oven production may be calculated according to the following formulas (Anon. 1963):

$$\text{Oven capacity (lb/hr)} = R \times N \times L \times \frac{1}{C} \times \frac{60}{T}$$

where R is rows of goods per ft of band length, N is rows of goods across band width, L is baking chamber length in ft, C is count of baked biscuits per lb, and T is baking time in min. To determine N and R, let W = actual product width in in., D = maximum baked size of cookie, S = spacing from center to center of adjacent rows, and X = clear band space at each edge, then

$$W = (N \times S) - S + D + 2X$$

or, $$N = \frac{(W + S - D - 2X)}{S} \qquad \text{and, } R = \frac{12}{S}$$

when all dimensions are in inches.

ELECTRONIC OVENS

Electronic ovens generate heat within the product as a result of vibrations set up in some of its molecules by absorption of electromagnetic radiation of high frequency. The radiation is created by electronic circuits resembling radio transmitters in a very general way. There are two principal types of electronic ovens, the dielectric heating variety using frequencies in the range of 30 to 40 megacycles, and the microwave or radar types using frequencies of 915, 2,450, 5,800, or 22,125 megacycles per sec. Microwave systems are the basis of the electronic ovens used in homes and restaurants for the rapid heating of miscellaneous foodstuffs. The installations in cracker and cookie plants have been principally of the dielectric heating type because the economics appear to be considerably better at high heat transfer rates. Some microwave baking units have apparently been used for bread and similar products.

According to several published reports, installation of electronic heating in existing ovens, though possible, has many disadvantages. The heat tends to break down insulation and the accumulation of deposits on the conductors and electrodes causes current leakage. Furthermore, any unevenness in the band, which must form part of the electronic system, causes nonuniformity in the current flow. The solution to these problems has been to place the electronic heating unit just beyond the terminus of the band oven, using a plastic conveyor.

Use of electronic heating during the early stages of the baking process seems to offer no advantage over more conventional techniques. In the first or development stage, the dense dough can be rapidly heated on the band. In the second stage when spring occurs, the application of electronic heat leads to an excessively vigorous evolution of gases with some distortion of the structure. It is in the final stage, where a delicate balance between coloration and moisture reduction must be maintained, that electronic heating is most valuable. Moisture can be reduced with little or no browning because the localized dehydration with rapid temperature rise which occurs during conductive, convective, or infrared heating does not occur. Regions with the most moisture are heated most and the heat input drops as the moisture approaches zero. By bringing the entire piece to very low moisture content, a slight internal coloration can be induced.

This relationship of the moisture content of the piece to energy input is a very important feature of electronic ovens. The wetter the dough the more heat is generated in it. Because of this relationship, the ovens are self-compensating for variations in moisture content. The final moisture contents of the baked cookies and crackers tend to be evened out by this effect. Not only can excess or uneven browning be alleviated, but checking can also be reduced.

DRYERS

Puffed snacks must be dried after extrusion in order to obtain moisture contents sufficiently low to insure good texture and storage stability. Several companies have designed and manufactured ovens for this purpose. The conveying means is rather generally a series of metal mesh conveyor belts which pass the product back and forth through an insulated chamber (see Fig. 18.1). Gas flames are often used for direct heat-

Courtesy of Food Engineering Corp.

FIG. 18.1.   LARGE MULTI-TIER SNACK DRYER

ing of the chamber, with fans to maintain a constant air flow. They have many obvious similarities to baking ovens, and the primary difference is that the dryers do not provide a baking surface on which the dough pieces may lie or a conveying means for pans. Also, the heat input is generally much less in the dryers.

Dryers intended specifically for puffed snack foods and similar foods are being made by the following manufacturers.

Food Engineering Corp.
2722 Fernbrook Lane
Minneapolis, Minn. 55441
phone: (612) 544-5055

Food Machinery Div.
Manley, Inc.
1920 Wyandotte Street
Kansas City, Mo. 64141

The Lanly Co.
26201 Tungsten Road
Cleveland, Ohio 44132

Krispy Kist Korn Machine Co.
120 So. Halsted Street
Chicago, Ill. 60606
phone: (312) 733-0900

The dryer manufactured by Food Engineering Corp. has automatic humidity-temperature controls for temperatures up to 300°F. It is recommended for drying flakes, granules, chips, pellets, etc. Air movement through the dryer is symmetrical from both sides to provide an even air flow throughout the product area. The product is fed onto the top belt and progresses through a series of up to 14 lower belts, finally emerging on the bottom belt (see Fig. 18.2). The amount of exhaust air and circulating air, and the heating capacity, are tailored to the specific needs of the user. Conveyor belt widths can be from 2 to 10 ft, while lengths of up to 100 ft are available.

The Lanly Co. supplies a snack food dryer with a capacity of 1,800 lb of corn puffs per hour. Moisture is removed at a rate of 130 lb per hour at an average temperature of 350°F and an average process time of 1.5 min. Maximum temperature is 500°F.

The 34 in. wide conveyor belt of wire mesh is arranged in a three deck configuration, with the corn puffs falling from one pass to another and finally on to the packaging conveyor. Conveyor speeds can be varied from 2.27 to 22.7 ft per min.

A continuous rotary drying oven is offered by Krisy Kist Korn Machine Co. It heats with natural, butane, or manufactured gas and has automatic gas and temperature controls with safety pilots. The variable speed drive is adaptable to capacities of up to 400 lb per hour.

Manley, Inc., baking and drying ovens have recirculating airflow heating systems using gas or electricity. They incorporate multi-tier (3, 4, or 5) continuous belts with variable speed drives and exterior bearings.

Courtesy of Food Engineering Corp.

FIG. 18.2.   SEVEN PASS SNACK DRYER—SIDE PANELS ARE REMOVED TO
SHOW CIRCULATING FANS AND HEATING COILS

FRYING

Block (1974) states that frying differs from other heat processing
methods in the following principal respects. (1) Cooking is accomplished
in a relatively short period of time, generally within 5 min, due to (a) a
great temperature difference between the heat source (fat) and the food,
and (b) the size of the individual food unit cooked is usually very small,
often being less than one ounce in weight. (2) The frying fat becomes a
significant component in the end product—varying from as little as 10%
by weight of the end product in breaded fish sticks up to 40% or more by
weight in potato chips. (3) Fried products are usually crisper on the
outer surface than other heat processed foods. (4) The heat transfer medium (frying fat) is subject to changes in composition and often in performance characteristics during its process life. (5) There are unique

mechanical problems involved in commercial frying operations. Although Block did not base his discussion primarily on snack frying operations, it is clear that most of his points apply with equal force to such products as potato chips.

Batch fryers are available, but they are now used almost entirely by restaurants and other retail food outlets. They are far too inefficient from a labor standpoint to be satisfactory for high volume snack food manufacturing. The batch fryers used by the original small potato chip manufacturers have evolved through a series of improvements and enlargements into completely continuous, high capacity, closely controlled automatic frying systems.

Most continuous frying systems consist of at least five more-or-less independent sets of equipment: (1) the tank or trough containing the frying medium and providing a means for directing drained fat back into the kettle, (2) a conveying means for moving the food into, through, and out of the fat, (3) the fat system which pumps and filters the frying medium and replenishes it as needed from a bulk supply, (4) a heat source with its control system for generating thermal energy as required and transferring it to the fat, and (5) an exhaust system for removing the hot vapors emerging from the vat (see Fig. 18.3). The fat itself is an essential and interacting part of the whole system, and its characteristics strongly affect the functioning of the equipment and the results of the operation as reflected in the quality of the finished product.

Conveying mechanisms for transferring the product from the entrance to the exit end of the trough include the following: (1) spacer bar conveying, in which the frying piece floats between transverse bars that push it over the surface at a rate adjusted so as to give the desired cook by the time it emerges from the end of the vat, (2) fat current conveying, in which the flow of the pumped fat carries the product through the length of the fryer, (3) drop plate conveying, in which the product rests on shelves moving below the fat surface until the pieces develop enough buoyancy to rise to the surface and contact the upper conveying means (older styles have stationary plates), (4) submerger conveying, in which a mesh band near the surface of the fat carries the floating product, (5) conveying nonfloating products by means of baskets, belts, or other holding devices traveling well below the surface of the fat, and (6) enclosed conveying, in which products are carried between two horizontal mesh belts.

Wright et al. (1973) describe a fryer said to improve on pre-existing equipment by reducing the tendency of chips to adhere to the equipment and promoting uniformity of product cook. The Wright et al. fryer is an elongated trough wherein a major portion of the cooking oil is introduced into the kettle at the food-receiving end over a large area in es-

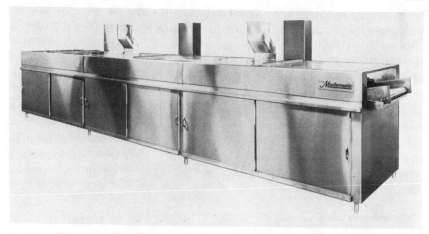

Courtesy of J. C. Pitman & Sons, Inc.

FIG. 18.3. MODEL 24–30 PITMAN MASTERMATIC AUTOMATIC CONTINUOUS FOOD FRYER

sentially counter-current contact to the food slices being brought into the kettle. The chips initially fall towards the bottom of the kettle but are caused to move gently upward by a rising current of hot cooking oil which also helps to separate the wet slices as a result of the violent boiling action. In a second zone of the heater an agitating device and oil issuing from nozzles positioned above the surface cause enough turbulence in the cooking medium to separate the slices and expose all surfaces to the oil. In this second zone, a portion of the cooking oil is withdrawn and passed to the third zone, where the final cooking is performed at temperatures most effective for completion of the cooking process. The concentration of slices in the cooking oil results in what may be called a slurry which is dense enough for essentially plug flow. It is claimed that cooking times may be reduced by as much as 33%, and greater uniformity of product results.

Control Systems

On-and-off and modulating control systems are available. The on-and-off control consists of a thermocouple activating a thermostat switch and can maintain the temperature within a range of about 10°F. The more expensive modulating thermal control systems can adjust the amount of gas flowing to the burner as required to maintain the temperature within ±2°F.

There are three types of controllers used for ovens, dryers, etc.: pneumatic, electric, and electronic. Pneumatic controllers receive a mechanical incoming signal and use a relatively complicated air pressure balancing system to produce a pneumatic force which operates the oven valves. The sensing element can be a liquid filled bulb connected by a capillary tube to the control instrument. A system of levers multiplies the slight motion of the liquid to an extent sufficient to activate the balancing section. An electric controller also receives a mechanical incoming signal, and uses a simple electric resistance slide wire balance to generate the electric impulse which operates the valves. The electronic controller receives an incoming electric voltage signal from a thermocouple and processes this through an electronic circuit to modulate the electric current which operates the valves. Most, if not all, modern equipment utilizes electronic controllers.

Exhaust Systems

Exhaust systems must collect and dispose of large volumes of hot vapor in a safe and sanitary manner. The fryer may be completely inclosed or overhead hood may be used. The inclosed design contributes very little total air movement to the room, but hoods exhaust large amounts of room air along with the steam and it may be necessary to provide make-up air to the area in order to compensate for the loss. All exhaust systems must be designed so as to prevent condensed materials such as fat from dropping back into the fryer.

Heat sources may be located within the kettle or they may heat fat in an outside reservoir for transfer to the frying area. Gas-fired tubes immersed in the fat and electric immersion heating coils have been used. Heat exchangers have been designed for direct and indirect heating of the fat. The fat can be circulated in a tube running through a chamber in which gas is burned, or indirect systems such as those utilizing chlorinated hydrocarbon heat transfer fluids can be used.

Suppliers of Snack Frying Equipment

There are several manufacturers of continuous frying equipment. Among these are:

Design + Process Engineering, Inc.
29 B Street
Burlington, Mass. 01803

Heat and Control, Inc.
225 Shaw Road
South San Francisco, Calif. 94080

J. C. Pitman and Sons, Inc.
Concord, N.H. 03301

The fryers made by Design + Process Engineering are direct electrically heated with an endless conveyor belt which travels just below the surface of the oil. A screw conveyor in a V-shaped trough at the bottom of the fryer continuously removes · food particles. The conveyor discharge material as well as skimmed surface oil is pumped from the discharge end of the fryer to a fine-mesh vibrating screen and then passed through a 5-micron filter.

Heat and Control manufactures high capacity potato chip fryers with continuous circulation heat exchangers, gas or oil fired. They have provision for continuous removal of fines from the oil, and a central control panel with a pneumatic proportioning temperature control. Standard sizes can fry from 1,400 to 3,200 lb per hour of potato chips.

Equipment offered by J. C. Pitman includes a continuous snack fryer which is adaptable to extruded, puffed, or popped products. Among the claimed features are "power jet" burners, a lift-out conveyor, constant time-temperature-motion regulation, and a "Cool Zone."

BIBLIOGRAPHY

ANON. 1963. Direct-fired and wire-mesh ovens. Greer Tech. Information Sheet *BT-101.* J. W. Greer Co., Wilmington, Mass.

BLOCK, Z. 1964. Frying. *In* Food Processing Operations, M. A. Joslyn, and J. L. Heid (Editors). Avi Publishing Co., Westport, Conn.

DAY, A. M., and DOWNING, G. D. 1972. Automatic popcorn popping method. U.S. Pat. 3,697,289. Oct. 10.

DERSCH, J. A. 1960. Bakery ovens. *In* Bakery Technology and Engineering, S. A. Matz (Editor). Avi Publishing Co., Westport, Conn.

DOWNS, D. E. 1958. Mechanized fried foods production. Bakers Dig. *32,* No. 2, 52.

KAISER, V. A. 1974. Modeling and simulation of a multi-zone band oven. Food Technol. *28,* No. 12, 50, 52–53.

LUGAR, T. R. 1962. Economic Biscuit and Cracker Baking. Thomas L. Green & Co., Indianapolis, Ind.

PIRRIE, P. G. 1934. The use of steam in bakers' ovens. Am. Soc. Bakery Engrs. Bull. *98.*

RITCHTER, O., JR. 1973. Operations of an oven. Proc. Am. Soc. Bakery Engrs. *1973,* 53–58.

SMITH, H. L., JR. 1960. Heat transfer in hot fat cooking. Food Technol. *14,* 84–88.

SMITH, W. H. 1966. What happens in the baking oven? Biscuit Maker Plant Baker *17,* 652–656.

SUAREZ, P. E. 1963. Oven bands—their use and maintenance. Biscuit and Cracker Baker *52,* No. 9, 74, 76.

WRIGHT, E. S., ANGSTADT, J. W., and GARROW, G. L. 1973. Apparatus for deep fat cooking. U.S. Pat. 3,754,468. Aug. 28.

CHAPTER 19

Specialized Equipment for Popcorn Processing

INTRODUCTION

Although some of the equipment described in other chapters may be used for certain phases of popcorn processing (e.g., packaging), there are many commercially available devices which have been specifically designed for such applications. I propose to discuss in this chapter the types of specialized equipment a food technologist would have available in setting up a line suitable for the large scale production of popcorn-based snacks, and to give specific examples, including the manufacturers' identification and some brief specifications thereof.

A typical fully automated dry popping line might include cleaning devices, a continuous dry popper, a sifter, a seasoning-coating reel, and the conveyors (pneumatic and mechanical) needed to get the ingredients and product from one processing unit to the next. If caramel corn is being made, the additional equipment required would be a caramel mixing kettle with automatic controls and metering pumps, film type caramel concentrator, corn coater, cooling separator tunnel, and a continuous belt-type cooling conveyor which delivers the finished product to the packaging equipment (de Muesy and Stinson 1971).

Havighorst (1970) described a large popcorn plant built to supply popped corn to concessions in a chain of 287 theaters. The corn is received in 100-lb bags and stored in a cool area to reduce moisture loss. When needed, the corn is dumped into a conveyor which delivers it to a series of hoppers, each supplying two batch-type popping machines. As directed by the automated system, a hopper will deliver 8 lb of corn through a flexible metal pipe to the popper. The popper has no direct temperature controls, but the stack reading during operation is about 650°F. Popping time is adjustable, but in this facility is claimed to be 4.5 min. Automatic controls reverse the perforated drum to allow removal of any unpopped kernels through a slot made by overlapping part of the drum surface.

Drums of coconut oil are liquified, a small amount of butter flavor added, and the mixture is pumped to a spray unit which coats each 100 lb of popped corn with 25 lb of oil. Salt of fine particle size is dusted on the oiled kernels.

Commercial popping units have developed through a series of evolutionary changes from the skillet or kettle used in home popping to very efficient continuous units which can deliver several hundred pounds per

hour of extremely uniform product. Home popping is mostly wet popping, in which some sort of vegetable oil is used as a heat transfer medium, but perforated baskets for dry popping over an open fire have received some use ever since colonial times. Commercial poppers were
originally wet poppers, too, but there are inefficiencies connected with
this process which led to the development of dry popping equipment.

POPPERS

There are two categories of popping equipment—wet and dry. Although dry poppers predominate in continuous production facilities
today, there is still some use of wet popping equipment (which is essentially batch type).

Popcorn can be wet popped in oil (usually coconut oil) with or without artificial color and flavor. Temperatures of about 480°F are required for maximum yield and expansion. A fully automated wet popping line called the Pop-O-Matic is offered by Manley, Inc. The popper
accepts a 5-lb charge of corn about every 3 min from a hopper holding
400 lb or more. Loading, popping, and dumping are controlled automatically in accordance with the temperature sequence. Popcorn, oil, and
salt are dispensed in measured amounts for each batch, and the quantities may be changed to suit different formulations. The complete plant
includes a seasoning mixer kettle, a conveyor assembly for moving the
popped kernels to a tumbler (for removing small pieces, unpopped corn,
etc.), a blower conveyor, and a remote control panel (see Fig. 19.1).

Cretors, Inc. produces straight line wet method popping plants consisting of conveying systems, mechanical oil feed, poppers, and a sifter.
They recommend that the popped corn storage bin be fabricated on site
to fit the space limitations of the customer's plant. Conveyors to the bin
are not included in the basic setup. The basic frame is made to accom-
modate three gas popping units, and Cretors will assemble as many of
these units as are required to give the needed output.

The framework of the plant is reinforced square 2-in. tubing painted
in sanitary white baked enamel. The conveyor is 10 in. wide sanitary
coated belting which travels inside a stainless steel trough attached to
the frame. The conveyor is driven by a ⅓ hp gearhead motor. A unique
recirculating oil system automatically delivers oil to each popping unit.

This system has a positive displacement pump which constantly re-
circulates the oil at low pressure past the discharge points for each popping unit. The oil charge is measured automatically in a volumetric tube
at each discharge point. A turn of the valve releases a measured oil
charge directly into the kettle. Returning the valve to the original position allows the volumetric tube to refill in preparation for the next
batch.

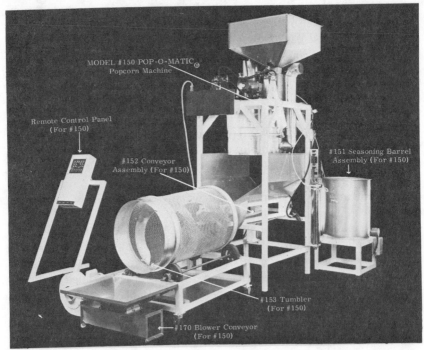

Courtesy of Manley, Inc.

FIG. 19.1.   A FULLY AUTOMATED WET POPPING LINE

Courtesy of Krispy Kist Korn Machine Company

FIG. 19.2.   A POPCORN LINE BASED ON FIVE SMALL "FRENCH-FRY" POPPERS

Krispy Kist Korn Machine Co. also manufactures wet poppers or "French fry" units which can be assembled in a series over a conveyor belt to give semi-continuous operation (see Fig. 19.2). The heavy cast aluminum kettles can pop about 3 lb of corn per cycle. Gas is supplied through a ball joint coupling allowing the kettle to be rotated for dumping the charge. A motorized stirrer is mounted in the cover.

There are two kinds of dry poppers, the batch style, which may be automated and assembled in groups to give essentially continuous operation, and the continuous style.

Batch style poppers consist essentially of a rotating perforated drum mounted in an insulated cylindrical housing in which gas is burned. The commercial dry popper (Model 685W) of Krispy Kist can deliver approximately 150 lb per hour of popped corn, or 10 to 12 lb per cycle. It is gas heated and manually loaded.

The Cretors "Flo Thru 200" automatic and continuous popper processes approximately 200 lb per hour of raw corn. The immersion-type gas burner consumes about 100 cu ft per hour of 1,000 Btu natural gas at 6 to 8 in. water pressure. Major items within the oven are a supply fan, gas burner, and popping drum. The inclosure is insulated with 2 in. of fiberglass.

The Manley continuous flow automatic dry popper (Fig. 19.3) is made in two sizes—a large unit having a capacity of 350 to 600 lb per hour and a smaller machine with a capacity of 200 to 250 lb per hour. An adjustable speed feeding device automatically delivers the raw kernels into the popping tumbler. Forced hot air at controlled temperature is directed against the tumbling kernels. As the kernels pop, they are immediately moved to the exit end of the drum, leaving only unpopped corn in the heated zone. The machine has a recirculating air flow system with an Eclipse fresh air burner. Gas controls are fully automatic and include safety devices. Controls include Partlow indicating temperature controller, Honeywell modulating controller, pilot pressure regulator, and Partlow high temperature cut-off (Middleton 1969).

The Krispy Kist Korn Machine Co. manufactures a continuous automatic dry popper.

SIFTERS

After the corn is popped, it should be sifted to remove unpopped kernels, small fragments of popped corn, charred debris, etc. All popped corn sifters consist of a rotating inclined drum made of wire screen or perforated metal (Fig. 19.4). The unwanted scrap pieces fall through the screen and are collected in a pan or on a conveyor underneath the drum. The pattern of the mesh weave and the size of the openings are critical

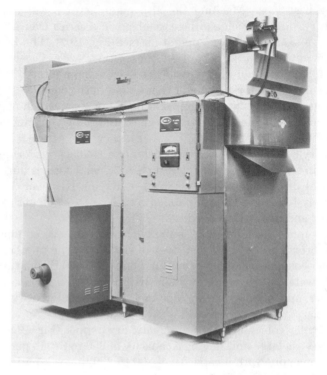

Courtesy of Manley, Inc.

FIG. 19.3.   CONTINUOUS DRY POPPER

in reducing clogging of the sieve by lodged kernels and it also has an effect on the percentage of kernels which are broken down by the tumbling and abrading action of the sifter. A helical metal ribbon is affixed to the inside of the cylinder to insure movement of the kernels to the exit end. Usually the rate of rotation is fixed. A typical throughput would be 750 lb per hour. The steel mesh cylinder may be totally inclosed by a metal housing, or the top half may be left exposed. Most firms manufacturing poppers also offer sifters to round out their line. Krispy Kist, Cretors, Manley, etc., make or distribute equipment of this type.

COATERS

Equipment for applying oil, salt, cheese, and the like is described in detail in Chapter 23. Candy coating equipment is described in the following section on Caramel Corn Plants.

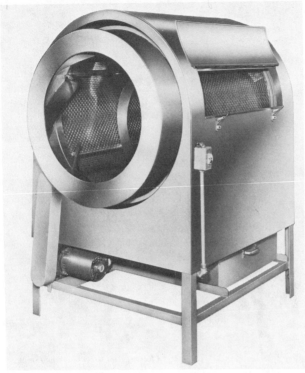

Courtesy of Krispy Kist Korn Machine Company

FIG. 19.4.   CONTINUOUS ROTARY POPCORN SIFTER

CARAMEL CORN PLANTS

A fairly large percentage of commercially popped corn is further processed into caramel corn. The range in size of these operations is extremely wide, varying from the small retail operations which may produce a few tens of pounds per hour by hand stirring, to fully automated lines putting out thousands of pounds per hour of nationally distributed confections.

Caramel corn can be produced, as it originally was, in gas heated copper or tinned pans of the Savage type, in which the operator stirs the contents of the kettle with a wooden or plastic paddle and adjusts the temperature by manually turning a gas valve (Fig. 19.5). Equipment of this general type is actually in current use producing some widely distributed gourmet confections. The drawbacks of such an operation are too obvious to require comment. It is possible, however, to produce good quality candy if the operator is sufficiently skilled and conscientious.

Courtesy of Krispy Kist Korn Machine Company

FIG. 19.5. BRICK-LINED CANDY STOVE WITH COPPER PAN
AND ACCESSORIES FOR THE BATCH PRODUCTION OF CARAMEL
CORN

The next step up in plant size would require equipment such as the hydraulic candy corn mixer (model 1785) made by Krispy Kist. This is said to be capable of producing up to 1,200 lb per hour of finished product. A cylindrical stainless steel kettle or mixing can fits on a rotating turntable which can be raised and lowered hydraulically. An auger-shaped agitator is mounted near one side of the turntable. A compound mixing action results from the placement of these elements. Popped corn and molten candy are manually scaled into the mixing can in the desired proportions.

Large-scale batch-type caramel corn plants are relatively efficient and can produce good quality relatively uniform product. Integrated systems for cooking and mixing, with capacities of 100 to 1,000 lb per hour and requiring only one person in attendance, are offered by Manley, Inc. They can be combined with product bins, soda dispenser, ingredient vats, dispensing devices, and a two-stage heat control to give completely automated plants. Operation would proceed as follows. An air

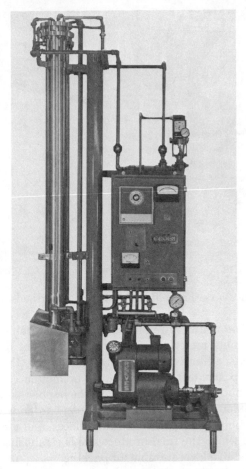

Courtesy of Cretors and Company

FIG. 19.6.　FILM-TYPE CONCENTRATOR FOR SYRUP

conveying system loads the popcorn hopper. A level control in the hopper shuts off the popcorn supply and starts a pump allowing a pre-measured amount of syrup to be delivered to the kettle. The heat is then turned on and agitation begins. At a first predetermined temperature level, butter and nuts are fed into the kettle. When a second temperature stage is reached, soda is added and the heat is shut off. Finally, the popcorn is added and blending continues for a selected length of time after which the kettle rotates to dump its contents and complete the cycle.

The hot, coated product must be passed through separating tumblers

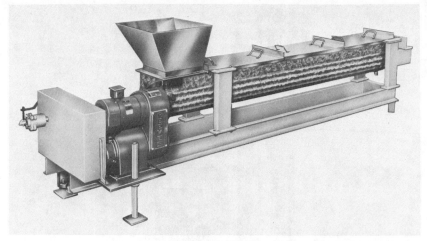

Courtesy of Cretors and Company

FIG. 19.7.   CONTINUOUS COATER FOR CARAMEL CORN

and cooling conveyors. These devices will be described in detail later in this section.

Completely automatic systems for producing caramel corn will include as a minimum metering and mixing equipment for blending the candy ingredients, a film type concentrator to reduce the moisture content of the syrup and bring it to a liquid or molten condition, a corn coater for distributing the hot syrup on to the popped kernels, a cooling-separating tumbler which allows the candy to harden while preventing excessive agglomeration of the individual pieces, and a cooling-spreading conveyor on which the final product size is established and the temperature reduced to a level suitable for packing.

Most of the concentrators used in the caramel corn industry are made by Groen Co. and are of the falling film type (Fig. 19.6). A typical caramel coating would consist of liquid sugar, corn syrup, coloring, and flavor continuously mixed and heated to 150 to 180°F in a steam jacketed kettle equipped with automatic temperature controls. The blended ingredients would be continuously pumped at a controlled rate into the concentrator, where high pressure steam heats a thin film, evaporating nearly all of the water content to produce an effluent syrup having a boiling point of about 290°F. Average residence time of the syrup in the concentrator is 4 to 10 sec. Dwell time can be varied to cause different extents of caramelization—color and flavor development. Salt and/or soda can be metered into the syrup as it emerges from the concentrator and falls into the collecting receptacle. Typical hourly capacity is 430 lb per hour per tube.

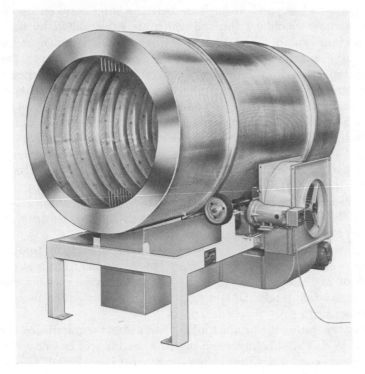

Courtesy of Cretors and Company

FIG. 19.8. COOLING AND SEPARATING TUMBLER

Continuous automatic caramel corn coaters generally take the form of heated screw conveyors which thoroughly mix the corn, nuts, and syrup (Fig. 19.7). In some cases, large rotating drums with helical flights are used to perform this function. Gentle action is needed in order to avoid excessive breakdown of the corn, and this dictates slow speeds, a minimum of tumbling, and no compaction. The action of the continuous units is generally more severe than given by the batch mixers. Cretors supplies a caramel corn coater which is a mixing screw conveyor with 700 lb per hour capacity. It is steam jacketed with hollow steam heated screw. Popped corn is dropped through a chute at the extreme end of the conveyor and the molten candy is fed in at an opening a short distance down stream. Total length of the unit is 6 ft, and the inside diameter is 8 in. The 0.5 hp motor has a variable speed drive.

Whether the corn is coated in batch- or continuous-type mixers, it must subsequently be cooled and separated to produce clusters of the appropriate size. Several firms distribute equipment for this purpose. In the Manley, Inc., unit, the hot confection first enters a conveyor consisting of a stainless steel woven wire belt traveling between upper and lower plenum chambers through which air is circulated by two fans. Ambient air is used, i.e., there is no refrigeration. The conveyor speed is variable. From the slightly elevated exit end of the conveyor, the still warm confection drops into a separator. The latter is a horizontally disposed drum with stainless steel ends and paddles, the circumference being made of 0.25 in. diameter stainless steel rods. The drum is rotated by a 0.5 hp motor operating through a variable speed drive. In a modified design, the drum has square mesh openings rather than slots.

The Cretors cooling and separating tumbler consists of a large cylinder of perforated metal with helical flights inside (Fig. 19.8). Its axis is slightly inclined to the horizontal plane. Power for rotation is applied by motorized, variable speed wheels to two tracks running around the exterior of the drum. Ambient air is forced by a fan into the drum from a plenum applied to the bottom of the cylinder.

Krispy Kist also makes a continuous cooling-separating tumbler for use in combination with their mixer. This equipment has a stated capacity of 2,000 lb per hour and consists of a long steel cylinder rotating on four external tracks. Operating principles are similar to those of the cooler-separators previously described.

After separating the product into clusters of the required size, it is desirable to chill it still further so that the coating will be nonsticky and the clusters completely firm before the product is packaged. This is generally accomplished by equipment consisting of conveyor belts passing through chambers in which ambient air is circulated. In a few operations, refrigerated air is used to speed cooling and increase the throughput. A typical ambient air unit is the Manley Model 323 caramel corn cooler. The warm clusters are fed on to a wire mesh conveyor belt which passes through the cooling chamber in three tiers. Fan forced air is filtered before it contacts the product.

BIBLIOGRAPHY

DE MUESY, E., and STINSON, W. 1971. One man makes 550 lb. caramel snacks per hour. Food Process. *32*, No. 1, 20–21.

HAVIGHORST, C. R. 1970. Automatic popcorn production. Food Engineering *42*, No. 12, 76, 78.

MIDDLETON, J. C. 1969. Personal communication. May 16. Manley, Inc., Kansas City, Mo.

CHAPTER 20

Specialized Equipment for Potato Chip Processing

INTRODUCTION

This chapter contains discussions of specialized potato chip equipment. More general types of equipment sometimes used in the potato chip industry, such as packaging equipment, certain types of fryers, salt applicators, etc., will be covered in other chapters.

The earliest equipment for potato chip processing was adapted from kitchen utensils originally intended for other uses. The tubers were often hand peeled, then sliced on meat slicers before being fried in wire baskets immersed in open kettles of hot oil. These batch operations with a high content of manual labor are obviously unsuitable for large volume production and have been replaced by automatic and mostly continuous equipment.

PEELERS

The most common equipment for peeling potatoes relies on abrasive removal of the surface layers. The effectiveness of these peelers depends upon maintaining the product in intimate contact with the abrasive surface and keeping the abrasive surface itself in effective condition. Practical difficulties have been experienced in bringing the potatoes from the center of the peeler into contact with the grinding area. Starch, fiber, and other potato constituents tend to fill in the spaces between the sharp particles reducing their effectiveness and, of course, the grit tends to wear down and lose its cutting edges.

One method of overcoming these problems has been described by Brady and Catalina (1973). Their invention provides an apparatus in which the abrasive surface is cleaned continuously during the peeling operation by the sweeping action of a solid stream or jet of water directed over it at a shallow angle.

Abrasive potato peelers are made or distributed by the following firms.

Allen Machinery Company
500 E. Illinois Street
Newberg, Ore. 97132

J. D. Ferry Division
Blaw Knox
1534 Fillmore Avenue
Buffalo, N.Y. 14240

Hobart Manufacturing Company
Troy, Ohio 45373

Vanmark Corporation
215 North Walnut Street
Creston, Iowa 50801

MacBeth Engineering Corporation
10th and Hanna Streets
Harrisburg, Penna. 17105

Magnuson Engineers, Inc.
1010 Timothy Drive
San Jose, Calif. 95105

Potatoes can also be peeled by a "dry" caustic process which breaks up and loosens the skin with an alkali treatment followed by infrared radiation at a temperature of about 1650°F. See Fig. 20.1, Huxsoll *et al.* (1973). Rotating rubber rollers with flexible fingers remove the loose debris from the surface of the tubers. In a commercial line operating on this principle, a metering bin regulates the flow of potatoes to a caustic immersion unit for a 30 to 60 sec exposure to the alkali solution. Following a holding period of about 3 min, the potatoes fall on to a roller conveyor which carries them through a gas-fired heater for about 2 min to concentrate by evaporation the lye solution adhering to the potatoes. Most of the softened skin and some of the underlying starchy tissue is removed by a mechanical scrubber having scrubbing rolls with soft rub-

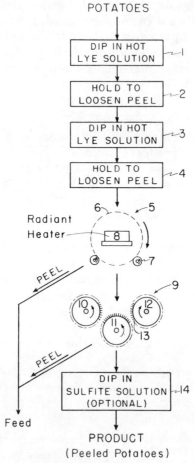

FIG. 20.1. FLOW SHEET FOR "DRY" CAUSTIC PEELING OF POTATOES

ber studs. Final cleaning is performed by a brushing washer which uses a small amount of water. Loss is said to be about 13% as opposed to perhaps 18% by conventional peeling methods. Because most of the waste is collected as a semi-solid material, the total volume of waste is reduced by more than 75%.

Peeling equipment of the dry caustic type is exemplified by the USDA-Magnuson Infrared Anti-Pollution Peeling Process developed jointly by the Western Regional Research Laboratory of the USDA and Magnuson Engineers, Inc.

Peeled and washed potatoes should be inspected to remove substandard tubers and those with minor visible defects should be trimmed. These operations are performed at an inspection conveyor leading from the peelers. The conveyor will vary in length and design according to the needs of each plant, but the standard size is about 10 ft long and is equipped with solid plastic belting of the sanitary type used in canneries.

Heat and Control, Inc. makes single- and double-lane trim tables in which the peeled tubers are transported on PVC rollers running on silicone rubber tracks which are inclosed at the sides of the bench. The double-lane tables have a center return belt for the culls. Various widths and 10 and 15 ft lengths are available. Variable speed drives are standard, and construction material is stainless steel.

SLICERS

Slicers are typified by the Urschel Model CC (Fig. 20.2), which is widely used in the chip industry. Peeled potatoes are dropped into a hopper leading into the bowl of a centrifugal slicing section. The bottom of the hopper is made of bars spaced far enough apart to permit small stones, nuts, bolts, and other foreign material to drop through. The centrifugal force acting on the potato as it is whirled around the bowl by an impeller causes the tuber to press against the inner surface with approximately seven times its normal weight. While in this position, it slides over a series of eight knives, which are adjacent to slots leading to the exterior (Fig. 20.3). The slices are ejected through these slots, first moving against a conical ring then dropping into a chute leading to a conveyor or collecting container. The replaceable blades are straight edged for ordinary potato chips and corrugated for crinkle slices. Knives can also be obtained for making Julienne strips and some other shapes. It is claimed that 80% of the slices will vary less than 0.004 in. in thickness and all will be within a range of 0.009 in. Standard heads and impellers are made to slice thicknesses up to 0.125 in. and diameters up to 4 in., but special parts are available to produce thicker slices. Capacity is on the order of 5,000 lb per hour.

Courtesy of Urschel Laboratories, Inc.

FIG. 20.2. SLICER FOR POTATO CHIPS

Following slicing, it is necessary (or, at least, very desirable) to re-move the loose surface starch. This is usually accomplished by washing the slices with hot or cold water, either by a tumbling immersion or by spraying.

SLICE WASHERS AND CONDITIONERS

Some processors feed the sliced potatoes directly into the fryer. Other manufacturers believe that a better product is obtained if the slices are first washed to remove the adhering starch which tends to cause the

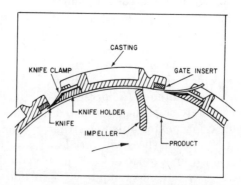

Courtesy of Urschel Laboratories, Inc.

FIG. 20.3. PRINCIPLE OF SLICING IN THE URSCHEL MODEL CC

chips to stick together and cook and color unevenly. In addition, it is sometimes desirable to leach out part of the reducing sugar content in improperly conditioned potatoes to prevent excessive darkening in the fryer.

Heat and Control Corp. offers a potato slice washer-conditioner which accepts pieces from the slicer, washes them, rinses them with top and bottom sprays, and removes the excess surface water with air blow-off jets and a shaker belt. Washing is accomplished by tumbling the slices in a perforated drum with internal spiral guide. The water can be maintained at any temperature up to 160°F and dwell time in the wash zone can be adjusted over the range of 0.5 to 4 min. Cold water is normally employed in the wash zone for properly conditioned (low sugar) potatoes, while hot water is used for removing excess reducing sugars. Slivers and small pieces are continuously removed from the wash water by a motorized fines catch box. The address of the manufacturer is Heat and Control, Inc., 225 Shaw Road, South San Francisco, Calif. 94080.

J. D. Ferry Div. of Blaw Knox, MacBeth Engineering Corp., and Salvo Machinery Corp. also make slice washers. Addresses of the first two companies have been given previously, and the latter can be reached at: Salvo Machinery Corp., 218 Shove Street, Fall River, Mass. 02724.

MICROWAVE DRYING

Coloration can be better controlled and output of the fryers substantially increased if chips are removed from the fryer with about 6 to 10% moisture content remaining, and the water brought down to 1 or 2% by microwave heating. Cryodry Corp. of San Ramon, Calif., was one of the leaders in promoting microwave equipment for finishing potato chips.

Frequencies used in test units have been 915 and 2,450 megacycles. It is not clear from the literature which is preferable, but commercial units seem to be based on the 915 megacycle frequency. Use of warm air currents through the drying chamber can double the rate obtained with microwave drying alone (Jeppson 1964).

Infrared heating ovens are also being used to finish the chips, i.e., reduce moisture content, facilitate oil drainage, and improve color. MacBeth Corp. is one of the manufacturers supplying such equipment.

FRYERS

Specialized fryers for cooking potato chips are made by several manufacturers; for example, Heat and Control, MacBeth Engineering, J. D. Ferry Div. of Blaw Knox, and Salvo Machinery Co. All large scale equipment made today is continuous, batch types being too inefficient from a labor standpoint and not capable of yielding as uniform a product.

MacBeth equipment includes fryers heated by gas-fired immersion tubes as well as units supplied by external heat exchangers, and with capacities of 750 to 3,000 lb per hour. Externally heated systems are composed of the frying machine, a stainless steel tube and shell heat exchanger, and a gas- or oil-fired remote heater. The heat exchanger is located adjacent to the fryer to minimize cooking oil volume. The oil is pumped around the fryer-exchanger circuit at a constant velocity. The heater, which can be located at any reasonable distance from the exchanger, utilizes a fire resistant heat transfer fluid at a maximum operating temperature of 600°F. This fluid is circulated through the shell side of the heat exchanger and back to the burner unit, while the cooking oil flows through the tubes of the exchanger. The units are available in heat generating capacities of 2 million to 18 million Btu. Characteristic of the heat exchanger types of fryers, the temperature of the cooking oil goes down immediately when the heater is turned off, and this has some advantages. An additional advantage claimed is that the cooking oil is never exposed to temperatures in excess of 600°F, while oil heated by immersion burners can reach much higher temperatures and thus is subjected to more deteriorative reactions.

MacBeth immersion heated fryers use suction, atmospheric-type burners with internally baffled firing tubes. The design is said to give better utilization of heat by reducing stack losses and to hold a greater amount of heat at the burner end where cold slices normally tend to reduce cooking oil temperature.

Heat and Control potato chip cookers utilize continuous circulation external heat exchangers, gas- or oil-fired. Low temperature differentials are achieved through high oil circulation rates. The units have central control panels with pneumatic proportioning and recording temperature control, and fuel flow rate recorders can be supplied as an option. There are provisions for continuously removing fines from the cooking oil. Standard sizes have capacities of 1,400, 2,100, 2,600, and 3,200 lb of chips per hour.

VACUUM FINISHING OF POTATO CHIPS

The excess browning and off-flavor development which may occur during the last stages of moisture removal by frying can be prevented by interrupting the usual frying process while a few percentages of moisture remain in the potato pieces and then subjecting the satisfactorily colored and flavored product to a vacuum-frying step which reduces the water content to a level commensurate with optimum texture and storage stability. Utilization of this process has been confined largely to Europe and South Africa, although at least one such plant has operated in the United States.

Equipment developed in Holland for vacuum finishing of chips which have been brought to 6 to 10% moisture by conventional frying conditions has a capacity of 1,500 to 2,000 lb of chips per hour. The vacuum frying oil bath is maintained at about 212°F by a heat exchanger. The color of chips processed by this equipment is noticeably lighter than that of chips produced in conventional fryers.

SORTING POTATO CHIPS BY SIZE

It is very difficult to separate potato chips into groups based on size because of the very friable nature of the product, its irregular conformation, and the presence of salt and other powdery materials. Allen Machinery Div. of Allen Fruit Co. makes a disc grader which has the capability of sizing potato chips as well as popcorn clusters and other irregular pieces with minimal breakage. Horizontally aligned interleaved arrays of vertically oriented rotating discs carry the larger chips over the top to one exit point while the smaller chips fall through the interstices and are transported to the other outlet. The size at which the separation is made can be varied by changing the distance between the axles of the arrays of rotating discs. Because of the method of operation, clogging or blinding does not occur as with vibrating screens. The equipment is available in 1,000, 2,000, and 4,000 lb per hour capacities.

Allen Machinery Div. also offers Spira-Lift reciprocating conveyors for cooling and conveying fragile snack products with minimal breakage. Outlet heights vary from 8 to 20 ft.

PROTECTING THE ENVIRONMENT

The increasing emphasis on environmental considerations by regulatory bodies makes it essential that snack food manufacturers, as well as all other businesses, evaluate the impact of their operations on the environment, especially when any change is anticipated in the amount or kind of pollutants discharged into water courses or into the atmosphere.

Considerable amounts of waste water containing starch and peel fragments result from conventional peeling, slicing, and washing operations. The starch has commercial value, in some cases justifying installation of treatment systems which can recover this material. One system (Shaw et al. 1973) passes a slurry containing about 3% starch into tanks equipped with baffles. From these flow control devices, the slurry moves through a Sweco screen where small pieces and peel particles are removed, and into a holding tank. A series of hydroclones subsequently separates the fluid into a 35% starch slurry and a much larger overflow fraction that is recycled back to the potato slicers and washers. The slurry is trucked away to a starch processing plant.

A more complete system for treating effluent from a potato chip and corn chip plant was designed by Heat and Control, Inc. (Anon. 1972A).

The effluent from this plant carries up to 3,000 to 4,000 ppm of colloidal waste. It is first pumped through a screen to take out the largest particles and then passed into a settling tank where it is held 5 min to permit partial precipitation of the suspended starch and other solids. The clearest fraction of the outflow from the tank is withdrawn and pumped to a bank of hydroclones which concentrate the suspended starch particles and pass the slurry back to the settling tanks. Thick starch paste is continuously withdrawn from the tanks by a cleated belt and deposited on a moving cloth vacuum filter belt. The clarified effluent is re-used in some of the chip operations, but since it includes alkaline corn steep water from the corn chip processing, it is not used for slice washing.

Fumes generated by the fryer can create an objectionable situation, if they are vented to the atmosphere without treatment. Some of the constituents have odors to which property owners in the vicinity might object, and there is a possibility of discharging detectible amounts of oil or particulate solids. Treatment of stack gases can be accomplished by providing a special furnace or afterburner for incinerating the combustible materials coming from above the fryer. A recent patent suggests directing the fumes back to the burners heating the cooking oil, where the organic material is converted to carbon dioxide and water by combustion (Newcomer 1973). The flow of gases from the cooking process can be controlled at a volume no greater than required for combustion air to insure that all of the fumes are incinerated. Condensers may be utilized in conjunction with such a system to remove most of the moisture carried by the fumes before they reach the burners, thereby maintaining the burners at a high level of efficiency.

BIBLIOGRAPHY

ANON. 1972A. Tom's cares about water quality too. Snack Food 63, No. 2, 46.

ANON. 1972B. Urschel Model CC Slicer. Urschel Laboratories, Inc., Valparaiso, Ind.

ANON. 1973. Starch recovery system cuts costs, adds income. Snack Food 64, No. 11, 32–32.

BRADY, K. J., and CATALINA, E. L. 1973. Method and apparatus for abrasive peeling. U.S.. Pat. 3,757,677. Sept. 11.

COPSON, D. A. 1975. Microwave Heating, 2nd Edition. Avi Publishing Company, Westport, Conn.

FOSTER, R. D. 1957. Report on various types of filters, their applications and limitations. Natl. Potato Chip Inst., Proc. Prod. and Tech. Div. Meeting 1957, 30.

JEPPSON, M. R. 1964. Consider microwaves. Food Engineering 46, No. 11, 18–20.

NEWCOMER, J. L. 1973. Elimination of cooking odors. U.S. Pat. 3,762,394. Oct. 2.

O'MEARA, J. 1966. Microwave drying of potato chips. Natl. Potato Chip Inst., Proc. Prod. and Tech. Div. Meeting 1966, 55–59.

SHAW, R., EVANS, C. D., MUNSON, S., LIST, G. R., and WARNER, K. 1973. Potato chip from unpeeled potatoes. Am. Potato J. 50, 424–430.

SMITH, O. 1975. Potato chips. In Potato Processing, 3rd Edition, W. F. Talburt, and O. Smith (Editors). Avi Publishing Co., Westport, Conn.

WHITEMAN, T. M. 1951. Improvement in color of potato chips and French fries by certain precooking treatments. Potato Chipper 11, No. 3, 24, 26, 28, 30, 32.

CHAPTER 21

Packaging Materials

INTRODUCTION

It is axiomatic to the food scientist that the product consists of the packaging material plus the foodstuffs. These two components interact at so many points that consideration of one without the other is almost certain to lead to costly and time-consuming errors. Development of the package should be undertaken at the same time as the foodstuff development begins, and both components should be regarded·as part of a single project with satisfactory completion of both an essential goal.

In a larger and perhaps more accurate but less familiar sense, the product consists of many more elements than the containers and the foodstuff. Some of these elements are abstract, incapable of precise measurement or manipulation by presently available techniques. The image projected to the consumer is one of those, and it consists, not only of product acceptability, package design, advertising approach, and company reputation, but of factors not accessible to the development scientist or marketing manager, such as the consumer's education, prejudices, financial condition, etc.

The package designer does have a great many variables to work with, however, and they can have a significant positive influence on the product's success. Common packaging films (printed paper/poly/foil/poly) may use from 25 to 40 components, all of which affect the performance in some way. A K-cello/ink/adhesive biaxial polymer may have from 20 to 30 components. By proper choice of materials and skillful design of the package (including the method by which it is formed and sealed), the food can be protected from damage and environmental contamination, the graphics artist can be presented with a suitable substrate for an appealing design, and the costs can be kept within a range allowing a satisfactory profit at a reasonable price.

TYPES OF CONTAINERS

Pouches

Flexible containers or pouches can be formed of plastic films, foil, or paper, but they usually consist of a composite structure in which two or more films are combined. A strip of this laminated web is mechanically formed into a tube, filled with the snack product, and heat sealed (usually). Some preformed bags are still being used, but the usual procedure is to form the envelope at the time the filling step occurs.

The relatively low cost of the packaging materials, the high speeds attainable in the filling process, and the protective features afforded by the container have combined to make pouches the most favored package for snack products.

Cartons

Ordinary fiber board boxes are unsatisfactory containers for most snacks because of the rapid transfer of gases, including moisture, through the material. In addition, soaking of oil into the board creates a large surface area and greatly accelerates the development of oxidative rancidity. Hermetically sealed bags within boxes have been used for years with complete satisfaction. The comments on pouches in another section of this chapter are applicable to bags in boxes. In these combination containers, the carton affords resistance to crushing damage and provides a relatively large flat surface for the graphics. It also makes it easier to display the container on a shelf. Disadvantages are the added costs for materials, labor, and equipment.

There are three basic kinds of bag-in-box packaging methods: (1) forming the pouch and the carton separately and dropping the filled and sealed bag into the box, (2) obtaining a pre-made bag-in-box in collapsed or flat form and filling and sealing on special equipment, and (3) making the bag and forming the box around the same mandrel thereby insuring a snug fit. Further descriptions of these procedures will be given in the next chapter.

Composite Cans

Composite cans is a term used to identify rigid containers of circular or approximately rectangular cross section consisting of a body made from laminated films (including foils and papers), mechanically bonded to metal ends. The laminated bodies may be formed by wrapping a strip of material at right angles to the axis of the tube (convolute), giving a straight seam up one side, or by helical winding of the strip around a mandrel, giving an angular seam up the body (spiral wound). The ends are generally either aluminum or tin plate, although unplated steel and plastic have also been used. Several kinds of easy-open features are available.

Such containers have been used for many kinds of snacks: corn curls, onion rings, pork rinds, caramel corn, simulated potato chips, and nuts, for example. They give excellent protection against crushing and can be good moisture barriers, if properly constructed. They are lighter than metal cans of the same size and are usually less expensive. All commercial structures give excellent protection against light damage to fats.

The bodies can be made of many different kinds of laminated films. Kraft paper and aluminum foil are common elements. Polyethylene and polypropylene as well as other polymer films and coatings are also frequently used. If foil is one of the components, good moisture protection can be obtained, even though some gas transfer does occur across the paper edges. Special lapping and folding procedures applied to the edges can reduce the moisture vapor migration to a very low rate. In general, it appears that spiral wound structures give better protection than convolute winding.

It is more difficult to design the graphics around a spiral seam than a convolute seam. Registration at the seam is never consistently good enough to allow the positioning of lettering or other important design features across this discontinuity, and the angular nature of the seam makes it difficult to avoid completely. On the other hand, straight convolute seams can be avoided quite readily in graphics design.

Although the operation used to affix the metal ends to the body is not the typical double-seaming operation which is employed in sealing sanitary tin cans, conventional can-closing equipment (suitably adjusted and modified) may be employed by the snack packer. Special lower-cost slower-speed equipment has also been made available by composite can manufacturers. None of the available cans has sufficient strength to maintain its shape under a moderate vacuum, and the seam is not reliably gas tight, so vacuum-packing is not practical. It is possible, however, to draw a vacuum in a chamber containing a composite can, inject nitrogen into the chamber to bring the pressure close to atmospheric, and then seal the can. Such methods are in commercial use today as a means of increasing the shelf-life of simulated potato chips.

Although a number of easy-open mechanisms are available, the snack packer may find that the particular pattern he desires is not available in the diameter he intends to use. Can manufacturers will not tool up for a new size without a guarantee of very large volume because of the tooling expenses involved. In such cases, the small or medium size packer will have to make a concession in size of container or type of opening feature. The snack manufacturer should insist that the can supplier affix the easy-open lids to the containers before they are shipped, while the packer applies the plain lid. More scrap is generated when the easy-open lid is seamed on, and its cost is considerably greater than the plain end.

If the snack is of the kind which requires the consumer to insert his hand into the can, protection against cutting his hand against the edge of the lid must be provided. Ridged ("necked in") bodies, plastic rings, and special conformation of the lids have been used.

Composite cans are seldom, if ever, made by the snack manufacturer in his plant. There may be some instances where the container manufacturer has a plant in close proximity to the snack producers and transfers the cans continuously as they are required at the packaging line. The efficiencies of this arrangement are too obvious to require further comment. The small to moderate-sized snack plant will receive his cans in large paper bags shipped in box cars or trucks. This may be of potential concern for the small packer due to the space lost for the storage of unfilled composite cans. In some cases, cans may be bundled on pallets or shipped in the corrugated cases that will be used as shippers for the finished product.

Suppliers of composite cans suitable for snacks include:

American Can Co.
Boise Cascade Composite Can Group
J. C. Clark Mfg. Co.
Clevepak Corp.
Container Corp. of America
Continental Can Co.
Stone Container

The first point to consider in evaluating the moisture vapor barrier properties of a package is, of course, the intrinsic permeability of the film or other material of which it is constructed. This can be readily measured by objective instrumental methods and can be expected to remain relatively constant for all material meeting the same specifications. It is not as simple to evaluate the effect on barrier properties of inadequate heat seals or the increase in water vapor transmission due to damage of the film by pouch-forming equipment. A range and average values for moisture absorption by product samples packed on the equipment during a normal production run are probably the most meaningful data which can be presented. These data can be obtained by exposing a fairly large number of samples to a controlled environment of high relative humidity and measuring the weight increase of the contents at regular intervals.

Rigid Plastic Containers

Thermoformed tubs of polystyrene or high-density polyethylene have been used for snacks for a long time. They are especially popular for caramel corn. Adequate protection for extended shelf-life is generally not possible since the necessary moisture vapor barrier is usually not achieved, because the closure is not designed for this purpose. If rapid turnover can be assured, and the humidity conditions are not extreme, the product may reach the consumer in acceptable condition. Taping the edges of the closure provides very little additional protection. Heat sealing a membrane of foil-containing laminate across the opening could give a satisfactory barrier against moisture vapor transfer, but it is diffi-cult to get a reliable seal all the way around a wide mouth container.

Polystyrene is not as satisfactory as polyethylene because of the greater moisture vapor transmission rate of the former resin. Polysty-rene is visually superior, however.

Rigid high-density polyethylene jars with air-tight screw closures should allow distribution of snacks through normal nationwide distribu-tion channels. A minimum wall thickness of about 0.030 in. is recom-mended. High-density polyethylene should be used. These containers will be translucent, not transparent. Costs per container should be ap-proximately competitive with glass, and freight costs will be much lower because of the lighter jar.

Metal Cans

Metal cans are excellent containers for snacks, especially from the standpoint of product protection. They have the disadvantage of rela-tively high cost, and some consumer inconvenience results from their use because of the opening feature, or lack of it. It is not possible to mer-chandise cans through some of the conventional snack outlets, such as racks or vending machines.

A snack packaged in a hermetically sealed can has complete protec-tion from crushing damage and environmental contamination, including insects. Light effects on fat are eliminated. Moisture transfer does not occur. If the snack is vacuum-packed, enhanced shelf-life will be ob-served as a result of decreased fat oxidation.

The opening feature presents a serious problem. For snacks to be con-sumed away from the home, the use of can openers is not feasible. The key wind opener is considered to be too inconvenient. The so-called ring pull full panel opening feature is preferred. It has at least three disad-vantages: (1) added cost of about 2¢ per container, (2) availability in a limited range of diameters, and (3) presence of sharp edges on the panel and around the opening. Various means of preventing consumer injury from the sharp edges have been devised. Each manufacturer has a dif-ferent patented method.

Can closing equipment can be leased from the can supplier. Cans are obtained in large paper bags or strapped on pallets. In some cases, it may be possible to obtain cans in the corrugated cases to be used for fin-ished product.

TYPES OF PACKAGING MATERIALS

Papers

Papers are matted cellulose fibers which have been treated with various chemicals and coated or filled with minerals, plastics, etc. They are used to add strength, textural qualities, appearance factors (including printing substrates), and other desired characteristics to laminated structures used for snack packages. When used alone, they have inadequate resistance to moisture vapor transfer and grease spotting.

Papers are designated by basis weight, that is, the weight in pounds of a ream (500 or, in some cases, 480 sheets) cut to a given size. Weights for most standard production papers are based on 500 sheets measuring 24 × 36 in. (3,000 sq ft). Bond base sheets are 17 × 22 in., lithos 25 × 38 in., and Kraft 24 × 36 in. Tissues base weights apply to 480 sheets 24 × 36 in.

Bond papers are uncoated sheets made of bleached chemical pulps in weights ranging from 20 to 70 lb. Special treatments are applied to give them wet strength, various finishes, tear strength, or ink compatibility. Because of the cost, bond papers are not often used for snack laminations.

Kraft papers are the unbleached versions of the bond specifications. Available in a basis weight range of 25 to 80 lb, they are classified as Northern Kraft or Southern Kraft. The former papers made from softwoods grown in the northern part of the United States are relatively uniform, less porous, and have higher tensile and burst strength values. Southern Krafts are comparatively porous, nonuniform, and rough, with high tear strength. Krafts are usually cheaper and stronger than other papers.

Tissues are lightweight papers made of fully or semibleached chemical pulps in basis weights of 8 to 20 lb. Special treatments add wet strength or mold resistance. Machine glazed bleached tissues can be used in laminates to give a bright white surface. Open and porous tissues are used in foil/wax/tissue where good wax strike-through is necessary.

Litho papers are available in basis weights from 29 to 60 lb and coated on one or both sides. Although they are not as strong as bonds, they afford a better printing surface.

Pouch papers are produced on glassine processing equipment and have a similar appearance and feel. Plasticizers added to pouch papers give them softness and machinability needed for flexible packaging.

Paperboard is similar in composition to paper, but has a thickness of $7\frac{1}{2}$ points as compared to 4 or 5 points for the heaviest paper. Boards are classified as either chemical pulp boards or wastepaper boards.

Glassine and greaseproof papers are produced by processes that beat and refine wood pulp until the fibers attain a high degree of hydration. As a result, the papers are resistant to wetting by grease, fat, and oil. Glassine is supercalendered (pressed between heated rollers) to give a smooth glossy surface, high density, and transparency. About 85% of the production is used for food packaging, but only relatively small quantities are utilized by the snack food industry.

Although their resistance to oil absorption prevents staining by potato chips and the like, unmodified grease proof and glassine papers have very poor resistance to water vapor transmission. Waxing, coating with polymers such as PFDC or laminating to moisture-resistant films are used to increase their suitability for snack packaging. Vacuum metallizing improves the appearance of glassine and is used extensively for candy bar wrappers.

Greaseproof paper, as such, is not used as an external ply for snack food packaging. The surface is not suitable for printing or application of heat-seal coatings. Greaseproofs are used extensively as liners of composite cans, i.e., the loose corrugated liner in Pringles.

Paperboard

Folding cartons are usually fabricated from either waste paperboards or chemical pulp boards. To be suitable for use as a material for folding cartons, a paperboard must be able to withstand a 180-degree fold without splitting.

Among the types of waste paperboards found in folding cartons are those called bending newsboard, bending chip, solid kraft lined chip, bogus kraft lined chip, white patent coated news, white day coated news, bleached manila lined news, manila lined news or chip, and white vat lined news or chip.

These boards can be calendered or coated to provide smooth white finishes. Box boards contain sufficient long cellulosic fibers on both sides of the sheet to prevent splitting. Since cylinder board is built up in layers on the paper machine, the center can be filled with lower grade fibers. Most news boards are filled with material from unsorted mixed waste papers. If the sheet is made entirely of these low grade materials, it is called chipboard.

Solid chemical pulp boards are usually made of 100% kraft fiber—bleached, semi-bleached, or natural. Semi-bleached board has a buff or manila color but in other quality factors is similar to fully bleached. Natural kraft boards are used where maximum strength and low cost is wanted. They can be laminated to white paper or white coated to improve appearance and printing qualities.

Cellophane

The most important advantages of cellophane are probably its transparency, glossiness, and crisp feel. Cellophane, however, by itself, is an almost unknown item in snack food packaging but it can be upgraded by laminations to foil or by coatings of various plastics. Probably 99½% of the cellophane used is a coated product. The coatings applied to the cellophane provide the packager with an item that can be heat sealed either to itself or a compatible heat-seal coating, has moisture barriers, can be printed on, can be laminated to, has gas barriers and various ranges of durability based upon specific in-use applications. The economic aspects of cellophane usage have varied remarkably in recent years. Generally speaking, polymer coated cellophanes are approximately equal in price to polymer-coated oriented polypropylenes, and gauge for gauge are two and three times more expensive than conventional polyethylene and unoriented polypropylene films. The key to selection of packaging material, either single or multi-ply is related to cost/barrier/function relationships. Price per pound of any packaging material is of incidental interest to the user as the barrier/function/cost relationships must be adequately and accurately defined for a package product to be successful in the market place.

Some of the characteristics of a cellophane film may be determined by examining the code symbol used by the manufacturer to identify it. For example, a du Pont cellophane might be identified as K 140 HB 13. The K designation signifies a polymer-coated cellophane; 140 indicates a yield of 14,000 sq in. per pound; HB means high barrier (WB = white high barrier, DB = high durability high barrier); the first digit of the last number refers to film series; and the final digit indicates the degree of surface treatment from 0 to 4 (with even numbers indicating both sides have been treated). Other manufacturers of film (Olin and Avisco/FMC) may have different notational systems (Doar 1973).

High barrier cellophanes consist of a cellulose base web which has been coated on both sides with a highly moisture-proof and heat-sealable layer of polyvinylidene chloride copolymer.

Plastic Films

Most plastic films, such as polyethylenes and polypropylenes are extruded from a single or homopolymer type of resin. Copolymer films are extruded from a resin that has been copolymerized with a resulting product being a material such as ethylene/vinyl acetate copolymers and ethylene/acrylic acid copolymers. There is another class of films that are seeing much use in packaging today and these are the mechanical mixtures of homo and copolymer resins. Films produced from these mixtures are generally multi-polymer films and are represented by classes of materials such as ethylene-propylene/ethylene vinyl acetate multipolymer films. Another class of films that is being used extensively in snack food packaging is co-extruded films. Co-extruded films use the previously mentioned polymers and other polymer materials such as nylon, polyvinylidene chloride as structural components. Co-extruded films are produced by extruding, through separate extruders, melts of desired polymeric materials and bringing these melts together generally in a multi-lipped die to form a film that has several layers, each layer location being determined by the composition of the melt within the extruder—as an example: co-extruder films of ethylene/vinyl acetate co-polymer as one of the plies and polypropylene as another ply. Three-ply structures can be produced that use low density polyethylene as an outer layer, nylon as an interior layer, and medium density polyethylene as the final exterior layer of the co-extruded composite material. Co-extruded structures are layered materials available from single pass operations where the film possesses a variety of physical and chemical characteristics tailored for each specific application.

Oriented materials are films that have been stretched during the manufacturing process to alter their physical characteristics. The most common oriented films are polypropylenes and polyesters. Oriented films are stretched across the web and/or in the direction of film flow. If the film is stretched equally in both directions it is said to be a balanced oriented or biaxially balanced oriented film.

Polyethylene.—High-density polyethylene and medium density polyethylene are often used for snack food processing in 1.0, 1.25, and 1.5 mil thicknesses as films, extrusion coatings, or extrusion laminations. Types with densities greater than 0.935 contribute good stiffness, low moisture vapor transfer rate (MVTR), and fairly good grease resistance. Lower density versions permit good sealing but are relatively low in oil and grease resistance. These materials tend to burn through, stick to the jaws, or seal at high temperatures, depending on their position in the lamination.

Polypropylene.—Polypropylene is a catalyst-polymerized olefin similar in many respects to polyethylene. It is tougher and somewhat more transparent than polyethylene but unoriented polypropylene lacks low-temperature durability. Up to now, it has always been more expensive than polyethylene but there is no certainty that this relationship will continue indefinitely. Some polypropylene films are "oriented"; the properties (such as resistance to stretching) measured down the roll are different from those measured across the web. Unbalanced oriented polypropylene is stretched 0.5 to 2 times in the machine direction and 8 to 10 times in the transfer direction. Balanced film, on the other hand is stretched 5 to 8 times in each direction. There are significant differences in behavior of the two types both in the converting or laminating process and on the form-fill seal equipment.

Because of its lack of low-temperature durability, most applications that require unoriented polymer use a copolymer or a multi-polymer of polypropylene.

Polypropylene is probably the best film to use when it is necessary to protect the bag or pouch from punctures by sharp snack pieces, such as corn chips. Its resistance to tearing can create some inconvenience for the customer who must open the pouch but this difficulty can be reduced by providing tear slits at the edge of the seal. A common snack food lamination is thermal stripe/oriented polypropylene (OPP)/adhesive/polymer coated cello. The thermal stripe is necessary to obtain the overlap back seal. Polypropylene has relatively low thermal stability. It will start to pucker at about 265°F and completely distorts at 300°F. This property dictates good temperature control at the sealing jaws.

Foil

For all practical purposes, foil means aluminum foil to the snack packager. At one time, steel foil was available for composite cans and the like, but it appears to be unused in snack containers at present. The chief advantage of aluminum foil is its moisture transmission rate, which approximates zero when the material is properly used. There is also a decorative potential which can be utilized in many container designs. Because of its opacity, foil affords complete protection against acceleration of fat oxidation by light, which is a major deterioration problem in potato chips and certain other snack products.

Aluminum foil has been defined as a continuous web of aluminum metal ranging from 0.00023 to 0.006 in. in thickness, but the snack food packager will be dealing mostly with gauges of 0.00025 to 0.00050 in. The pure metal is not used, alloys being primarily those covered by specifications #1145, #1100, or #1235 of the Aluminum Association.

Questions as to the effect of pinholes on the barrier properties of foil are frequently asked. The incidence of pinholes (very small irregular discontinuities penetrating the foil) increases very rapidly as thickness decreases, especially below about 0.00025 or 0.00030 in. It appears that very small foreign particles either bound in the metal itself or accumulated in the rolling mill process create punctures in the film. In addition, the plastic limit of the metal may be exceeded as it is being rolled, causing small tears.

Good quality foil thicker than about 0.00035 in. will normally have fewer than 20 pinholes per square foot. The barrier properties are affected very little by these holes. If the foil is laminated to a plastic film, the effect of the pinhole count on the moisture vapor transmission rate becomes negligible, for all practical purposes.

For heat-sealing purposes, it is necessary to coat the foil with nitrocellulose, ethylene vinyl acetate, polyvinyl chloride copolymers, etc., or to laminate it with a suitable plastic film. Apparently, polyvinylidene chloride (PVDC) can cause corrosion, so it is not generally used in direct contact with aluminum.

Although 0.00035 in. foil is adequate for most snack packaging applications, the use of vacuum or gas flushing in a pouch may dictate the use of thicker webs, up to 0.0007 (Gayner 1974).

Laminating Films

Most films used for making snack pouches are combinations of two or more materials laminated together so that they function as a single structure on the filling machines.

There are four basic methods used to make the laminations: (1) thermal lamination, (2) dry bond lamination, (3) coating one film with a dissolved adhesive, driving off the solvent by oven treatment, then bonding it to another film by applying heat to activate the adhesive, and (4) extrusion lamination.

In thermal laminating, two webs are brought together between two heated rollers (Fig. 21.1). The coatings melt, and pressure of the rollers

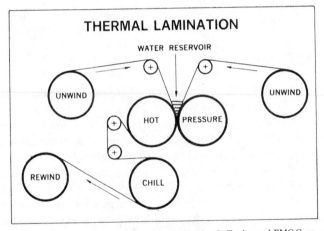

Courtesy of Charles C. Taylor and FMC Corp.

FIG. 21.1. COMBINING TWO WEBS BY THE THERMAL LAMINATION METHOD; IN THE METHOD SHOWN HERE, A WATER RESERVOIR HAS BEEN ADDED TO PREVENT AIR ENTRAPMENT BETWEEN THE FILMS

causes the inside coatings to fuse together. The coatings on the two materials being combined act as the bonding agent or adhesive. Examples of films combined in this way are two webs of cellophane coated with polyvinylidene chloride, and PVDC-coated cello with oriented polypropylene.

The dry bond method starts with a web of flexible packaging material which has been coated with an adhesive dissolved in a volatile liquid (Fig. 21.2). The solvents are driven off in a heated tunnel. The adhesive-coated web is then combined with another film when they pass between a pair of heated rollers.

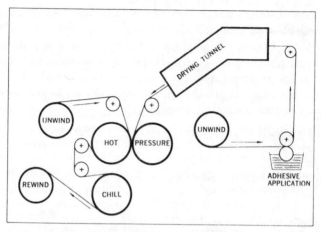

Courtesy of Charles C. Taylor and FMC Corp.

Fig. 21.2.   Combining Two Webs by the Dry Bond Lamination Process

The third method is really a variant of thermal bonding. One web is coated with a dissolved adhesive at the last station of the printing press. The solvent is evaporated in a drying tunnel. The printed and adhesive-coated web is rolled up and stored for lamination at some later time. In the laminating process, the prepared film is bonded to another ply on a thermal laminator. This procedure is used primarily for cello/cello and cello/polypropylene combinations.

In extrusion lamination, the plies are bonded by a layer of molten material, usually low-density polyethylene, which is extruded from a die mounted above the nip (aperture between the rollers). One of the films may be "primed" by previous treatment with a very thin layer of dissolved adhesive. The roll set in this equipment generally consists of a cooled polished metal cylinder and a rubber pressure roll. Fairly complex structures can be formed by extrusion bonding, oriented polypropylene/polyethylene/coated cello and coated cello/print/polyethylene/foil/heat seal coating being typical examples (Taylor 1971).

With some laminations in which the outer ply carries no coating or a coating that is not compatible with the heat-seal coating on the inner ply, it is necessary to apply a thermal reactive strip on the uncoated side of the web so as to enable the lap-back seals to be made on form-fill-seal machines.

SPECIAL FEATURES REQUIRED IN CERTAIN SNACK PACKAGES

Concealing Crumbs

In packages of snacks which break up readily, and this includes most snacks, a collection of small pieces gradually accumulates at the bottom of the container. Because it is believed that the crumbs are unsightly to the consumer, pouches used to package these snacks are sometimes made of opaque material. Alternatively, the pouch may be made of transparent film which is printed with a "crumb line" to give an opaque lower section to hide the settled fragments.

Protection Against Light-induced Damage

Light accelerates the development of rancidity in most, if not all, snacks. Fluorescent light is regarded as particularly effective in this regard. The bad effects are most apparent in potato chips. Opaque packaging materials are generally effective countermeasures although some films which are opaque to visible light allow ultraviolet radiation to penetrate. Foil laminates are the protective materials of choice for pouches, but many pigmented plastics have also been found to be satisfactory. Testing of films for transmittance of rancidity-accelerating radiation has been entirely empirical—bags of product are exposed to sunlight (actual or simulated) and/or fluorescent light for varying periods and then examined chemically or organoleptically for indications of rancidity. It should be possible to devise equipment which would measure electronically the amount of radiation transmitted through a given area of film, and relate these data to development of rancidity.

Preventing Destruction by Insects

Almost all packages except hermetically sealed glass jars and metal cans are subject to penetration by some species of insect. Plastic jars with uniformly thick sidewalls are also protective under nearly all conditions, however. Pouches, composite cans, boxes, and similar containers will be attacked at rates depending upon: (1) the species of insect, (2) the number of contacts by insects, (3) the temperature, (4) the presence of attractive material such as food or deterrent substances such as pyrethrins on the exterior of the container, (5) the inherent resistance of the package structural material to penetration, (6) adequacy of the seals, and (7) damage to the structure by sealing equipment or other agencies.

The first two factors can be minimized by shipping and storing under sanitary conditions. Inspection of boxcars, warehouses, etc., and the application of residual insecticides where necessary should be routine duties of the physical distribution department or the quality control personnel.

Refrigerated storage (40°F) will very effectively control all forms of insects which attack food directly. Air conditioning (70°F) is much better than ambient summer temperatures but will still allow product contamination.

Under practical distribution conditions, the relative resistance of the package itself to breaching by insects (larvae or adults) is a very important factor affecting the number of contaminations observed by the ultimate consumer. Although there is no film which is completely resistant to penetration by all species of insects, polypropylene in thicknesses of 0.9 mils and up approaches the ideal very closely. The package must be well sealed (polypropylene to polypropylene) without damage to the laminate if the protective feature is to be retained. Failure, when it occurs, is usually at the ends of corrugations caused by the sealing jaws.

TESTING FILMS

The technical evaluation of a film consists in evaluating, not only its protective properties, but also its machining properties on the pouch-forming equipment. Porner (1974) described the process used at Frito-Lay to screen and approve new structures proposed for packaging potato chips. First, the structure is tested for its ability to perform satisfactorily on production form-fill-seal equipment. He determines: (1) Is the slip correct? (2) Are the end and back seals strong enough? (3) Is the sealing temperature range narrow or wide? (4) What is the optimum speed at which the film can be processed? If these results are satisfacto-

TABLE 21.1

COMPARISON OF WATER VAPOR TRANSMISSION RATES
OF FILMS USED IN SNACK POUCH STRUCTURES

Film	Water Vapor Transmission Range[1]
Aluminum foil (0.00035-in.)	0.00–0.02
Oriented polypropylene	0.2–0.4
Medium density polyethylene	0.7–1.0
Nitrocellulose-coated cellophane	0.5–1.0
Polymer-coated cellophane	0.5–0.8
Unoriented polypropylene	0.6–0.9
Low density polyethylene	1.0–2.0
Polymer-coated oriented polypropylene	0.3–0.4

[1]Grams per 100 sq in. per 24 hr per mil thickness.

ry, the structure is tested in the laboratory for: (1) water vapor transmittance, (2) resistance to breaking after exposure to low temperatures, (3) impact strength, (4) coefficient of friction, (5) residual solvents, (6) component weights, (7) ink scuffing, if surface printed, (8) tear strength, (9) bond tensile strength, (10) film tensile strength, and (11) bond strength of the thermal stripe.

Following the laboratory tests, fresh product is packed into bags made of the material. The bags are placed into regular shipping cases and subjected to shipping tests after which the product is examined for breakage.

Product shelf-life is determined as bagged product held at 85°F and 80% RH. Moisture pickup is determined gravimetrically at intervals until it reaches a level at which the product is stale from an organoleptic and texture standpoint.

Access to a reliably controlled high-humidity cabinet or room is absolutely essential for the packaging technologist dealing with snack products. No other equipment except a balance will supply as much pertinent information.

LEGAL CONSIDERATIONS

The laws and regulations governing packaging of foodstuffs are becoming increasingly restrictive. All package designers must keep abreast of these restrictions if they are to function effectively. It is no longer safe to rely entirely on supplier advice as to the legality of a particular food and package combination, especially if the supplier ordinarily deals with nonfood packagers. Salesman's comments such as "This film is GRAS." or "This material is in the process of being approved by the FDA." are not a safe basis of package development. After much time has been spent in design, it may be discovered the salesman has been overly optimistic in his interpretations of an off-hand comment by some technical person in his own organization or that of a competitive packer.

Furthermore, sources of supply may be suddenly cut off because the supplier's plant is suddenly found to be violative of some safety or environmental regulation and can no longer manufacture the needed material. Packages regarded as safe and acceptable by virtue of long usage without consumer harm may become legal pariahs because some politically influential fanatic imagines they are harmful to the environment or are economically unsound. Many of these actions are unpredictable by logical means and can be dealt with only after they occur.

It is clear enough that food-contacting materials must be composed only of substances which have FDA approval, since some migration to or contamination of the foodstuff by the material is almost certain to occur in normal manufacture and distribution. If the legal and technical staffs of the packager are not adequate to interpret the regulations, then the supplier will have to be relied upon. This means, as a minimum, the obtaining from the supplier unequivocal guarantees that the specific material being supplied conforms to all applicable Federal and local regulations for food packaging.

BIBLIOGRAPHY

ANON. 1960. The Canning of Potato Sticks. Technical Service Dept., American Can Co., Barrington, Ill.

ANON. 1963. The Canning of Nut Meats. Technical Service Dept., American Can Co., Barrington, Ill.

ANON. 1974. Living with packaging shortages: a look at paper. Food Prod. Develop. 8, No. 7, 88–91.

CAVALETTO, C. G., and YAMAMOTO, H. Y. 1968. Criteria for selection of packaging materials for roasted macadamia kernels. J. Food Sci. 22, 97–101.

DOAR, L. H. 1973. Cellophane—still very much alive and kicking. Package Eng. 18, No. 2, 19.

FIRMAN, E. F. 1973. Effects of light on snack food packaging. Snack Food 62, No. 1, 70, 72.

GAYNER, H. 1974. Personal communication. Alcoa Technical Center, Alcoa Center, Pa.

HELLER, J. 1972. Packaging basics. Packaging Develop. 1974, (Nov./Dec.) 2–6, 8–10.

HICKMAN, J. J. 1973. New generation of high-barrier cellophanes. Snack Food 62, No. 5, 38–39.

PORNER, F. E. 1974. Potato chip packaging now and tomorrow. Snack Food 63, No. 2, 32, 34, 36.

PORNER, F. E., and WHITE, O. L. 1972. Save time and money by running tests in proper sequence. Package Eng. 17, No. 9, 55–57.

QUAST, D. G., and KAREL, M. 1973. Simulating shelf life. Mod. Packaging 46, No. 3, 50, 52–54, 56.

SACHAROW, S. 1974. Packaging requirements for snack foods. Food Eng. 46, No. 10, 70–73.

SOUTHWICK, C. A., Jr. 1973. Simulating shelf life. Mod. Packaging 46, No. 3, 50, 52–54, 56.

TALWAR, R. 1974. Plastics in packaging: gas and vapor permeation. Package Develop. 1974, (Sept./Oct.), 29–32, 49–52.

TAYLOR, C. C. 1971. Flexible packaging for snacks. Snack Food 60, No. 1, 1–3.

CHAPTER 22

Packaging Equipment

INTRODUCTION

Many types of equipment have been used and are being used for packaging snacks, but the generally fragile and bulky nature of the products and the advantage of being able to apply inexpensive packaging materials have led to the predominance of pouch-forming equipment in this industry. The measuring process is an essential part of the packaging system, and from hand-filling with volumetric scoops it has advanced through an evolutionary series of changes to automatic weighing and filling on very sophisticated high-speed pouch formers.

Although equipment designed for packaging other products such as candy, cookies, granular materials, etc., has been adapted for snacks and is still utilized quite successfully for certain of these items, the extremely variable shapes and densities of snacks such as potato chips and popcorn has necessitated the development of specialized equipment. Within the space limitations of the chapter, only a few of the most important alternatives can be discussed. For more extensive treatments of general lines of packaging equipment, the reader should consult *Food Packaging* (Sacharow and Griffin 1970) or *Principles of Package Development* (Griffin and Sacharow 1972).

FOLDING CARTONS

Folding cartons made of paperboard are used for many baked snacks although they do not offer sufficient protection against moisture vapor protection to be satisfactory for the more hygroscopic products. For the latter items, cartons can be used to inclose hermetically sealed pouches of moisture-resistant films, giving an improved appearance and considerable protection against crushing.

Folding cartons can be classified as either tray-style or tube-style. The tray-style cartons are formed from paperboard with the sides folded at right angles and locked, glued, or stitched together at the corners. Combination top and bottom units can be formed from two trays, or one panel can be extended and folded to form a cover.

The tube-style cartons are made from a sheet of paperboard which is folded over and then glued to form a rectangular sleeve. The panels which form the ends may be fixed in place by applying adhesive, tucking extensions into folds, or interlocking slit tabs.

It is generally recognized that the tray style offers greater strength

Courtesy of Wright Machinery Company

FIG. 22.1. EQUIPMENT FOR SETTING UP PRE-FORMED BOXES AND FILLING THEM WITH A WEIGHED AMOUNT OF PRODUCT

than the tube version. It can also be loaded from the top, which is often more convenient than the end loading of the tube style. Many variations of design are possible.

In almost every case, the cartons are delivered to the user in knocked down form. Setting up, filling, and sealing can be done manually or with automatic equipment in any desired combination of functions (Fig. 22.1).

PREFORMED POUCHES

Some snacks are packed in prefabricated bags or pouches made of heat-sealable films. Preformed bags are delivered to the user in a continuous roll, bound on wickets or stakes, or loose. A number of manufacturers make equipment for dispensing, filling, and sealing preformed bags. These generally consist of assemblies of devices to dispense a single bag, hold it in a properly oriented position, inject an air blast to open it, dispense a measured amount of product into the opening, and then heat seal it. Companies providing this type of equipment are Tele-Sonic Packaging Division, Automated Packaging Systems, Inc., and Errich Packaging Machinery Co.

Among the suppliers of stock bags for snack products are Milprint, Inc., which makes stock potato chip bags in a wide range of materials and sizes, Tower Products, Inc., and Automated Packaging Systems,

Courtesy of Bodolay Packaging Machinery, Inc.

FIG. 22.2. MODEL 51-CK PACKAGING SYSTEM FORMS A BOTTOM-GUSSETED BAG FROM ROLL STOCK FILM, FILLS THE BAG WITH A SNACK PRODUCT, CLOSES THE BAG WITH A TWIST TIE, AND DISCHARGES THE COMPLETED BAG INTO A REGISTERED MOVING LINE OF SEALRIGHT CHIPKAN PAPERBOARD CANISTERS

Inc. The latter sells plastic bags in a continuous roll for use in their equipment.

In a few cases, relatively large bags of thin gauge polyethylene are used as inner containers in cylindrical paper containers or tins with friction fit closures. The bags are usually sealed with wire ties or plastic clips. Such containers are suitable for limited shelf-life applications, but obviously will not give much protection against moisture vapor transfer. The equipment shown in Fig. 22.2 automatically forms, fills, ties, and places bags in paperboard outer containers. This is the model 51-CK System of Bodolay Packaging Machinery (Springfield, Mass.), which automatically forms a side-weld, bottom gussetted bag from roll stock film, fills the bag from an automatic weigher, twist ties the bag, and discharges the completed bag into a coordinated moving line of fiber canisters. The system has been used for packaging potato chips, pretzels, cookies, and other free-flowing snack goods. Bag size capabilities range upward to 14 × 22 in. Most smaller snack producers will perform these functions by manual operations.

Centerfold or J-fold roll stock film can be formed into pouches by relatively simple sealing equipment. Most pouches used in the snack industry today, however, are formed from flat webs on vertical or horizontal form-fill-seal devices. The speeds and labor economies which can be achieved are some orders of magnitude greater than those attainable with the manual filling of pre-made bags. This equipment will be described in the following section.

FORM-FILL-SEAL EQUIPMENT

Most of these machines fall into two categories: the vertical type, which generally forms a tube by making a continuous seal down the middle of the back side of the moving film, and the horizontal, which usually forms a trough or U-shaped cavity of film that is then separated into compartments by vertical heat seals before it is filled. The former is much more common in the snack industry, although horizontal equipment is used for nuts and similar dense products. Other types of pouch formers are also used occasionally for snacks.

In vertical form-fill-seal equipment, the film is drawn from a roll under tension and carried over a "collar" of unique geometry which shapes it into a cylinder. The film—at this point still open at the edges—travels down and around a metal tube through which the product will be dropped. The back seal can be made by pressing a hot sealing iron, often bar-shaped, against the apposed edges of the film using the tube as a back-up surface. This gives an overlap seal. Alternate arrangements can give a face-to-face or fin seal. Two webs can be used, in which case heat seals are made on four sides (see Fig. 22.3).

The motive power for the web movement is provided by horizontal sealing jaws which pull the film downward the length of one pouch in each sequence. These jaws simultaneously form the end seals (top seal of the bottom pouch, bottom seal of the top pouch of the set). The film may be severed by the action of a knife in the heat-seal bars, or the cutoff may be made by a separate device.

Among the best known form-fill-seal equipment are Package Machinery Company's Transwrap, FMC Corporation's Stokeswrap, Hayssen's Compak, Triangle's Flexitron, Mira-Pak's Mira-Former or Mira-

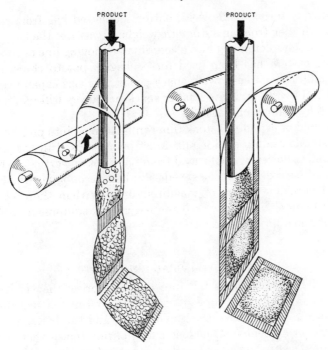

Courtesy of Thomas E. Walmsley

FIG. 22.3. TWO METHODS OF FORMING POUCHES IN VERTICAL FORM-FILL-SEAL MACHINES

Wrap (Fig. 22.4), Woodman's Pacemaker, and Wright Machinery Company's Mon-O-Bag units (Fig. 22.5).

After these steps are completed, the draw bar opens and is returned to the top of the next pouch. During this return movement, the product is dropped through the metal tube into the open mouth of the half-formed pouch.

The hot-tack seal at the bottom of the tube must form very rapidly and be strong enough to resist the weight of the product attempting to force it open. Many packaging materials such as glassine and uncoated cellophane will not form such a seal, and coatings of nitrocellulose, rubber polymers, PVDC, etc., must be applied to give the necessary properties. Under certain conditions seals can be made to polyolefins and some other films.

The heat-seal material may be applied as an overall coating or layer or as a thermal stripe limited to the area in which the seal is to be formed.

Courtesy of Mira-Pak, Inc.

FIG. 22.4. MIRA-WRAP SERIES 10000 MODEL H, A TWIN TUBE FORM-FILL-SEAL MACHINE

Conformations of the sealing jaws, temperatures, dwell times, and some other factors will be different for each different sealant combination.

The tube and former assemblies must be changed when bag width is altered. Sometimes a change in the film composition will also necessitate use of a different assembly. Dimensions, shape, materials of construction, surface finish, and other characteristics of the forming head have marked effects on the efficiency of pouch forming and on the damage done to the film. Some heads which give excellent results with cellophane will not function properly with polypropylene or paper, etc. The factors governing the film and former interaction seem to be rather inadequately understood, even by the engineers of the packaging machine companies.

Courtesy of Wright Machinery Company

FIG. 22.5. WRIGHT MODEL "ES" MON-O-BAG SYSTEM WITH BAG INCLINE AND ROTARY PACK-OFF TABLE

Many of the available machines have unique design features intended to give superior performance, at least for certain limited applications. Examples are:

(1) Hayssen equipment can use powered rollers to exert additional force in pulling film from the roll.

(2) The Rovema filler distributed by Pneumatic Scale Corp. uses rubber wheels to push the pouch material down the tube and past fixed sealing and cut-off jaws. Gusseted pouches can be made on the Rovema.

(3) Mira-Pak equipment eases the fall of product by carrying it down to the pouch in a moving cup. Breakdown of fragile materials such as potato chips is minimized.

(4) In Woodman equipment, the tube is slanted so as to reduce product momentum when it falls into the pouch.

(5) The Wright Multi-Pak system forms, fills, seals, and attaches pouches to a card in shingle style. Possible variations include multi-packs on a strip of cardboard for case packing or rack sale, and single or twin strips of multi-packs in a master bag.

Product Measuring Devices

The equipment for measuring the amount of product to be deposited

Courtesy of Wright Machinery Company

FIG. 22.6. MODEL AC COUNTING FILLER WITH SEVEN HEADS

in each bag is generally separate from the form-fill-seal machine, although manufacturers of the latter devices will usually supply integrated systems. Auger, bucket, volumetric, and net weighers as well as counting devices can be used to measure out the portions.

An advanced type of counter, the Wright Model AC counter with seven heads, is shown in Fig. 22.6. The electronic counting system consists of a receiving hopper serving a multiple-head counter of modular design with individual vibrator feeds for each counting head, and remote control station. Product is supplied into a channelized hopper, from each channel of which the individual pieces are fed through a series of vibratory feeders where the flow is automatically modulated and the pieces are oriented into single file. At this point, there is a very narrow section designed to spill off any pieces that may be going through side-by-side or piggyback. The pieces to be counted, now flowing in single file at close intervals, ride over a rotating wheel which accelerates them, putting enough space between them to allow them to be counted with a light beam and a photoelectric cell. The counted pieces drop directly into the lower compartment of a batch accumulator or bucket. When the desired count is reached, an angular-shaped butterfly gate shifts to create a temporary upper compartment in the bucket, and the counter resets to zero and begins counting and accumulating a succeed-

Courtesy of Wright Machinery Company

FIG. 22.7. AN IN-LINE SYSTEM FOR WEIGHING AND FILLING FREE-FLOWING PRODUCTS INTO RIGID CONTAINERS SUCH AS COMPOSITE CANS

ing batch into the upper compartment. A trap door opens in the bottom of the bucket, dropping the counted batch into a carton or other container. As the trap door closes, the butterfly gate shifts back to its original position allowing the accumulated pieces to drop into the lower compartment. To synchronize the dumping of the precounted batches into containers being supplied at a uniform rate, the dumping of each batch is delayed slightly when necessary to allow the container to move into the proper position, and the feeding of pieces to the counter is interrupted for the same amount of time. The solid state counter can be set for loads of up to 99 pieces, and counting rates up to 300 pieces per minute per counting head are possible. Equipment of this type is suitable for counting relatively uniform pieces of intermediate size.

The Wright Model FA inline modular weigher system (Fig. 22.7) can

Courtesy of Wright Machinery Company

FIG. 22.8. A VERTICAL FORM-FILL-SEAL MACHINE WITH AUGER
FEED FOR POWDERED MATERIALS

be used for weighing and filling free flowing products into rigid containers. The system consists of 2 to 6 Electroflex scales and a package indexing conveyor for handling cartons, jars, and cans. The product is received from a continuous processing conveyor or low-level feeder into a
specially designed balanced channelized surge hopper. A separate product flow control system at each transfer point provides a smooth and
continuous flow to each weigh bucket. The bulk product feed system is
equipped with vibrators to separately control bulk- and finish-fill feeders. The finish fill feeder is V-shaped to single file products for the final
weight.

The electroflex scales are said to have such features as cross flexure
pivots, zero pendularity, zero stress in beam at the point of measurement, and balanced beam with mechanical application. Detection is
achieved by contact points on a lever, which activate a solid state relay.
Wright claims that the solid state correct weight detector allows instan-

taneous finish fill control that remains stable throughout a wide temperature range, during line voltage changes, or under adverse environmental conditions.

The weighed product enters individual product chutes which are mechanically lowered into the package, eliminating spilling and allowing the product to settle prior to discharge. The package indexing conveyor is equipped with a product settling device, package flow controls, and dual indexing mechanism.

Rotary net weighers provide relatively high rates of measuring product. The Wright Rotary Net Weighers can be made with up to 18 heads mounted on the rotating turret. Motion is continuous—while the turret is revolving, the product is fed continuously into each of the weigh buckets. About 95% of the desired weight is fed by bulk flow, the remaining 5% is fed by finish-fill. Speeds up to 180 fills per minute can be attained. In addition to snacks, cookies, crackers, and cereals are being packaged on these lines. Provision can be made for blending of two components (e.g., popcorn and nuts) at the weighing unit.

The Wright Mon-O-Bag standard net weighing system is used for free flowing snacks requiring finish fill. Product is received into a specially designed two- or three-channel surge hopper. Balanced channelizing of the product to each weigher is closely controlled at transfer points to ensure a smooth and continuous flow into the weigh bucket. A vibrator continues the flow into separate bulk and finish feeders.

Auger feeders are not much used in the snack industry. They are more suitable for powdered materials such as beverage bases or soup mixes. In these devices, a rotating screw or auger is fitted into a tube leading from the bottom of a cone-shaped hopper into the container (Fig. 22.8). The number of rotations of the auger is adjusted so as to deliver the desired volume of product.

Tumble fillers have been suggested for use in packaging snacks, although they are perhaps more suitable for irregularly shaped fruits and vegetables. The equipment shown in Fig. 22.9, manufactured by Solbern Corp., carries cans or other rigid containers down the axis of a rotating drum containing a quantity of the material. Pieces of the product are carried up the sides of the drum and fall in and around the container. By adjusting the speeds of drum and conveyor, accurate weights are said to be obtained.

Effects of Environmental Conditions on Packaging Efficiency

It is generally agreed that the stiffer the film, the better it will run on packaging equipment. Polyethylene film stiffness, measured either as the modulus of elasticity in flexure or as tensile modulus is directly proportional to cost, since, for example, increasing the modulus of flexure

FIG. 22.9. A TUMBLE FILLER FOR PLACING IRREGULARLY SHAPED PIECES
INTO RIGID CONTAINERS

can be accomplished by using a thicker gauge film while a higher tensile
modulus could be obtained by using denser types of resins (which cost
more than lower density resins). As temperatures at time of use increase, these films become more flexible and more difficult to run
(Membrino 1974). The coefficient of friction increases, but this can be
offset to some extent by using slip additives or dusting with starch.

Static Electricity.—Plastic films are nonconductive. As a result of
unwinding and rubbing across machine parts, the web can become elec-
trostatically charged. Clinging and distortion of film as well as hang-ups
of food particles and dust can result. Tinsel brushes, polonium strips,
and electrical ionizing equipment can be used to overcome these problems. Conductive coatings can also be applied to the film. Cellophane,
aluminum foil, and other conductive layers will generally prevent the
accumulation of static electricity.

INNER-LINED CONTAINERS AND ROTOSEAL MACHINES

These packages and the equipment for filling and sealing them are
distributed by the Interstate Folding Box Co. of Middletown, Ohio. Pre-
lined bag-in-box containers are delivered flat, and erected, the bag heat
sealed, and the box closed on Rotoseal equipment. Weighing and dump-

ing is done on separate equipment not sold by this firm. As delivered by the supplier, the bag material is sealed at the side and open at the top and bottom. It is also glued to the inside of the box so that the bag opens as the box is squared up. All seals are fin or face-to-face seals, and the liners are made of laminates of films and foils. Capacities are from 15 to 420 cu. in.

The basic Rotoseal equipment produces up to 30 finished packages a minute, and this can be doubled by adding an automatic infeed-transfer unit. A second Rotoseal and auxiliary equipment can be used to increase the line speed to 120 units per minute. One or two operators are said to be required. Height changes within a certain range can be made by simple adjustment of the machinery, but alterations in other dimensions require change parts.

Package design options offered by the Interstate Folding Box Co. include boxes with handles on a gable top, push-pull pour spouts, two sectioned inner bags, and dispenser packs.

AUTOMATIC CASE PACKAGING FOR FLEXIBLE BAGS

From one dozen to 100 pouches, depending on size and retailer requirements are ordinarily packed in corrugated shippers before they enter the distribution system. An extra labor charge is involved which manufacturers would like to minimize. Neodyne Industries, Inc., has developed equipment for the automatic performance of this operation.

Their equipment check weighs each pouch, assembles a set of pouches in the configuration needed for one layer, then picks up the layer with vacuum cups and deposits it in a shipping case. Adjustments can be made for size of pouch, number of pouches in a layer, number of layers in a case, and speed of delivery.

Neodyne claims their equipment can permit about 10% more output from a form-fill-seal machine by eliminating the human constraints on case packing.

The Standard-Knapp Flexible Bag Packer performs a similar function although design features are different. Bags are taken directly from the form-fill-seal machine and each bag is checked individually for weight. The seal on each bag can also be checked, if desired, and any package which fails the weight or seal check is automatically rejected. The accepted bags are accumulated and then placed into an empty corrugated case which has been automatically set up and moved into loading position. Speed varies in the range of 25 to 110 bags per minute, depending on the model.

MILK-CARTON TYPES

Many kinds of snacks can be packed in plastic-coated board cartons of the type used to package fluid dairy products such as milk and cream. The limited size of the opening in the conventional design makes it difficult to dispense potato chips and clusters of caramel corn, but nuts, single kernel popcorn, and other snacks in this size range should pour out readily. Several companies make equipment for setting up the blanks, filling them with a measured amount of product, and heat sealing the container.

PACKAGING NUT MEATS

For maximum shelf-life of a nut meat product, it has been shown that vacuum or vacuum-gas packing in hermetically sealed containers is required. Their high content of unsaturated oil causes nuts to be particularly subject to the development of rancidity when stored in contact with oxygen. Rancidity can be effectively delayed to provide a useful shelf-life of a year or more for many kinds of nuts merely by removing most of the air from the can at the time of packaging.

Although there is no sharp dividing line between a satisfactory and an unsatisfactory vacuum level for canned nuts, it is generally recognized that a vacuum of 27 in. or better is preferred. About 90% of the air and oxygen is removed at this vacuum level. U.S. military specifications require an oxygen concentration of 2% or less in the head space of canned nuts. Most of the major can companies will lease equipment for vacuumizing or gas flushing metal cans and sealing them.

BIBLIOGRAPHY

BALL, C. O. 1963. Flexible packaging in food processing. *In* Food Processing Operations, J. L. Heid and M. A. Joslyn, Avi Publishing Co., Westport, Conn.

BRICKMAN, C. L. 1957. Evaluating the packaging requirements of a product. Package Eng. *2*, No. 7, 19.

GRIFFIN, R. C., JR., and SACHAROW, S. 1972. Principles of Package Development. Avi Publishing Co., Westport, Conn.

KRIEG, H. R. 1973. Data analysis for packaging operations. Snack Food *62*, No. 5, 42–44.

LEFFLER, W. H. 1974. Part II. Evaluating existing machinery systems, justifying the new. Package Develop. *1974*, (Jan./Feb.) 23–30.

MEMBRINO, J. N. 1974. Do you know the impacts temperature and humidity changes make on polyethylene films? Package Eng. *19*, No. 12, 52–55.

SACHAROW, S. 1971. Packaging machines—what type for your product. Food Eng. *43*, No. 3, 70–74, 77–80.

SACHAROW, S., and GRIFFIN, R. C., JR. 1970. Food Packaging. Avi Publishing Co., Westport, Conn.

WALMSLEY, T. E. 1973. Packaging of variety cake products. Proc. Am. Soc. Bakery Engrs. *1973*, 169–174.

CHAPTER 23

Miscellaneous Equipment

INTRODUCTION

A fundamental problem in the attempted categorization of equipment for such a diverse industry as snack food manufacturing is the large number of classes composed of only a few examples which do not seem to fit well into any of the major categories, or at least have sufficient points of difference to make their placement in separate groupings more meaningful. One possible approach, and it is the one followed in this book, is to establish a catch-all class of "Miscellaneous" equipment in which all of the difficult-to-categorize equipment is included.

Among the kinds of equipment covered in this chapter are flavor applicators, nut processing machines, measuring devices, and bulk handling equipment. The coverage of equipment in this chapter and the preceding chapters of this section is not intended to be exhaustive—representative rather than complete listings are included. It is obvious that every type of equipment used in snack factories cannot be covered in a volume of convenient size. In a more encyclopedic work, exhaustive coverage might be aspired to.

NUT PROCESSING EQUIPMENT

Most nut processing facilities will perform the minimum steps of receiving, dumping, storing, shelling, cleaning, blanching, roasting, salting, testing, packaging, and shipping. Some of these operations are conducted with generalized equipment suitable for use with many different kinds of particulate foodstuffs. The devices to be discussed under this heading are used predominantly for processing nuts.

Sorters

Nuts can be sorted and cleaned by visual inspection, screens, rotary pocket separators, specific gravity devices, and photoelectric sorters. Different combinations of these methods will be used on a line depending upon the type of nut being processed, the form in which it will reach the consumer, and the manufacturer's assessment of the importance of quality factors. Additionally, magnetic separators are usually applied at several points along the line to remove ferromagnetic debris.

The Sortex Air Separator (Fig. 23.1) is a relatively simple and inexpensive machine for cleaning particles in any shape up to 0.75 in. in the largest dimension. It separates the infeed into 2 or 3 groups on the prin-

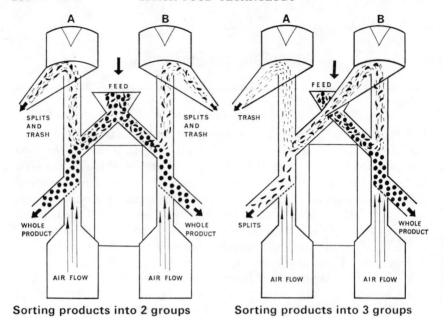

Sorting products into 2 groups   **Sorting products into 3 groups**

Courtesy of Gunson's Sortex Ltd.

FIG. 23.1. DIAGRAMS OF EQUIPMENT WHICH SORTS NUT MEATS AND THE LIKE INTO TWO OR THREE GROUPS ACCORDING TO DENSITY

Sorting Products into Two Groups: (1) Whole products. (2) Lighter particles including splits, broken pieces, chaff, trash, shells, and other rubbish.

Uncleaned product is fed by vibrator into two feed channels which simultaneously serve Separators A and B. A continuous stream of air in the cleaning chutes lifts the lighter particles to the top of the separators for reject collection.

Sorting Products into Three Groups: (1) Whole products. (2) Splits. (3) Lighter particles, broken pieces, chaff, trash, shells and other rubbish.

The head of Separator B is swiveled around to feed into Separator A and the vibrator feed into Separator A is closed. Uncleaned product is then fed into Separator B where heavy, whole products in the continuous air flow fall through the lower outlet. Lighter particles are lifted to the top of the chute and are directed into Separator A. The Separator A air flow is regulated so that the heavier splits fall through the lower outlet and the lighter particles (broken pieces, chaff, trash, shells, etc.) are lifted up the chute and emerge for separate reject collection.

ciple that the desired fraction is either lighter or heavier than the other particles. Uncleaned material is floated through a rising stream of air. The lighter particles are lifted to the top of the cleaning chute for collection while the heavier pieces flow through the lower outlet. Throughputs of up to 1,500 lb per hour are obtainable, depending on the product.

The Bauer specific gravity separator uses a combination of mechanical and pneumatic action to remove loose skins, stones, immature peanuts, skins, and other foreign objects.

The Sortex Electronic Color Sorting Machine (Model 512) is recom-

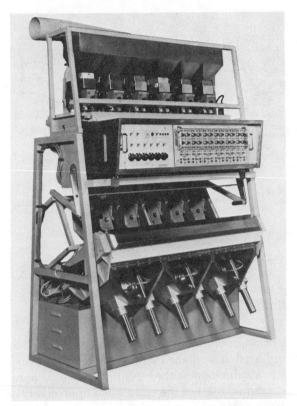

Courtesy of Sortex Company of North America

FIG. 23.2. SORTEX 964 OPTICAL COLOR SORTER

mended by the manufacturer for separating in-shell peanuts, walnuts, pecans, etc. The basic principle involves aligning the nuts in single file prior to dropping them through a scanning arrangement of light beams and photoelectric cells which sense the reflectance properties of individual nuts. Those kernels which have a nonconforming reflectance pattern are diverted into a reject chute by a blast of air. Similar equipment is available for sorting shelled peanuts, walnut and pecan pieces, granules, etc. See Fig. 23.2.

Blanchers

Peanuts may be oven heated to about 225°F for 1 hr to loosen the red skins prior to mechanical removal. Although dry heat is used in this step of the process, it should be regarded as a dehydration procedure rather than as cooking or roasting. Nuts may also be wet blanched in a hot water bath.

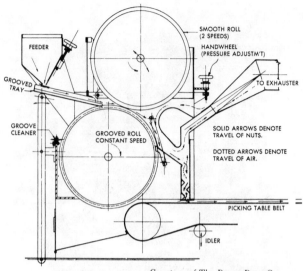

Courtesy of The Bauer Bros. Company

FIG. 23.3. METHOD OF OPERATION OF A WHOLE NUT BLANCHER

In oven blanching, equipment of the same type as used for roasting is commonly employed. The blanching process requires bringing the nuts to about 250°F, at which point the moisture content is reduced to about 1.5% from the original 6%, but the color is not appreciably changed. Following heat treatment and cooling, the nuts are transferred to mechanical units which strip off the skins.

The Bauer whole nut blancher (Fig. 23.3) passes the kernels between resilient rubber rollers and draws off the loosened skins in a stream of air. A vibrating slotted screen removes split nuts and hearts. Blanched nuts fall on to an illuminated inspection belt which is part of the machine. Capacity is 350 to 450 lb per hour. The Bauer split nut blancher also employs rubber elements to remove skins efficiently with minimum production of meal. Cyclone collector and piping are furnished for the pneumatic collection of skins.

Roasters

Nuts can be roasted in oil (sometimes called French frying) or dry roasted by hot air currents or infrared radiation. Batch and continuous roasters are available for either process. The Bauer Roaster is a rotary cylinder with an infrared heat source mounted inside (Fig. 23.4). The heater can be either oil or gas fired. Since it is a batch roaster, adjustments can be made easily in the conditions used for each lot of nuts. If

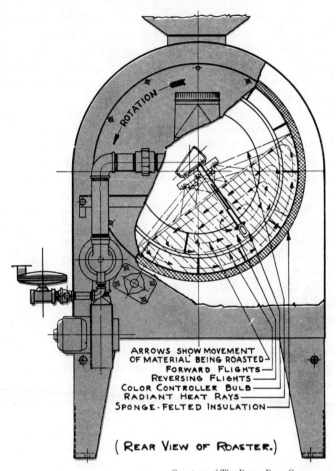

ROTATION

ARROWS SHOW MOVEMENT
OF MATERIAL BEING ROASTED
FORWARD FLIGHTS
REVERSING FLIGHTS
COLOR CONTROLLER BULB
RADIANT HEAT RAYS
SPONGE-FELTED INSULATION

(REAR VIEW OF ROASTER.)

Courtesy of The Bauer Bros. Company

FIG. 23.4. A RADIANT ROASTER FOR NUTS

intended for peanut butter, the contents of six bags or 720 lb of nuts can be brought up to 285 to 290°F within 20 min. If the nuts are of the Virginia variety and intended for salting, they may be blanched in this equipment—i.e., given a "white roast" as it is called. The batch is constantly mixed by forward and reversing flights through the radiant zone. A stainless steel sensing bulb immersed in the peanuts automatically records the temperature on a disc chart. It is claimed that old or new crop, cold or warm, or cold storage peanuts can be roasted to identical coloration when brought to equal control settings, the only variable being the time required to bring the nuts up to finish temperature.

When used for inshell peanuts, the load is brought to about the same temperature used for roasting salted nuts, but the heat is then turned off and the product is allowed to tumble within the hot cylinder for an additional 5 min to give a heat rise of about 15°F. This treatment allows for maximum development of the roasted kernel without any noticeable darkening of the shell (Bubp 1972).

The Jetzone® fluidized bed machine can be used for continuously roasting peanuts. In this equipment, hot air is forced through an assemblage of vertical tubes down into a bed of nuts traveling on an oscillating steel solid pan conveyor. Heated air is forced by a fan into a pressure plenum area, then passed down through tubes onto an oscillating conveyor, creating a bath of turbulent air under and around the nuts. Air rising from the conveyor passes around the tubes and is directed into return ducts at the sides. Returned air is carried into cyclones where airborne particles are removed before it passes into the heater.

Manley, Inc. offers a 5-tier stainless steel peanut drying and roasting oven based on a recirculating air flow heating system and continuous product flow. Gas or electrically heated versions are available.

Several firms make fryers which can be used for oil roasting of nuts. The Bauer fryer and cooler combination cooks, cools, glazes, and salts nuts in one continuous operation. A stainless steel hopper and spreader distributes salt evenly over the full width of the conveyor.

J. C. Pitman and Sons offer several models of their Mastermatic® continuous automatic nut fryers which automatically cook, cool, glaze, and salt the product. Capacities range from 3,000 to 5,580 lb per hour of Virginia jumbo peanuts, with other varieties requiring different rates. Btu inputs are from 1.4 to 2.75 million. Cooker and cooler belts are of type 18-8 stainless steel wire mesh. Frying temperature is controlled by means of two indicating thermostats, one for each end of the oil bath. Heat is applied by immersion tubes. There is a variable speed drive unit for providing holding times of 45 sec to 7.5 min. A stainless steel hopper and electronically controlled salt spreader distributes a curtain of seasoning evenly over the full width of the conveyor, and a rotary agitator gently mixes the nuts after oil and salt are applied.

Coolers

After nuts are blanched or roasted, they should be quickly cooled to halt the thermally induced reactions leading to color and flavor development. The Jetzone® cooler is a fluidized bed unit operating on the same general principle as their roaster, but using ambient rather than heated air to create the fluidized bed.

The Bauer vertical automatic cooler is particularly designed for fast uniform cooling of peanut butter stock. This company also offers two

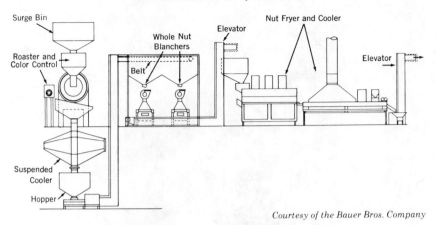

Surge Bin

Whole Nut
Blanchers

Elevator

Nut Fryer and Cooler

Roaster and
Color Control

Belt

Elevator

Suspended
Cooler

Hopper

Courtesy of the Bauer Bros. Company

FIG. 23.5.   TYPICAL LAYOUT OF A SALTED NUT-PROCESSING SYSTEM

models of suspended nut coolers. The smaller size has a removable cooling basket of 3 bag (peanuts) capacity, while the larger is of 6-bag capacity. Exhauster (up to 10 hp) and cyclone collector are optional elements.

Integration of the foregoing equipment into a complete salted nut processing line is shown in Fig. 23.5.

OIL, POWDER, AND GRANULE APPLICATORS

Oils are usually applied as sprays directed on to a layer of product traveling on a mesh conveyor belt, but they can also be applied by spraying or dripping inside a rotating cylinder. Powders and granules are usually added by blowing or sifting the seasoning in to a rotating cylinder through which a continuous stream of product is passing (Fig. 23.6), but tumbling drums can also be used. In some cases, cheese-based seasonings are suspended in oil which is then sprayed or dripped on to the product. Salt is often added by sprinkling on to a traveling bed of product.

Oil Sprayers

Krispy Kist Korn Machine Co. sells a sprayer and melting tank combination suitable for applying oil and cheese mixtures to snacks. It includes a 120-lb capacity stainless steel kettle provided with a jacket for circulating gas-heated water and a propeller-type mixer. A bronze pump receives the oil mixture from an adjustable faucet and delivers it through a pressure hose to the spray jet.

Powder Dispensers

Powdered condiments are fed into the product area in three ways: by

Courtesy of Cretors and Company

Fig. 23.6. A Seasoning Coater for Continuous Flow of Product

gravity, by forced air currents, and by spinning cones or discs. The gravity feed devices are suitable for dispensing salt from a hopper on to a bed of product passing beneath. Variations in the method of controlling the rate differentiate the equipment of different manufacturers. Flavor Dispenser of Heat and Control Co. uses a special mesh belt with a variable speed drive to draw flavoring materials from bottom of hopper at a desired rate. Reciprocating wiping blades above a screen form the controlling element in other dispensers. Vibrating hoppers can be used with some very free flowing condiments, the rate being proportional to the frequency of the vibration.

Manley, Inc., offers a series of spice applicators consisting essentially of a fan forcing air past the bottom of a hopper into which a vibrating pan feeds material at a constant rate. A long tube feeds the air current into the coating drum. Another version uses a variable speed auger to feed the seasoning powder to the blower tube. Heat and Control also offers a more elaborate cabinet mounted pneumatic salter which feeds the entrained seasoning through a tube to a dispensing unit the width of the product conveyor belt.

Another design of pneumatic applicator for salt and spices is offered by American Food Machinery Corp. This system consists essentially of a blower with inlet filter, a pressurized hopper preceded by an air proportioning valve, a variable speed agitator valve in the hopper, and a delivery line to a spreader through which the powdered material is dispensed. The air proportioning valve operates in conjunction with the clean-sweep agitator to draw the salt or spice from the pressurized hopper on a continuous or intermittent basis. The ingredient is then carried to the spreader which dissipates the air and allows the material to fall on the food product as it passes along a conveyor channel. Discharge may also be made through a variety of nozzles directly into tumblers or batch mixing vats. The dispenser can be controlled manually or automatically by photocell, timer, or other switching device. The unique valve provides instant material flow at a predetermined rate, and complete cut off.

AMF recommends this dispenser for use with any hygroscopic or hard-to-handle powdery substance, specifically salt, seasonings, and spices as used in the snack food industry. Flow rates of 0 to 10 lb per minute are obtainable, depending upon the material.

The Be-Mo Machine Co. (Ft. Lauderdale, Fla.) makes a Spinner-Spreader based on the use of centrifugal force to distribute flavoring powders.

A simple type of gravity feed seasoning applicator is made by Leon C. Osborn Co., Inc., Harlingen, Texas.

Electrostatic Salters

The electrostatic salter introduced by Morton Salt Co. in 1965 has been widely accepted for surface salting of chips, crackers, and pretzels. The new ESB Model has a frame and hopper fabricated from a cellulose/lignin laminate having good electrical insulating properties. Electrostatic dispensing is accomplished by inducing a high-density negative static charge on the surface of rotating induction/dispensing cylinder and a corresponding positive charge on the opposing dispersion rod (Fig. 23.7). When polar material such as salt passes over the roll, a negative static charge is induced on the particle surfaces. When adequate potential develops, the particles are pulled off the rotating cylinder by the attraction of the positively charged dispersion rod, but the attraction is

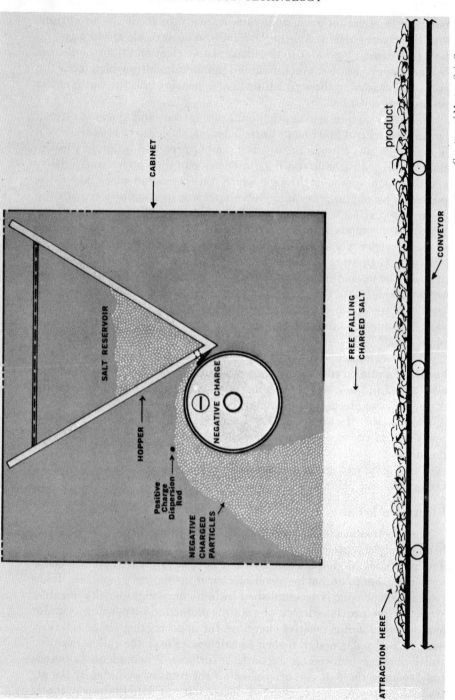

FIG. 22.7. PRINCIPLE OF THE ELECTROSTATIC SALTER

not sufficiently strong to draw the particles across the 2-in. gap and the salt falls toward the dispensing area. While they are falling, the negatively charged particles mutually repel one another to maintain uniform separation, and fan out into a uniform pattern before striking the target surface. The static charge is quickly dissipated after the salt strikes the product or conveyor. There is some evidence that initial salt adherence on oil-fried snacks can be significantly improved when the product is salted on a solid metallic surface, such as a vibrating conveyor, that is adequately grounded. According to the manufacturer, dispersion uniformity can be controlled to less than a 10% coefficient of variability which is equivalent to a standard deviation of ±0.15% for a 1.5% salt level on potato chips. Any grade of chip salt may be used effectively in this equipment if it contains a small amount of conditioner (0.1 to 0.2%) to keep the crystal sufficiently dry for inducing an electrostatic charge (Strietelmeier 1972).

The capacity for inducing an electrostatic charge density on the induction roll is limited by the potential of the transformer which is 25,000 volts. The voltage is regulated by a rheostat control dial, usually in the range of 70 to 100% of maximum potential for high polarity materials such as salt, antioxidant salt, and vitamin and mineral fortification. This potential is insufficient to achieve an adequately uniform dispensing pattern with less polar materials such as sugar and various seasonings, but the pattern for these materials can be improved by moving the dispersion rod a fraction of an inch closer to the induction/dispensing roll.

There are three sizes of ESB Model Electrostatic Salters covering a total dispensing range of 3 to 430 lb per hour of salt.

Watkins (1970) described another type of electrostatic applicator for seasonings. Food products such as cookies, popcorn, crackers, or pretzels are dropped into a flavoring chamber through which air containing ionized condiment particles is being circulated. The particles are ionized by being blown through a tube containing an electrode. The food pieces attract a coating of the condiment and are collected in a basket in the closed chamber of the machine. Batches are dumped when coating has been completed.

Coating Tumblers

The Cretors Flo-Thru seasoner coater consists of a baffled rotating cylinder (horizontal axis) of stainless steel. Integral pumping and spraying apparatus or salters are available. Manley, Inc., offers somewhat similar equipment for treating 300 lb of popcorn or 500 lb of colletts per hour. Inlet and outlet flights are used in conjunction with tilt angle (adjusted by raising or lowering legs) and variable speed drive to control flow rate. The continuous coater of Krispy Kist Korn Machine Co. operates on the same principle. It includes a stainless steel drum, variable speed drive, frame with adjustable legs, a vibrator feed blower for salting, and a stainless steel pipe with sprayhead adapters for oil. A batch-type coating tumbler made by the same company treats 10 lb of popped corn or 12 lb of corn curls in 4 min.

TRANSFER AND STORAGE EQUIPMENT

General Considerations

Bulk storage facilities fall into two categories based on the state of the product—whether it is liquid or solid (i.e., granular). Cornmeal or flour will, of course, always be handled as powders, and it is usual to handle sucrose, corn sugar, and salt in this form. Fats, corn syrups, and invert sugar are frequently handled as liquids in bulk systems. Bulk shipping, receiving, and handling of potatoes is assuming increasing importance in the chip industry, but apparently there is no bulk handling system for nuts.

There are some general principles applicable to the design of all bulk handling facilities. Sanitation should be a primary consideration in the design. The ingredient contacting surfaces should be nonporous and resistant to corrosion and abrasion. They should be nontoxic and should not transfer odor, taste, color, or particles to the ingredient. They must not accelerate deteriorative changes in the food material. Physical changes (as in particle size) during transfer or storage should be held to a minimum consistent with necessary design limitations. Tubes carrying powders should be grounded to prevent the buildup of static electricity.

Sampling and testing of bulk loads present some problems. The receiving department is generally anxious to unload the shipment quickly in order to avoid demurrage and other costs and complications, while the quality control department will not want to risk contamination of tanks or old stocks of ingredients with unsatisfactory material and will insist on completing the necessary tests before unloading is started. There is usually no provision for returning unsatisfactory material which has already been taken into the system, an additional reason for caution. The interests of the two departments necessitates close coordination of effort and maximum speed in sampling and testing incoming shipments.

Sampling of liquid loads for physical and chemical tests can be performed with a simple bottle-on-a-stick type of device inserted through the loading hatch or by collecting fluid from an appropriate port. Sampling of dry materials such as sugar and flour is accomplished by inserting a long trier or specialized sampling equipment through the opened manhole at the top of the car, and withdrawing and combining portions taken from several directions. There are automatic sampling devices which divert a certain portion of the ingredient as it flows into the plant. The sampling problems resulting from stratification of either liquid or powder loads must be kept in mind constantly in order to avoid unpleasant surprises on the production line.

Assurance should be obtained that an adequate supply of delivery conveyances will always be available. Shortages of trucks and cars are commonplace and are more common in some parts of the country than in others. Sources of supply are more restricted for producers who must rely on bulk delivery.

Scheduling of deliveries is much more difficult for bulk receiving because the storage space is absolutely limited. It is nearly always possible to find a place to put a few hundred bags of sugar, but, if the syrup tank is nearly full, nothing can be done to remedy the situation. "Running out" is also more serious, since it is often quite difficult to introduce bagged or drummed material into bulk systems.

Bulk Handling of Ingredients

Each of the ingredient types used in snack food manufacturing involves special handling problems. Many ingredients are still received in bags, boxes, or drums because of the unresolved engineering difficulties or the high cost of bulk transfer systems.

Considerable progress has been made in the handling of potatoes for large chipping operations. Vanmark Co. engineers and builds integrated lines for unloading, storing, washing, and peeling potatoes. The heart of their one-track handling system is the bulk storage bin using an angle-of-repose loading principle which reduces product bruising and crushing. Tempered air can be fed into the bottom of the bin to increase storage life.

Large manufacturers find economies in the use of bulk vegetable oil installations for receiving, storing, and dispensing oil for frying and coating snacks. Bulk oil facilities consist of pump, pipes, and tanks with auxiliary equipment of various kinds. Shipments are received from 60,000 lb tank cars or from 20,000 or 30,000 lb tank trucks.

Both round and rectangular tanks are used. A convenient arrangement consists of 2 or more tanks of 65,000 to 70,000 lb capacity for each type of shortening. Since each tank will hold a full tank car of oil, it will not be necessary to mix fresh oil with residues left in the tank from previous shipments. Tanks of 35,000-lb capacity are more convenient for truck deliveries.

Stainless steel, type 302 or 304, is the preferred material for constructing storage tanks. These metals have no adverse effect on fats and oils and are not affected by common cleaning solutions, but they are high in cost and difficult to fabricate. Consequently, most tanks are made of mild steel plate or even of black iron. The configuration of the tank and the location of the outlet should be such that complete drainage of the contents can be obtained. Vertical tanks can be made with dished or cone-shaped bottoms and rectangular tanks can be slightly tilted toward the outlet to achieve this result. The tanks must be equipped with gauges, manholes, etc. The inlet pipe should be brought to within a few inches of the bottom to minimize splashing.

Liquefied fats should be held at about 10°F above the AOCS capillary closed-tube melting point (not the Wiley melting point). Lard will be completely liquid at 120°F, but some vegetable shortenings or oleo may require slightly higher temperatures. The fat can be heated by internal coils of pipe carrying hot water or steam or by external coils or jacket. Hot water heating is preferred since steam may cause localized overheating with damage to the fat. Nonaerating agitators should be installed if steam is to be used for heating. Side-entering propeller-type agitators or recirculating pumps are satisfactory. Thermostatic controls must be provided. Tanks may also be installed in a heated room or a separate heated enclosure, in which case no additional temperature-adjusting equipment is necessary.

Liquid level gauges can either be sight tubes or float type indicators. The latter can be arranged to operate signals or controls. Sight tubes are not suitable for use with high temperature installations. The installer should calibrate the tank in terms of depth versus volume (or weight) of contents.

Some suppliers recommend installation of stainless steel screens at the inlet, others indicate this is not required. Lines and valves may be of standard iron or steel. No copper-containing alloys such as brass, Monel, or bronze should ever be allowed to contact the fat. Gate-type valves are recommended. Welded construction is highly desirable since few pipe compounds and gaskets are compatible with edible fats and oils.

Transfer pipes within the plant are almost always heated and insulated. A convenient method of heating is by means of an electrical heating tape wrapped around the pipe. Steam tracing or hot water tracing is also used. Unloading lines are not heated. They are freed of fat after use by draining or by blowing nitrogen or air through them.

It is good practice to incorporate in the system a holding tank located close to the use points. A float switch in this tank is connected to a pump which maintains the liquid level within a narrow range. The temperature is thermostatically controlled. Maintaining the head and temperature constant increases the accuracy of the dispensing meters and confers other advantages.

Recommended turnover is a tank every 2 to 3 weeks. Many operators take four weeks to use a tank. The tanks should be thoroughly cleaned at least every six months or at the first sign of sediment or off-odors suggestive of rancidity. A new shipment of shortening should never be dumped on top of the residue from a preceding shipment. Antioxidants should be used up to the legal limit.

Nitrogen blanketing is sometimes recommended as a means of increasing the storage life of bulk shortening. Melted shortening will absorb oxygen during manufacture and transport to the user's plant. Blanketing the shortening under a layer of nitrogen while in the user's tank will retard further uptake of oxygen, but the amount already present can cause sufficient deterioration to make the shortening unusable. Oxidation at the surface is small compared to that due to entrained and dissolved oxygen. The nitrogen blanket is not a sure preventive of rancidity, but it is a further precaution which, in combination with other good practices including prompt turnover of stock, will assure a constant supply of shortening having excellent organoleptic qualities. Cost of the equipment for nitrogen blanketing is only a few hundred dollars while the expenditure for gas should not exceed $10 per month per tank. To hold down condensation in the head spaces of tanks held in a room where the temperature fluctuates, a worthwhile precaution is to blow air across the surface of the liquid with a fan of about 90 cfm capacity.

Bubbling nitrogen through fat in short bursts is a good method for preventing stratification or settling of higher melting point fractions. This is probably not needed for vegetable oils, but may be helpful for maintaining uniformity in hydrogenated shortening, oleo, or lard held near the melting point or if the required temperature is not being maintained in all regions of the tank.

Some varieties of gear-type single or double-lobe pumps have been used to dispense liquid shortening. However, these pumps have a high degree of slip and do not accurately measure low viscosity liquids if any fluctuation in head occurs. Positive displacement pumps can be used. There are pumps expressly designed for this application. According to Abbott (1959), a packed type, ball check, reciprocating pump will give good accuracy with low viscosity liquids.

The fluid shortenings consisting of oils with added stearine flakes and emulsifiers do not need to be heated and, in fact, heating above the melting point of the stearine can lead to separation on subsequent cooling. These fluids should be kept under continuous agitation with a slow moving blade that circulates all the way around the outside of the tank. At temperatures within the range of 70 to 87°F, the shortening should retain good odor and flavor for at least 3 to 4 weeks.

Pumps, pipes, valves, and nozzles should be fabricated of stainless steel, if possible. Food grade plastics are also suitable materials for some liquid bulk handling applications. In any case, copper and copper-containing alloys should never be permitted to contact fats or oils intended for use in snacks, since even minute traces of copper will greatly accelerate the development of oxidative rancidity. Some installations rely on iron piping. Although not the material of choice, iron is marginally acceptable where economy is the over-riding consideration.

Bulk Handling of Products

Transfer of snack products poses special problems because of their fragility, nonuniformity, and low density. Pneumatic conveying avoids some of the physical abuse inherent in many mechanical transfer methods, but can cause significant amounts of breakdown, especially where the pipeline undergoes several turns. Manley, Inc., offers "Air-Veying" systems. These are low-pressure positive effect systems including Teflon-coated hoppers, Venturi type blowing unit with stationary tongue, air intake with washable stainless steel filter, adjustable damper to control air velocity, and sanitary couplings with flush gaskets. They are designed for use with 6-in. diameter stainless steel or aluminum tubing.

Manley, Inc., also sells live bottom bins for accumulating and temporary storage of product. These are generally rectangular in profile, and deliver product on a conveyor belt running along a trough at the bottom. A sliding gate controls the size of the exit aperture and rate of product delivery. This company also supplies a product moving auger lift to elevate snacks from a floor level hopper (200 to 800 lb capacity) to a processing intake about 10 ft high.

MEASURING AND WEIGHING EQUIPMENT

Accurate weighing and metering of both ingredients and products are critical, not only to the economic health of the snack producer, but even to the continued existence of his business. Accuracy of ingredient measurement is related directly to the quality of the finished product and it also affects costs. Since composition of the product is regulated by Federal, State, and local laws in the case of standardized items and must in any case conform to the label declaration of ingredients, management must be certain that the amount of each component found in the finished food will fall within acceptable tolerances.

The hand scaling of individual ingredients has been replaced at an ever increasing rate in recent years by automatic weighing and metering. Some of the advantages which may result from such changes are greater reproducibility (not necessarily greater accuracy), lower manpower requirement, and improved sanitation. In addition, automatic measuring is almost essential in a bulk-handling system.

Designers of automatic portioning and metering equipment for the snack industry have borrowed freely from other industries. Water meters of various kinds have proved to be adaptable, with changes of varying degrees of complexity, to the measuring of other liquid ingredients. Weighing devices originally used for other kinds of foods or in nonfood applications have been modified for use by the snack manufacturer. A great deal of ingenuity and expertise has sometimes gone into these adaptations and modifications.

Powders and granular materials are generally measured by gravimetric techniques while liquids are more commonly measured volumetrically. Volumetric equipment is the least expensive, as a rule, but is also inherently less accurate and dependable since the feed rate necessarily depends upon the density and flow characteristics of the material being measured, and these factors cannot be expected to be perfectly uniform, especially in powdered materials. As a practical matter, volumetric feeders function perfectly satisfactorily in many bakery applications for sugar and other relatively free-flowing ingredients. Gravimetric measurement is inherently more difficult to automate but is theoretically capable of greater accuracy, at least for powdered and granular materials which may vary considerably in density.

The next two sections of this chapter describe the equipment which has been developed for the automatic weighing and the automatic metering (i.e., volumetric measuring) of ingredients.

WEIGHING

The two major categories of weighing devices are balances and force-deflection systems. The simplest weight, or force, measuring system is the ordinary equal arm balance. It operates on the principle of moment comparison. The moment produced by the unknown weight is compared with that produced by a known weight. If the 2 arms are equal in length, the 2 weights will be equal when a null balance is obtained. Obviously, such equipment is unsuitable for most practical applications. Some industrial equipment is based on the principle of an uneven-arm scale, of which the ancient steelyard is the simplest example. An unknown weight on a short moment arm is balanced by moving a poise of known weight along a calibrated long moment arm. Multiple beam industrial scales introduce additional levers between the unknown weight and the beam to increase the ratio of unknown to poise. Small capacity bench balances are examples of the uneven arm scale while many platform scales and suspended hopper scales make use of the multiple beam construction.

Pendulum balanced scales use one or more "pendulums" (weights mounted on one end of a rigid rod which rotates about a pivot at its other end) to balance increasing loads as they move from vertical to horizontal. Because the weight which the pendulum will balance varies as the sine of the displacement angle, cams are used to give linear movement of the scale pointer. Metal tapes transmit the motion from pendulum to platform. These scales are commonly used where a dial indication of weight is required. They are available in recording and printing models, and with electric cutoff.

Because the deflection of a spring or any other elastic element is directly proportional to the applied force, within the elastic limit, a calibrated spring, tube, rod, or plate can serve as a weighing device. This principle is applied in the spring scale and the torsion balance. They are subject to hysteresis, fatigue, and temperature errors, but when properly designed and used can be valuable weighing devices.

The torsion principle of weighing is incorporated in the Thayer Flexure-Plate Leverage System. The weighbeam is connected to the sensor or indicator by a series of levers and steel plates which are calibrated to flex the desired extent. When the weighbeam is in the null position, all flexure plates are plumb and in tension and transmit no movement within the system. When the beam moves out of the balance position, the flexure plates are very slightly deflected and tend to urge the system into a balanced position. Advantages claimed for these systems include (1) no knife-edge or other friction surface to wear, (2) shockproof, (3) require no maintenance, and (4) not affected by dust, dirt, or vibration.

Other force-sensing elements adaptable to weight measurement include strain-gage load cells and pneumatic pressure cells. In a sense, strain gages are deflection scales, but the deformation is measured as a change in an electrical signal rather than directly as a change in dimension.

Continuous and automatic operating scales utilize the same weight-comparison principles used in manually operated unequal arm balances. Sensing means are used to energize the mechanisms that start and stop the flow of material into the weighing hopper. For example, weighing of flour in most modern bakeries is accomplished by a special automatic flour scale located immediately above the dough mixer. In some cases, the hopper is on a trolley, and can be moved over several mixers, as required. The usual model consists of a conical steel hopper mounted on four-point suspension bearings of knife-edged pivots made from case-hardened steel. Once the scale beam is set for the necessary amount of flour and the switch is activated, operation is completely automatic. A mercury switch activated by the scale beam shuts off the flour input when the preset weight has been reached. At the proper time, the operator discharges the contents of the hopper directly into the mixer. Capacities range from 200 to 1,000 lb of flour. Transfer operations are dustless, the outlet of the hopper being connected to the mixer inlet by a sliding sleeve. Air displaced as the flour goes into the mixer is conducted to the upper section of the hopper by a venting tube. The scale must be adjusted by the operator or automatically compensated for conveyor overrun to obtain satisfactory accuracy.

In another type of automatic scale, a balancing weight is positioned by a reversible electric motor. Deflection of the beam makes an electrical contact which causes the motor to move the weight in the direction necessary to restore balance, and the final balance position is translated by means of a potentiometer or digital encoding disc into a signal which can be used for recording or control purposes.

Automatic batch scales are adaptable to continuous flows of liquids, granular materials, or powders. If dry materials are being measured, the ingredient flows from a feed hopper through an adjustable gate into the scale hopper until the preselected weight is reached and a trip mechanism closes the gate and opens the outlet. When the scale hopper is empty, the weight of the tare forces the door closed, resets the trip, and opens the gate for another cycle. A dribble feed, resulting from a partial closing of the gate as the set weight is approached, reduces the rate of inflow so that the extent of overshooting the mark is reduced. A counter can be used to record the number of cycles and the total amount of material which has been dumped.

For systems in which continuous addition of dry ingredients is required, automatic gravimetric feeders are preferred.

Weighers Using Moving Belts

The simplest forms of this type of measuring equipment weigh the delivered ingredient but do not adjust the delivery rate to maintain a constant flow (i.e., no feedback). They are useful in automatic batching, but are not readily adaptable to continuous processing where the rate of delivery and not the total weight is important. In one semicontinuous model, the scale belt runs continuously but is fed intermittently with loads that do not extend its full length so that a definite increment is being weighted at any given time. Since the increments are deposited at short intervals, they approximate continuous flow. The weigh belt is mounted on a scale mechanism which records and sums the amounts removed.

A second type, the continuous conveyor scale, uses a scale-supported section of a belt conveyor to totalize the load which is being constantly deposited on it. The forms of weighing devices used in such equipment include spring-balanced beams, strain gages, and pneumatic load cells.

Fully automated belt weighers with feedback to control rate of delivery are useful in continuous or intermittent weighing.

The most common type utilizes a conveyor belt balanced on a weigh-beam. When the belt is driven at constant speed and the total weight of the belt, material, and associated mechanisms is held constant, the rate the material comes off the end of the scale is also constant and total weight for any known time interval can be computed. Imbalance of the beam actuates a change of the rate of material deposit onto the belt in the direction of restoring balance, by mechanical adjustment of the feed gate or by varying the speed of a belt or screw feeding the weighing conveyor. Accuracies are said to be as high as 0.0001% for certain products and equipment, but a more likely figure is about 0.1% for continuous measuring scales.

In another design of automatic belt gravimetric feeder, variation in the amplitude of vibration of a feed tray is used as a means of controlling the rate of delivery to the belt. A cam-operated mechanism driven off the belt transmission oscillates a driving plate at constant frequency in the direction of an opposing receiving plate connected to the feed tray. A rubber control wedge suspended from the scale beam transmits the vibration from the driving plate to the receiving plate. The amplitude of vibration is thus regulated by the position of the wedge between the plates. If the weight on the belt is excessive, the wedge is raised, the vibration is diminished, and less material is fed, and vice versa.

Loss-in-weight Feeders

In a typical design, the entire feed hopper is mounted on a scale which controls the rate of removal of material from the hopper. The hopper may be discharged, for example, by a rotating valve driven by a variable speed motor. If the scale beam counterpoise is retracted continuously by a constant speed drive, the rate of delivery will be constant and the equipment will be suitable for continuous processes.

Loss-in-weight feeders are of several design types. Vibratory, screw, pneumatic, and belt methods may be used to remove material from the hopper which is mounted on a multiple-beam scale mechanism. The hopper is filled and the scale beam balanced by manually adjusting the poise. The control dial on the rate setter is set to the desired rate of feed in pounds per hour. The setter, operated by a synchronous motor, retracts the poise by a lead screw at the exact feed rate desired. As long as the feeder delivers material from the hopper, causing it to lose weight at the same rate the poise is being retracted, the scale beam will stay in balance. If the feeder delivers too much or too little, causing loss in weight faster or slower than the poise is retracted, the beam will tip and operate controls which cause the feeder to correct its rate. Only a very small beam movement is required for control, and the scales are so sensitive that the beam is essentially in balance at all times (Anon. 1968).

Where the loss-in-weight feeder is electronically controlled, it can be coupled to the flow from another loss-in-weight feeder or from variable liquid flows through a Venturi tube, orifice plate, or other flow measuring device. The metering can be continuous or intermittent, and in the latter case, the sensing of a predetermined amount of one ingredient leads to the operation of the feeder for a fixed number of seconds.

Principles of Control of Gravimetric Feeders

Belt gravimetric feeders can be mechanically, pneumatically, or electronically controlled. Operation of belt gravimetric feeders is based on the principal of maintaining a constant weight of material on the moving belt by actuation of a positive-acting gate on the feed hopper. The rate of feed may be changed by varying the belt load and/or changing belt speed. Net weight of belt load may be sensed directly by a force balance pneumatic transmitter and load cell or by a differential transformer which transmit any variation from set load to the solid-state electronic controller. The feedback system actuates the control gate to maintain belt load at the set point and weigh platforms in a null position. Electronically controlled belt gravimetric feeders are available with capacities from about 0.1 lb per min to over 60,000 lb per hour and accuracies of ±1%. The belt type with preset cutoff uses a scale and tachometer signal to provide an instantaneous indication of rate of material flow by weight per unit of time. The resulting analog signal is converted into digital form through an integrator. A pulse counter records the digital information to provide visual indications of the total flow of material. The pulse counter can also be provided with a stop-point setting which will cause a contact to close and halt the conveyor when the prescribed amount of material has been transferred.

The proportional feed belt feeder also relies upon a scale and tachometer output to provide instantaneous rate of flow measurement. In this case, however, the resultant signal is electronically compared with that from a percent-setter to regulate conveyor speeds. The system can be designed to have a stop-point similar to the arrangement described above. In belt type systems, one of the chief factors affecting overall accuracy is the location of the sensing device in relation to the discharge point; the closer together, the greater the accuracy.

Hopper weighing systems can be provided with either analog or digital controls. The analog system employs a scale and a conventional sensing arrangement. Other types of weighing devices such as load cells can be used as sensors, provided their output is electrical. The amplified signal is compared with the present input through the balance detector. When the signals balance, indicating that hopper weight is at the desired level, the solenoid cutoff closes the feeder. The analog indicator (e.g., a large scale head graduated in 1000 increments) continually supplies progressive weight information.

The digital system employs a mechanical weighbeam and lever arrangement and a conventional scale head. The basic sensor can assume a variety of configurations and use several kinds of components, but a pulse signal giving a number of pulses which are proportioned to the material weight must be obtained. The remaining portion of the system features a preset counter with digital weight indication.

Fuller (1970) discussed the importance of receiving scales. He points out that without checking all ingredient receipts for weight (including bulk shipments), no control can be exerted over the shipper's errors or intentional fraud. Certified public scales and railroad scales are usually accurate but do not allow positive and direct control of the weighing operations by the purchaser. According to Fuller, many suppliers do not even have shipping scales and do not weigh their shipments, but only estimate them. Some of the weighing equipment used for receiving are railroad track scales, motor truck scales, bulk handling scales, bag check weighers, and dock platform scales. By installing an automatic bulk handling receiving scale in the unloading line leading from the truck or rail car to the storage bins and by installing automatic tank scales under the liquid storage tanks, the shipping weights can be verified and inventory of bulk materials can be closely controlled. In one type of bulk receiving scale, the material is air-conveyed to an overhead surge hopper, and then fed through a weigh hopper on a scale until a preset weight is reached at which time the fill feeder stops and the weight is recorded. The weigh hopper then discharges the lot of material to a selected storage hopper. When the weigh hopper is empty, the residual weight is recorded, and the cycle begins again. The computer adds each batch weight and subtracts each residual weight so that the contents of the shipping vehicle are known exactly at the termination of the receiving process.

Metering

A volumetric measuring device in its simplest form is a container of known volume, such as a tank, barrel, pipet, or volumetric flask. The container may be calibrated at various points, as in the burette, graduated cylinder, or some storage tanks. Volumetric measurements can be applied to solid materials as well as liquids; for example, in home cooking procedures where nearly all proportioning is done volumetrically in spoons and cups. But large scale manufacturing processes do not customarily include volumetric measurements for powders and the like because of difficulty in controlling density.

Location of the liquid surface, and hence the amount, of liquid in a tank, can be determined by several methods. Simplest of these is the tape or chain connected to an indicator and attached to a float on the liquid surface. A calibrated or plain glass tube attached to taps at the top and bottom of the tank can also be used for direct indicating of liquid level. The dielectric, conducting, or absorption properties of the liquid can be the basis of measurement. If the liquid rises between two plates of a condenser, a capacitance change will be produced proportional to the depth. The absorption of radiation between a radioactive source and a sensor located above the liquid and at the bottom, respectively, will be proportional to the depth of the liquid.

Much engineering work has gone into the construction of water meters, and virtually all fluid control and measuring systems are based on instruments initially designed for water. Although we rarely think of it in this way, water is a classical example of an ingredient handled and dispensed by bulk transfer methods.

Several accurate versions of flow meters are commercially available. These have been classified as (1) inferential meters, in which the liquid actuates a screw, a vane, or some other inertia-dependent mechanism, and (2) the positive displacement type in which a definite volume of water is allowed to pass during each complete cycle of the mechanism. Both types are accurate to a few percentage points of the total reading at high rates of flow. At low rates of flow, the displacement meters are generally more reliable.

Head meters (orifice plates, Venturis) relate the flow rate through a constant area to a variable pressure difference. It is also possible to relate the flow rate to the area change needed to obtain a constant pressure drop. Rotameters, a variety of inferential meter, are the most common type of variable area meters in the food industry. They consist of a gradually tapered vertical tube, usually of glass, with the fluid flowing upward through it, and a float capable of unrestricted movement up and down the tube. The float will assume an equilibrium position such that fluid drag and buoyancy just equal the downward force of gravity. The fluid drag is related to the area of flow between the float and the sides of the tube. If the flow rate increases, the bob moves upward to increase the area and keep the drag constant. The pressure drop through the instrument remains almost independent of the flow rate.

The calibration will vary with float dimensions, tube taper, and fluid properties such as viscosity and density. Special float designs are available which are relatively insensitive to viscosity effects; but in the bakery these meters will normally be used for one ingredient only, so viscosity fluctuations should not be a major problem. Means for compensating fluid density changes can be obtained. The mathematical treatment of fluid flow measurement is detailed in many texts on chemical engineering and hydraulics, but a good summary of this aspect of metering has been given by Rothfus (1968). By affixing a magnet or armature to the float and placing a sensing device outside the tube, rotameter readings can be transmitted for recording and control purposes.

Turbine meters (another type of inferential meter) have come into use for measuring such liquids as water, invert syrup, and ammonium bicarbonate solution (Hagedorn 1965). Their small size and weight permit installation at any convenient point. These meters consist essentially of a rotor located in a tubular housing (inserted as part of the pipe line) and a sensing device.

The flowing liquid impinges upon the turbine blades which freely rotate about an axis along the center line of the surrounding tube. The angular velocity of the turbine rotor is directly proportional to the fluid velocity through the turbine. The angle of the rotor blades to the stream governs the rotor velocity. Blade angles are usually between 20° and 40° because larger angles result in excessive end thrust and bearing friction while smaller angles cause undesirably low angular velocity and loss of repeatability.

Sensing of the rotation is done by magnetic interaction between the blade tip and a pickup coil located outside the tubing. The rotor blades are made of paramagnetic material and the pickup coil contains a permanent alnico magnet. Because the surrounding pipe is stainless steel, it does not interfere with the interaction. Frequency of the magnetic pulses is proportional to the flow rate. Pulses per unit volume may be varied in meters of small capacity by using some nonmagnetic blades. In large diameter meters, resolution is increased by installing a large number of small magnetic buttons on a rim which rotates with the turbine.

In addition to the wide range through which accurate measurements can be made (on the order of 10:1), a major advantage of the turbine meter is that each electrical pulse is proportional to a small incremental volume of flow. This incremental output is digital in nature, and so can be totalized with a maximum error of one pulse regardless of the volume measured. Accuracies of ±0.5% of the actual flow rate or ±0.25% over selected flow ranges have been claimed for these systems.

Continuously flowing streams of material can be measured volumetrically by displacement meters. There are several types, based on nutating discs, reciprocating pistons, rotating vanes, etc.

The piston meter is like a piston pump operated backwards and it is capable of accuracy to 0.1%.

For precise volume measurements, corrections for temperature must be made because of its effect on the density of the material being measured and the dimensions of the volumetric device. Pressure is generally not a factor since common liquids are noncompressible under the conditions of measurement.

The principle of nutating meters is as follows. A disc piston fits approximately horizontally in a chamber defined by truncated cones top and bottom (the apexes facing each other). A vertical diaphragm also separates the chamber. Liquid flows into the chamber on one side of the vertical diaphragm and pushes the disc up and down with a nodding or nutating motion (it does not rotate). At each complete cycle, the piston discharges a volume of liquid equal to the capacity of the measuring chamber. A spindle affixed to the ball-like bearing which supports the piston describes a circle as a result of the disc's nutation, and this motion is transmitted through a gear train to a register. The only moving part in the measuring chamber is the piston (Anon. 1963).

The meters previously discussed measure the volume of a liquid under a head of pressure applied at some other point. A calibrated pump can also serve as a measuring and dispensing device for supplying ingredients to batch or continuous processes.

The positive-displacement metering pump is electrically actuated by a controller which has a synchronous motor drive unit. The drive moves the hands on an indicator which can be preset for the quantity desired. The metering cycle continues until the synchronous motor in the controller has counted the predetermined number of revolutions of the metering pump, after which the controller stops the two motors at the same time. The metering pump is also driven at a constant speed by a synchronous motor, and, since both the controller drive and the pump motor are energized by current of the same frequency, their motions will always be proportional. Control of metering can also be accomplished by micro-switches activated by digital counters operated by cams on the pump shaft.

When the shut-off point is reached, the meter automatically resets itself to the original amount, ready to dispense the ingredient for another batch. Accuracy of 0.1% has been claimed for positive displacement flow meters utilizing a system of oval gear wheels to measure the liquid.

Metering pumps must be fed at a pressure (rate) sufficient to avoid cavitation. Piping must be assembled so that the meter does not "pump air." If these conditions are not met, accuracy of any degree cannot be relied upon.

Volumetric feeders for pulverulent materials can be classified on a basis of their action as: (1) a belt, disc, roller, or screw moves the ingredient through a gate which is usually adjustable in height, (2) a helix moves the material through a tube at a rate governed by the speed or "on time" of the screw, or (3) pockets are filled with the powder and then rotated to dump their contents, the speed of movement of the pockets past the release point governing the rate of delivery.

Some volumetric dispensers rely on helical feed devices. A typical feeder of this type would have two concentrically mounted and independently driven augers rotate in the same direction but at dissimilar speeds. The larger outer helix tends by its slower rotation to create constant motion in a zone of material surrounding the faster speed smaller inner auger (or metering auger). An optimum ratio of the two speeds is selected to give a constant uniform density in the material surrounding the metering screw. If product characteristics require it, a reverse helix can be added to further stabilize the flow pattern. Since the metering screw operates in an environment of uniform density, it can deliver a constant rate of ingredient at any given speed. Metering accuracy of ±1% of set rate is claimed for most materials.

A Sterwin feeder uses volumetric principles and is useful for low addition rates, such as vitamins and flavors. The feeding range can be varied from 4 oz to 60 lb per hour. The principle is that a vertically movable slide or gate controls the depth of powder on a horizontally rotating feed disc which is drawing the material from a hopper. The powder, after it is removed from the disc by a screw, falls down a chute to the mixing area. Addition rates are controlled by varying the speed of the disc or the height of the opening in the hopper. Disc speed can be varied by changing the gears or the speed of the motor. The gate is adjusted by a micrometer screw to give a 1:20 variation in the height of the opening. As in all equipment of this type, feed rate is not necessarily directly related to the dimensions of the opening or the speed of the disc, and it is important to check the settings by weighing the material discharged during a given time interval.

BIBLIOGRAPHY

ABBOTT, J. A. 1959. Weighing and mixing procedures. *In* Bakery Technology and Engineering, S. A. Matz (Editor). Avi Publishing Co., Westport, Conn.

ANON. 1957. Streamlined processing of nut meats. Food Eng. *29*, No. 6, 64, 120–123.

ANON. 1962. Split skins spun off in new nut blancher. Food Eng. *34*, No. 8, 51.

ANON. 1963. Neptune meters. Neptune Meter Co., Long Island City, N.Y.

ANON. 1968. Omega gravimetric dry materials feeders and weighers. BIF, Providence, R.I.

BUBP, A. D. 1972. Personal communication. Springfield, Ohio. Nov. 3.

FULLER, W. S. 1970. Automatic weighing and dispensing of wet and dry sundry ingredients. Proc. Am. Soc. Bakery Engrs. *1970*, 145–151.

MILLER, R. C., and DEVLIN, K. A. 1948. Processing Filbert Nuts. Oregon State Coll. Agr. Expt. Sta. Tech. Bull. *15*.

ROTHFUS, R. R. 1968. Working concepts of fluid flow. V. Flow measurement. Instr. Control Systems *41*, No. 7, 105–108.

STRIETELMEIER, D. 1972. Personal communication. Chicago, Ill. Sept. 15.

WATKINS, H. E. 1970. Apparatus for flavoring of snack foods and the like. U.S. Pat. 3,536,035. Oct. 27.

WRIGHT, R. C. 1941. Investigations on the Storage of Nuts. U.S. Dept. Agr. Tech. Bull. *770*.

Technical Functions

CHAPTER 24

Product Development

INTRODUCTION

The snack food industry has benefited from a substantial amount of creative product development over the last several years, most of it conducted by the large, diversified food companies, but also contributed to by the technical service staffs of suppliers of ingredients, equipment, and packaging materials. Although small manufacturers do little or no innovative product or processing research within their own organizations, they are often quite adept at copying the achievements of their competitors, frequently adding refinements which improve the acceptability or profitability of the item.

Copying is at best a risky proposition, however. The market can be saturated with a new product by the time the copy is ready for sale, patents may prevent duplication, and operating know-how essential for successful production may not be achievable without very extensive and costly engineering studies. If a company is to achieve growth or even maintain status quo in this rapidly changing industry, it is necessary to have some sort of product development program, though it may be limited to the part-time efforts of one man. Companies who imagine that they can continue to exist by manufacturing commodity type items at a cheaper cost than competition simply do not realize the contribution of technical advancements to production efficiencies and consumer acceptability.

MISSION AND OBJECTIVES

Every company should supply a statement of mission to its Research and Development or Product Development group. The laboratory director can then draw up a set of objectives which will describe the activities he plans to undertake in order to achieve the mission.

Desrosier and Desrosier (1971) quote a typical research laboratory mission statement, as follows.

(1) Develop and keep abreast of scientific and technical knowledge in areas of vested corporate interests.
(2) Apply this technical knowledge by developing new and improved products and processes for the corporation.
(3) Establish, issue, and maintain standards and specifications for raw materials and finished packaged products bought and/or sold by the corporation.
(4) Audit adherence to the approved quality standards and recommend corrective action as required.
(5) Represent the corporation in matters involving laboratory sciences and technologies.

Some managers would regard this list of duties as being too specific for a statement of mission, which should be brief, general, broad, and timeless.

According to Heid (1963), the objectives of applied research and development in the food industry are: (1) to produce, maintain, and increase sales by (a) improving processes, products, and packages, (b) developing new uses for products, and (c) developing new products, processes, and packages, plus testing acceptability of these products; (2) to eliminate losses and hazards of losses (by troubleshooting and by developing profitable utilization of surpluses, residues, and wastes, and by avoiding nuisances, damage suits, etc.); (3) to minimize dependence on raw material in inadequate or undependable supply (as the development of synthetics, improvement in tin-plating procedures, etc.); (4) to act as technical consultants to top management in evaluating proposed new equipment and processes, keeping management informed of developments of potential interest and importance (in nuclear energy, etc.); (5) to increase the profit on invested capital and on unit sales volume (by ensuring better, more uniform products at lowest possible cost); (6) to deal with governmental agencies in relation to technical matters including labeling, standards of identity and quality, waste disposal, etc.

Other authors regard some of these functions as more appropriate for a quality control or technical service department.

Stability in staffing of the R&D department and the existence of a detailed plan in which the long-term goals of the company are described are among the factors necessary for maximizing the usefulness of R&D. Only a few of the factors contributing to the successful operation of an R&D department can be discussed in this brief chapter. More thorough treatments can be found in the books by Walters (1965) and Stanley and White (1965).

ESTABLISHING A PRODUCT DEVELOPMENT POLICY

To evaluate the effect new products could have in your company's future, existing and expected conditions must be analyzed. The following steps, slightly modified from Rives (1964), can be helpful in this evaluation. (1) Prepare a long-range (5 to 10 yr) forecast of sales of existing product lines. (2) Prepare a long-range profit plan for the company based on existing product lines. (3) Prepare an inventory of the company's capabilities and resources, including but not restricted to (a) means of distribution, (b) geographical limitations, (c) financial limitations, (d) technical staffing limitations, (e) legal limitations, such as antitrust requirements, and (f) existing equipment. (4) Determine the most favorable market opportunities for new products based on company's capabilities and resources. (5) Prepare a modified long-range profit plan, incorporating the probable effect of new products. (6) Prepare a statement of the new company objectives including the intent of expanding sales and profits through the introduction of new products. (7) Assign new product responsibility and formulate procedures for establishing projects. (8) Provide for evaluation of new product performance.

ADMINISTRATION OF PRODUCT DEVELOPMENT

Position of Product Development in the Table of Organization

Product development functions generally fall in a broader department called Research and Development, or, sometimes, Technical Services. The former title implies the performance of some basic investigations, while a Technical Services Department will often include the Quality Assurance group as well as applications services performed for customers. We will first consider the more common organizational arrangement, in which product development is performed by a Research and Development Department.

Relationship of R&D to Other Departments

Because the success of the R&D department is essential to the health of the company and because its goals and methods of operation are unlike those of other departments, it should report directly to the president. In this way its influence can be felt throughout the company and support at the highest level will be continually evident. The filtering of information through a nonscientific channel, such as will occur if R&D is given a more subordinate position, will inevitably result in much slippage and distortion. By making the director of R&D a member of the executive committee, the president and other members can be expected to derive some benefit from the advice and example of a scientist.

There must be a single administrator responsible for all R&D functions. Fragmentation of the individual activities among other departments such as engineering or marketing is extremely inefficient and inappropriate. Process development, process engineering, and package design (with the exception of graphics), as well as product development, must be directed by one administrator in order that the close coordination essential to success in these activities will be possible.

The use of terms such as "research" or "product development" in other than their scientific connotation, as, for example, the use of "research" to denote consumer survey studies conducted by marketing, or "product development" to mean the sales program involved in the introduction of new or modified products, is inexact and may tend to confuse top management. Unless the R&D director is both positive and persistent in making clear to the chief executive the difference between scientific activities and other activities similarly named, confusion will inevitably develop as to the proper role of marketing and R&D in new product development.

It is usually found desirable to restrict the official contacts of the personnel of other departments with R&D personnel to formal channels of communication. Direct contacts at lower echelons can lead to dissemination of erroneous and sometimes harmful information. Many breaches of security in new product development can be traced directly to informal discussions in the laboratory (Wade 1965). Free food also has an overwhelming attraction for many people, and its availability in the R&D area can lead to a stream of distracting visitors unless access to the area is rigidly controlled.

Interior Organization

It is customary to separate the R&D department into a research section, dealing with fundamental problems not directly connected with a marketable product, a development section responsible for product and process development, and an engineering section responsible for implementing on a commercial scale the laboratory or pilot plant processes of the development group. Lines of demarcation are not always clear and there should be no attempt to emphasize the difference between research, development, and engineering lest valuable contributions be ignored because of their unorthodox origin. In smaller companies it may be best to eliminate administrative separation of these functions.

Line organization within the development staff can be formed in at least two distinct ways. Personnel can be divided into administrative groups on the basis of the products or processes dealt with or on the basis of the disciplines involved. By the former method, all persons working with vegetables would be supervised by one person while those concerned with fruits would be in a different group. The latter method would place chemists in one section and bacteriologists in another. There are refinements of these systems which need not concern us at this time. Each method has its advantages and disadvantages, but, in general, the product-oriented scheme is best suited for relatively small companies interested in short-range development projects while the discipline-directed system is better for larger companies with a wide range of interests and a long-term development policy.

Some companies may find that a completely different type of organization is better for them. The system should at least be consistent and it should be predetermined. Allowing groups to form and break up on the basis of short-term projects is a very poor policy.

An important requirement for any organizational scheme is that it must provide an environment which facilitates the performance of creative work. There are many ideas about how this can be accomplished, but proof as to the merits of the plans seems to be lacking.

According to Koning (1974), a good environment for research should include the following factors: (1) a scientist or group of scientists should be assigned an actual and significant problem that (a) is related directly to a specific goal of the organization, and (b) allows the scientist to develop his discipline or specialty; (2) within a problem area, scientists should be allowed the independence to develop studies necessary to solve the problem; (3) scientists should be supervised by knowledgeable scientists and administrators who show continuing interest in the work, who attempt to fulfill the individual needs of the scientists, and who make a sincere effort to provide all the physical and technical support the scientists require; (4) completely open communication should exist throughout the organization; and (5) scientists should be given generous recognition for work well done, and, conversely should be constructively criticized when work is poorly done. Although the complete application of these tenets would doubtless be possible only in some highly idealistic organization staffed by completely unselfish personnel (a situation not known to be represented by an actual example in government, religion, education, or industry), they at least represent a series of desirable goals.

Budgeting

The establishment of a realistic budget and then holding expenditures within this budget are essential parts of the procedure by which top management is assured that the R&D department is being operated in an efficient and predictable manner. The amount to be allocated to R&D, either in toto or by project, should be settled by conferences in which the desired output is related to company goals and resources.

In practice, allocations for R&D have often been based on a percentage of the previous year's sales, or whatever the competition is spending multiplied by some arbitrary factor. Research differs from many other company activities in that there may be no significant achievements for long periods of time and there is great difficulty in assigning a monetary value to the discoveries which have been made and described. A few science administrators have been extraordinarily successful in creating a kind of mystique which led to such things as the publication of numerous articles recommending a "hands off" style of management and a science-centered administration as the ideals. Tolerance of top management and stockholders of this situation has caused some research directors to reach the mistaken conclusion that no measure of the contribution of research to their company's financial welfare was necessary or even desirable. When hard times come and cost-cutting is required in order to survive, as it was for many companies in 1970 and 1971, such departments were the first to be decimated. Very few companies can afford to support a group of dilettantes, scientific or otherwise, in the current economic climate. Furthermore, the mystique has now been largely taken over by marketing.

Development has been in a somewhat better position. The personnel concerned with development have been expected to produce because their results are highly visible. Development projects are funded with the idea that a payoff will occur within some finite time period. Even here, however, much stricter budgetary controls can be expected in the future with more accurate attribution of development costs to specific products. However unpleasant and unscientific it may seem, the science administrator is going to find that a knowledge of budgeting strategy is essential to the proper performance of his job.

Continuity of the research effort over a period of years is necessary if it is to be efficient. Research expenditures cannot be alternately expanded and contracted without destroying a large part of the department's effectiveness. Shifting goals make for budget uncertainty, and major changes in management objectives during the year should always be accompanied by an immediate R&D assessment (distributed to top management) of the effect the changes will have on expenditures. It is probably better strategy to accept minor changes without attempting to modify the budget.

The budgetary controls involved in R&D can be regarded as being of two basic kinds. Policy controls are concerned with the nature, scope, and direction of the total effort, while operational controls are concerned with planning and evaluating specific projects during their life cycles. Policy, or company strategy, will dictate how much money is to be allowed for the overall R&D activity including administration, depreciation, etc. Out of this total will be allocated amounts for individual projects. It seems that R&D efforts tend to diverge from the predicted path more often than nonresearch activities; and a decision must be made whether to demand a strict accounting in the frame of project goals for money grants, or to allow considerable freedom in fund dispersal as long as progress is being made in the general direction of company goals. There is no single answer which will apply to all firms, but experience seems to indicate the desirability of moderately rigid guidelines, always understanding that technological breakthrough may suddenly establish new requirements which cannot be anticipated.

There are several details which experience has shown to be important. Production time and materials used for experimental runs in the plant should be charged to an R&D account. Allocations should be made in the budget for these runs. Generally, it will be found that some or all of the experimental product can either be recycled or packed off as saleable product, in which case R&D should be credited with the equivalent yield. Indirect costs which are difficult to associate with specific projects include library, telephone, secretarial and clerical salaries, administration, utilities, building depreciation, etc. If charges are to be made for occupancy or similar overhead items, the exact amount should be agreed upon at the beginning of the budgeting period.

To be effective, project controls must alert the manager to any deviation from the plan in achievement, cost, or time. Budget variance provides a comparison between the work which has actually been accomplished and the amount which was planned. A monthly report of expenditures by account is essential to the planning of the department's activities. A budgeting system is inadequate if it does not deliver such reports promptly. The R&D manager should insist upon receiving timely, accurate reports from the accounting department. He should promptly verify all charges by consulting vouchers, payrolls, orders, etc. Any unexplained items should be resolved by consulting with the comptroller.

SYSTEMATIZING NEW PRODUCT DEVELOPMENT

Before development projects can be intelligently assigned, it is necessary that a company plan be set up describing the areas of interest for the present and future. These should be made as narrow as possible to avoid wasted effort. A statement such as "the company is interested in any product that will make a profit" is evidence of an abdication of top management's responsibility to provide necessary guidance. Top management should know the limitations of marketing capabilities, available finance, legal restrictions, plant capacity, and other factors determining the type of product which the company can successfully make and distribute. The more clearly these limitations are set forth, the less is the chance that abortive projects will be initiated.

Within the fields of interest, market research should be asked to identify rather broad areas of marketing opportunities. Generally speaking, it is of doubtful value to ask market research to specify a definite product which is wanted. Desires of the consumer are likely to be amorphous and poorly defined until crystallized by the offer of a finished product. Furthermore, specifications arrived at by consumer research may be impossible of achievement in practice. The better approach is to design a product on the basis of broad guidelines developed by market research. When a superior or different product becomes possible because of advances in technology, it is perfectly reasonable for R&D to develop a prototype and then request marketing to determine consumer acceptability of that specific item.

In some cases, it is necessary to reproduce a known item in order to meet competition. Market research will probably have little part in this decision. The originator of the demand will be sales who reports that a competitive product is steadily encroaching on the volume of an existing company item. Sometimes, a strikingly novel and apt idea is suddenly brought forth and management immediately realizes that it represents a substantial opportunity. In other cases, R&D will realize that a new development in technology makes possible the production of an article devoid of some of the objectionable features of existing counterparts. Each of these new product ideas must be fully coordinated with all other departments before a research project is initiated.

There are nearly always more projects recommended than can be adequately funded. Souder (1970) lists four kinds of techniques which have been used, or suggested, for selecting from the list of proposed projects those which should be supported. These are capital budgeting formulas, cost prediction formulas, scoring and ranking methods, and resource allocation methods.

The capital budgeting approaches are standard return on investment and discounted rate of return calculations modified by the inclusion of risk parameters such as the probability of commercial and technical success. A computed value for each project is discounted for the risk. Projects are ranked in order of the resulting values, and successively less promising projects are selected until the total allocation of funds is exhausted. Uncertainties result from the evolutionary nature of a development project which provides constantly changing prospects of success or failure, as well as the effect of competitive R&D activities which may suddenly make the whole project obsolete.

Cost prediction formulas are based on the premise that historical relations exist between the cost of a project and the sales of the resulting product. This is at best a vague relationship suitable for use only when more accurate means are not available.

Scoring and ranking criteria are often used for selecting projects. Numerical values are assigned to several factors, such as product life, patentability, market size, competition, project cost and duration, etc. The raw values or values weighted by some previously chosen factor are summed, and the projects ranked in order of their total scores. The obvious difficulty in these methods is the uncertainty in the assigned values.

Resource allocation techniques are intended to provide bases on which to decide how much to spend for a given project out of a series of projects taking into consideration the total company situation, now and as projected. Data collected from all levels of the organization are submitted to computer analysis to determine the relative value of competing projects funded at various levels.

The level of difficulty of a proposed project is always an important consideration in deciding whether or not it should be initiated.

Desrosier and Desrosier (1971) assigned technological level of difficulty to four categories of new food products. In decreasing order of difficulty, these are (1) new factory, new technology, and new equipment are required—i.e., raw materials, processing equipment, packaging, warehousing, and distribution systems all differ from those the company is now using and familiar with; (2) the existing factory can be used, but new types of technology and equipment will be required—i.e., the process method and/or the packaging equipment are different from those currently available; (3) the existing factory can be used but some minor new technology and equipment will have to be implemented—i.e., a new step in processing or packaging is necessary; and (4) the present factory, equipment, and technology can be used with a new formulation, processing sequence, etc. This arrangement formalizes the intuitive recognition that projects requiring the most complex and expensive changes in an existing operation will be the most difficult to implement successfully.

Sources of Ideas

In a survey of over 150 American companies, Booz, Allen, and Hamilton found that 88% of new product ideas came from within the companies. Marketing together with research and development personnel contributed 60%. Sources of ideas, according to Kill (1965) might be as follows. (A) Internal: (1) sales, advertising, and merchandising staff; (2) R&D personnel; (3) production and engineering staff; (4) employee suggestions. (B) External: (1) consumers' suggestions; (2) trade suggestions (wholesalers, retailers); (3) suppliers' suggestions; (4) competitive products (domestic and foreign); (5) technical and trade publications; (6) patents.

Some companies like to distribute suggestion forms among all employees, in the hope that a startling new idea will be elicited. There should be a requirement that the submitter give the reasons why he believes the product or process would be worthwhile.

Ammerman (1963) lists the following possible sources of new product ideas: (1) company suggestion plans, (2) request new product ideas from the sales staff and provide a place for them on sales reports, (3) contests both within and outside the company, (4) government publications such as the *Product List Circular,* published by the Small Business Administration, (5) permanent product scouts who travel abroad as well as in the United States looking for new product ideas, (6) industrial research firms, (7) market research firm reports, (8) advertising agencies, (9) journals and other publications, (10) consultation with suppliers, (11) trade fairs, (12) plant tours.

Brainstorming or group-think sessions with deferred evaluation has had a certain vogue. The validity of such approaches to idea generation has been questioned in a number of experimental research studies. Results seem to indicate that individuals working alone are more effective than groups in producing truly novel but practical ideas. Tauber (1971) described an idea-generating method suitable for use by individuals. He assumed that creativity is a special form of problem-solving in which combinations of words represent product ideas. A list of words representing some important characteristics or concepts associated with existing products is set up in grid or matrix form and all cross-classifications evaluated for desirability or practicality.

The compensation, if any, for a successful idea should be clearly spelled out on the suggestion form. An idea does not become company property merely because the originator has described it to a company representative, even though the submitter may be an employee. Unless the employee has been hired specifically to invent, his inventions do not automatically become company property. If he has used company time and company property to reduce the invention to practice, his employer may acquire a royalty-free license or some other equity. Situations have occurred where a suggestion is made in an apparently off-hand manner by an employee or outsider, and then a claim for compensation presented when the idea is put into practice. For this reason, a clear understanding of the relative rights of each party should be reached at an early stage.

The least productive and most dangerous source of ideas is the outsider who submits an unsolicited suggestion. These are best returned without a review of any kind. It has been suggested that incoming correspondence which might include such suggestions should be screened by a person without technical training who will immediately seal up and return any communication as soon as he becomes aware it contains suggestions for a new product. The reason for these precautions is that a legal liability to the submitter could result from apparent utilization of a suggestion even though a similar idea had earlier occurred to a company agent. Often these suggestions will describe products which are already under development and this creates a very sticky situation indeed. Experience has shown that rarely, if ever, is a useful product developed from an idea originating outside the company. Further discussions of this point can be found in the article by Auber (1965).

Controlling Product Development Projects

When a new product idea has been approved for development, a series of events is initiated which must be carefully planned, closely controlled, and efficiently managed to maximize the probability of technical and marketing success.

Outlines of two different plans for controlling new product development step by step will be presented in this section. Each plan has different advantages and disadvantages depending on company size and organization, and other factors.

The key operating element in plan No. 1 is a Product Planning Committee (PPC) composed of the president of the company, the director of research and development, the head of marketing, the chief finance officer, the manager of production, and such other executives as are thought to have an immediate interest in and responsibility for new product development. The committee appoints a secretary, who issues agenda, takes notes of the transactions and issues minutes, and maintains records of the individual projects.

In the first plan for controlling new product development, entry of an idea into the system occurs as a result of the submittal to the secretary of the Product Planning Committee of a new product idea. A standard form which is made available to all employees can be used. Normally, the initiator will fill out the designated blanks himself, but in some cases the secretary will make these initial entries. The secretary assigns a number to each idea received. His function is strictly clerical, i.e., he performs no screening operation. If necessary, additional information is solicited from the submitter in order to clarify vague or inadequate terminology. The proposal, with such additional information as he has been able to accumulate, is read by the secretary to the PPC at the next meeting.

Experience has shown that most of the suggestions entering the system either will be close matches to others which have been considered earlier, will fall outside the company objectives, or will be clearly nonfeasible for some other reason. If the proposal has any merit, a followup representative such as the secretary is assigned to extract from the submitter and other sources as much information as is necessary to allow a reasonably accurate evaluation of the proposal. The PPC, on the basis of the completed information, either abandons the idea or assigns an analysis team, usually consisting of one person from marketing and another from R&D to further process the suggestion.

The function of the analysis team is to make a study of the marketing and technical aspects of the problem sufficiently extensive in scope to allow the PPC to determine if (1) a marketing opportunity exists, (2) the company is in a position to take advantage of the marketing opportunity, (3) development of the product is feasible, and (4) capital costs of getting into production can be afforded. The data at this stage will necessarily be crude and embody a considerable degree of uncertainty.

After the analysis team has completed its study, a brief but comprehensive report is submitted to each member of the PPC. At the next meeting, the committee decides to (1) request additional work, (2) abandon the idea, or (3) advance the proposed product to the level of a priority project.

In the latter case, the analysis team is disbanded, the project is given a name and assigned a place on the priority list (all priorities are temporary and can be changed at any meeting of the PPC), and a product team assigned. Unless there are very cogent reasons for a different composition, the product team will include representatives from marketing, R&D, and production (engineering may be included in place of production if it is a separate department).

The product teams are delegates of the PPC and thus of the president, and operate with his authority to gain action on necessary development problems and to coordinate the activities of the several departments. Team members will not ordinarily be department heads. If their efforts are successful, their final act will be the submitting to the PPC of a "Proposal to Manufacture and Market," a document which will recommend that the committee authorize expenditure of the funds required to produce and sell the new product. More will be said about this document later. Alternatively, the team may at any time recommend that the project be abandoned because of technical nonfeasibility or new marketing information.

In the interim between the assignment of a product team and their issuance of a "Proposal to Manufacture and Market," the necessary research, development, engineering, and marketing studies will have been completed. Occasional reports (not necessarily on any fixed schedule) are made to the PPC, and informal minutes are kept of the team meetings. Each project has an account to which expenditures for time, materials, and services are charged.

Each member of the team is responsible for the successful participation of his department in the development work necessary to produce the new item. When necessary, he enlists the aid and authority of his department head. Problems in securing cooperation of other departments, plans for future work, estimates of phase completion dates, fund requirements, and chances for success are discussed at team meetings.

From time to time, the composition of the team may be changed by the PPC as need changes for the participation of different departments. Prototypes of the new product (including package designs) plus market research data are presented to the PPC as they become available. Based on these exhibits and the reports of the product teams (presented orally or in written form), the PPC decides at each meeting whether to abandon the project, continue it as planned, or to alter direction. In the normal course of events, at each meeting some of the teams will present additional data of importance and the project will ordinarily be continued in such cases. If work is to continue, a priority is assigned or continued.

During the development process, if it is successful, equipment designs will be completed, product costs will be finalized, ingredient sources will be determined, and all arrangements brought to the point where successful production can be assured. The results of these studies will be set forth in the "Plan to Manufacture."

As stated previously, the "Plan to Manufacture" is essentially a proposal to top management that expenditures necessary to production and marketing of a new product be authorized. An important feature of the plan is a complete specification for the new product which gives firm, unequivocal requirements for each important characteristic of the product and of the manufacturing process. None of these requirements is to be changed without the prior concurrence of all interested parties, and production is charged with the responsibility of seeing that the quality assurance provisions are met. In case of disputes over the provisions of the plan, e.g., when production expresses doubt about its ability to consistently meet the quality standards deemed necessary by marketing, arbitration is performed by the PPC. Acceptance of the details of the plan by each responsible department head results in the transformation of this document into a specification which is used for the commercial product.

The second plan for controlling development projects for new products is better suited for small companies. Consideration of a new product is initiated by the submittal to the president of the company (or his representative) of a brief description of a proposed new item together with a statement of its supposed advantages over other available foods of the same type. The proposal can originate from any of the division heads within the company, from an employee suggestion plan, or from a source outside of the company. Unless the president finds that the suggestion appears to have no merit, he forwards the proposal to marketing, together with a directive to ascertain the marketing potential of the proposed item. If marketing concludes that the proposed new product has inadequate sales potential, their recommendation for abandonment of the project is sent to the president for final disposition. This recommendation is accompanied by a detailed justification for the adverse decision. The president bases his further actions on marketing's memorandum plus other advice and information.

Should marketing, on the basis of their tests and surveys, deem the proposed new product to have a reasonable chance of success, they establish certain broad and tentative specifications of a preliminary nature including (1) maximum acceptable unit sales price and minimum permissible margins, (2) product (package) size, (3) minimum storage life at some definite temperature with the method for determining the end point clearly specified, (4) production capacity to be targeted, (5) general characteristics of the product in terms of appearance, flavor, and texture, (6) limitations of the home preparation requirements (equipment, time).

These preliminary specifications, for which marketing assumes the responsibility, are incorporated into a memorandum. The original is forwarded to R&D and is accompanied by both the initial proposal and the directive received from the president. A copy of the memorandum is forwarded to the president. After an analysis of marketing's recommendation, the R&D staff submits to marketing either (1) a proposed schedule in which probable success of the proposed development is stated percentagewise, within the framework of total man-hours required, total research costs (including labor) and duration of the project,[1] or (2) indicates to marketing that the product as specified is nonfeasible due to technological problems or to some unrealistic limitations established by marketing and suggests changes, if appropriate. If the nonfeasibility is due to the establishment of limitations which seem to be impossible to meet, marketing and R&D confer in an attempt to reconcile their differences of opinion. If agreement is not reached in this conference, the file on the matter is submitted to the office of the president for final action.

After having studied the plan of R&D, marketing submits a memorandum to the president in which it is recommended either (1) that the schedule of time and costs for development be accepted, or (2) that consideration of the project be terminated because of excessive research

---

[1] An example of such a schedule statement is: "The project you have described would require approximately 2 man-years of labor, and $25,000 in total costs. It would be completed in 1 calendar year with a 75% probability of success."

costs and the limited probability of success. In either case, written concurrence of the president is required before the project can be considered terminated.

If the president concurs with marketing that R&D should perform the work they have outlined, a formal research project is initiated and appropriate funds to support it are allocated. Simultaneously, a committee consisting of one representative each from marketing, R&D, and production is set up to coordinate and expedite all activities directed toward getting the new product on the market. This committee has primary responsibility for the achievement of successful production. The members of the committee are appointed by the various divisions, with due notice to and concurrence by the president, and will ordinarily include persons having close and continuing contact with the actual work being performed. For example, R&D might appoint the technologist responsible for work on the project as its representative.

Each member of the committee is responsible for coordinating the activities of his department with the work being done by the other departments. Each member receives copies of the R&D progress reports on their project, and the representatives of marketing and production are responsible for informing the other members of the committee of all pertinent actions taken by their respective department. This information must be relayed to the other members within the shortest practicable time after the action is taken. Written concurrence of all committee members must be obtained before major actions, such as the purchase of production equipment, are undertaken. Information can be exchanged and concurrence obtained during committee meetings if desired, and in this case, the transactions are entered in the minutes of the meeting.

All decisions made in committee meetings are set down in the minutes, copies of which are circulated to all interested parties, including the president. The original of the minutes is retained in the file held by R&D. The representative of R&D on the committee acts as the secretary and is responsible for transcribing and reproducing the records of the meeting and for circulating them to the other members as soon as possible after each meeting. If either of the other committee members objects to the accuracy or completeness of the minutes or wishes to make additions or corrections for other reasons, he files in the record a memorandum setting forth his opinions. Copies of this memorandum are also sent to the president.

The final aim of the committee is production and marketing of the new product and its work does not stop prior to that stage unless the project is terminated by the authority of the president. When the product is being successfully produced, the committee will disband upon mutual agreement that their work has been completed. In this type of system, the final goal of the committee is an operating production line and distribution system rather than the plan described for the preceding system.

CONDUCTING DEVELOPMENT WORK

According to Bass (1967), the natural sequence of steps in a technical undertaking is (1) a search for a new idea within the field of interest, (2) the merits of the idea are examined by some type of evaluation—initially by mental processes and then often by preliminary experimentation, (3) if the conclusions are favorable, a tentative goal is set, (4) a plan is made that can be used to achieve the objective, (5) criteria are adopted for evaluating success, (6) work is undertaken and results accumulated, and (7) when analysis reveals the goal has been achieved, the project is concluded. The goal may have to be redefined many times before success is attained.

The initial step in any development project is a thorough literature search to determine the existing state of the art. The search should encompass not only scientific and technological periodicals and reference books, but also the patent literature, technical bulletins of ingredient and equipment suppliers, and the company files which may contain reports of previous work along similar lines.

Results of the initial literature search should be given in a formal report. This important step in the development project is frequently neglected, with the result that changes in personnel or other interruptions in continuity lead to the necessity for doing the work all over again. If time and funds are available, the literature search should be supplemented by discussions with workers in the field and visits to equipment installations.

The techniques of experimental design, conducting the experiment, and statistical evaluation of results are presumably known to the reader and will not be discussed here. In the following paragraphs will be described some special approaches which might fit particular situations and yet be overlooked by the investigator.

Use of Consultants

The use of consultants and contract research institutions to provide specialized assistance during a development project which seems to be beyond the available capacity of a company's internal staff is often economically justifiable. Countless hours of literature search and much fruitless experimentation can frequently be saved by resorting to the consultant's expertise. Such assistance is particularly valuable when a company is considering entry into new fields. At the start of a project a consultant who is an expert in the particular technology can be called in to advise on possible approaches. If competent, he will often be able to point out deficiencies in some of the initial proposals drawn up by R&D. He may have specific knowledge of how similar problems were solved in other companies. Since the consultant presumably is acquainted with the history of litigation on the type of product being considered, legal considerations, and especially the patent situation, can be analyzed.

In some cases, a consultant may be hired to work alongside the R&D staff on a particularly difficult problem. The wisdom of this approach is questionable; if the staff is adequate in number and skills it should be able to handle the details of development work. If not, the preferred solution is to hire more and better permanent staff, not to rely indefinitely on consultants. The consultant's time is generally considerably more expensive than that of the salaried technologist. Consultation during the course of an investigation, and particularly if a stalemate seems to have been reached, is often worthwhile, in order to resolve internal disputes.

Contract research organizations can be used for projects requiring special talent or equipment and particularly in areas outside the client company's usual fields of interest. Contract research is performed by profit-making corporations, nonprofit institutes, and universities. By contracting for outside development work, new investments in staff and facilities are reduced. Some of the values which have been attributed to outside research are: provides new viewpoints, resolves internal conflicts, relieves heavy in-house work loads, gains services of specialists, and saves time. Multiclient studies in which several companies jointly sponsor a program on some major problem of mutual interest are becoming popular.

The cost of a contract research project may be some negotiated firm figure, but is more often set as a maximum figure with periodic billings of overhead (usually a fixed percentage of salaries), direct salaries, rebilled charges (for travel, chemicals, etc.), and profit. Administration of the project is best assigned to a single individual in the client company who will make periodic audits of progress and expenditures. Conclusions of value are then transmitted by this project officer to interested parties in his organization.

In setting up an outside contract for research it is very important to specify the goals as clearly as possible. Much wasted effort can result from failure to give adequate direction to the investigators. Frequent detailed reports should be insisted upon and these should be supplemented by visits to the research facility for informal discussions with the technologists doing the work.

Operations Research

Operations research is a formal means of analyzing business problems. The traditional instrument of the technique is a team made up of scientists from various disciplines with an operations research specialist as the leader. Each member contributes the specialized knowledge of his field as well as any experience he has had with theories and models that might help solve the problem. Operations research has been used in planning the development and introduction of new products, although evidently not to a great extent in the food industry.

One widely-used operations research tool is "critical path analysis," or "network analysis," which can be used in project planning and control to determine which activities must be done in series and which can be done in parallel. Characteristic of these systems is a diagram or chart showing the temporal relationship of all the different tasks. These may be called PERT (Program Evaluation and Review Techniques) diagrams, Gantt charts, etc. The information contained in the charts can also be compiled in tabular form for computer analysis. Using these charts, or computer models, the analyst can find the shortest time through the network. This path determines the time expected to be required for completion of the project when all parts are scheduled in the proper sequence, and it is called the critical path. An example of the use of charting methods for controlling new product development is Raytheon's "Product Planning Monitor Schedule" which is essentially a Gantt chart giving target and performance dates for all the steps involved. The chart is revised on a monthly basis using the most recent inputs of information.

Although some dramatic successes have been claimed for this procedure, it does have certain defects. For example, all steps are considered as indispensable, or of equal weight, when, in fact, it is well recognized that some stages or components can be eliminated or condensed if management is willing to take certain risks or accept an inferior product, i.e., critical path analysis does not necessarily identify critical operations. Furthermore, all time estimates will be in error by an amount related to the predictive ability of the estimator, and these errors tend to be cumulative. Technical projections cannot be accurately made until some preliminary work has been done, in many instances.

As long as its faults are recognized, critical path analysis can be a worthwhile tool in the planning of development projects. The research administrator should have at least a superficial knowledge of these techniques not only because they may be helpful in guiding research projects and integrating them into the overall company activities, but also because they may be suddenly introduced by management with some disruptive effects on R&D customs. A review book of considerable value is Critical Path Analysis (Lang 1970).

Recording Data

In addition to the usual reasons for keeping notes on the experiments performed during a development project, the necessity for establishing priority and reduction to practice for patent purposes dictates use of a formal recording system. A haphazard collection of loose data sheets is of little or no legal value.

The research notebook should be permanently bound, not loose-leaf. Pages should be numbered and a record kept of the individual to whom it is issued. When the book is completed or its use terminated for other reasons, a brief index of its contents should be made and filed. The index can be prepared by photocopying the table of contents prepared by the experimenter during his use of the book. The book itself should then be stored in a safe or some other secure place.

Entries should be made in ink or indelible pencil. The date the experiment was performed should be placed on each page. A witness who understands the substance of the experiment but who has not participated in it should affix his signature and the date of signing to each page. No marks, including corrections, should be made on the page after the witness has signed.

The following outline for the description of the experiment has been recommended (Anon. 1963B): (1) reference number of the experiment and subject or title of project; (2) object of the experiment—what is intended to be accomplished by it; (3) apparatus and procedures—references can be made to published descriptions or procedures, or to those described in other notebooks; anything new should be fully described; (4) starting materials—source and grade of ingredients; any special treatment that has been applied to them before the experiment; age and storage conditions; (5) results: (a) products—details of all essential characteristics, (b) processes—refer to engineering drawings or photographs where possible, and (c) samples—identify any samples retained or set out for analysis. The number assigned may be the combined research book number, page number, and a sample designation mentioned on that page.

Alternative procedures will no doubt suggest themselves to the reader.

Evolutionary Operations

The technique of evolutionary operations (EVOP) is of special value for product improvement, cost reduction studies, and other investigations where radical changes are not anticipated. The advantage of this approach is that regular production lines are used. The difficulty in translating laboratory findings to production practice, which seems to be especially great in bakery formulations can be reduced.

The basic principle of EVOP is the making of changes during ordinary production which do not alter the finished product sufficiently to make it unmarketable. The changes are made in a systematic and predetermined way, and the results are carefully recorded and analyzed statistically, thus differentiating this method of investigation from the more-or-less intuitive changes made by line personnel to improve machining characteristics or product quality.

EVOP can be applied readily to a two-variable situation although the statistical procedure on which it is based, mapping of response surfaces, can be applied to situations of any degree of complexity. However, at any given time or place, EVOP should be limited to optimizing 1 or 2 variables.

EVOP procedure facilitates a constant probing in all directions to find a set of operating conditions or combination of ingredients that will result in an improved product, less scrap, greater production, etc. If the time and temperature of processing are factors under consideration, a continuous study can be undertaken with the present time and temperature as the central point and variations made in all directions. This naturally leads to a design where the number of time-temperature variations tried are five: (1) current procedure, (2) higher temperature and shorter time, (3) lower temperature and shorter time, (4) lower temperature and longer time, and (5) higher temperature and longer time. Formula changes can be made in a similar pattern, and such studies are particularly useful in finding the lowest possible raw materials cost for a product of given acceptability characteristics. Of course, all changes in finished products offered for sale must be within the constraints imposed by label statements and other product claims.

A word of caution is appropriate at this point. It is possible to progress from an article with a high level of acceptability to an article of very poor acceptability by a series of steps each of which causes a change that is imperceptible by any known or practicable means of sensory testing. Each of a series of changes can be selected so as to yield a product which the expert or consumer panel cannot differentiate at a significant level from the preceding modification, but the total effect of several of the changes will often result in a final version which is not well received by the consumer. Formula alterations for the sake of cost reductions can result in loss of market by this sequence of events, if the results of each successive change are compared with the previous slightly modified product. For this reason, it is very important to use the original product as the standard rather than to compare a new version with the immediately preceding modification.

The EVOP technique has been described in considerable detail by Kramer (1965). He implies that EVOP studies are best conducted by the quality control section while other authorities incline to the view that the product development group is more suitable for this work.

EXAMPLES OF PRODUCT DEVELOPMENT SYSTEMS

Actual examples of product development procedures are rather infrequently described because of security reasons, but it may be instructive to read what has been published about the procedures at some of the larger food companies.

Kill (1965) described a system which can be assumed to resemble the sequence of events at Nabisco.

(1) The idea is submitted on a form addressed to the R&D laboratory. It includes, besides administrative and control data, a project title or description, the ultimate objectives of the project, potential company benefits, and estimated cost of implementation.

(2) Industrial intelligence, such as availability of competitive items, potential demand, etc., is collected.

(3) A profit study with projected sales volume is made.

(4) The research department will calculate cost of man power, materials, and equipment needed to develop a prototype acceptable to the sales and merchandising departments.

(5) The production manager lists requirements for his department as he sees them.

(6) The product development committee considers a wide variety of aspects of the contemplated product as it affects the company's future before it passes on the desirability of the project. The committee may be composed of some or all of the following: members of the Board of Directors, owners, general executives, president, vice president, secretary, treasurer, or comptroller.

(7) If approved, the project is assigned a priority and time limit.

(8) Assignment to a product development group is followed by the production of prototypes. Progress is communicated to and discussed with the production manager, who can commence preliminary design work or express opinions as to the feasibility of the process. Liaison with marketing is maintained in a similar manner. Sometimes these liaison functions are performed by the product manager, usually assigned to the marketing department. Packaging development must continue simultaneously.

(9) A detailed description of all production costs are given to the fi-

nance department for a profitability study. If, as is often the case, adequate profits cannot be projected using the costs and prices at this stage, the various departments are asked to modify their requirements and adjust their calculations.

Hormel uses a client-for-hire system by which research costs are charged to marketing for specific projects. New product ideas are first evaluated by marketing. Those deemed desirable are discussed with the director of R&D to ascertain their technical feasibility before a project is set up. If it is decided to proceed, the following sequential steps are taken in what can be called Phase I. (1) The objective is defined. What is the projected expected to accomplish? (2) A justification is provided. How will the company benefit? (3) Preliminary specifications are written. The product or process is described as accurately as possible, including the physical, chemical, and organoleptic properties, package design, cost limitations, etc. (4) Priority, timing, work required, and other administrative aspects of the project are established. (5) Supporting data are supplied to facilitate the determination of scientific feasibility and the design of experiments.

When the above steps have been completed, a project leader is selected. He searches the literature, decides upon an approach, outlines the method of attack, and estimates requirements of time, money, and skills. Marketing decides whether or not to continue, based on this information. This is Phase II.

The project is readied for final review in Phase III. After presentations by all responsible functions, the project receives final approval or rejection from a group vice-president, the R&D director, and either the corporate vice president or a divisional manager.

In Phase IV, R&D personnel take over the responsibility for the project, and a team is selected to work for the leader. The team outlines the experimental approach and issues weekly reports covering money and time expended as well as monthly reports on technical progress.

According to Alleman (1967), these are the key points in Kroger's bakery product development program. (1) Have a definite goal or objective for each project. (2) Follow a set plan. (3) Keep eyes and ears open for new ideas. (4) Be flexible and adjust to changing times. (5) Be aggressive, don't be afraid to try new ideas. (6) See that all work is done carefully and accurately. (7) Be sure you know the cost of new products. (8) Get best possible consumer evaluations before going too far. (9) Accept help from any source. Don't be too proud. (10) Make every project justify itself. Don't do useless work. (11) Have enthusiasm for all new developments. If you are not impressed with your own work don't expect enthusiasm from others. (12) Have a positive attitude at all times and drive the projects to the goal of goals—more profits.

S. C. Johnson Co. has sponsor groups for new product development. These groups include (1) the man who had the idea, (2) the R&D man who is going to supervise the lab work, (3) the man who has to sell the product when it is completed, (4) financial and production personnel, and (5) a member of the new products department.

Wolf (1959) described the product development system established at Arnold Bakeries: (1) Ideas are obtained from employees, executives, and outside sources. (2) Market Research evaluates the idea on the basis of concept tests with consumer panels. (3) Management selects those ideas which can be manufactured profitably and assigns priority. (4) Development work is started. (5) Final product characteristics are specified by management and market research after examining the initial prototypes. (6) Experimental plans are set up by the laboratory. (7) Test bakes are made and evaluated by R&D personnel on the basis of the specifications established by management. (8) Prototypes which appear to meet the requirements are evaluated for acceptability by the market research department using a consumer panel. (9) Acceptable items are coordinated with production to develop suitable processing methods. Time studies enable predictions to be made of labor costs.

The management of product development in a large diversified international company with substantial food interests was described by Cotton (1970). National requirements vary so much that a single research, development, and engineering facility for the entire company was not considered feasible. Therefore, decentralized profit centers and laboratories have been set up in marketing regions unified by common requirements while a world headquarters technical group takes responsibility for overall policy guidance, information transfer, and budgeting. Maximum self sufficiency in the units is encouraged. Projects they work on are developed by the national research department in close cooperation with top marketing and production executives. The World Headquarters Technical Department receives copies of project reports and maintains liaison with the national laboratories at appropriate management levels.

In setting up a project, the first step taken by the national company is to collaborate with marketing personnel in defining the objectives. The second step is setting up a three man team generally composed of representatives from production, marketing, and R&D. These development teams can cut across normal lines of command to achieve their objectives. Each week a progress report is made to top management, and assistance in obtaining action is requested, if necessary. Research supervisors meet every three months with the director of research, the associate director of research, and the manager of research administration to compare results with planned schedules. The national companies also have research committees consisting of the chairman of the board, the president, and the vice president of research, which review programs and progress every month.

New Venture Groups

Some larger companies use the venture group approach to guide the development, production, and introduction of new products and lines of products. The characterizing feature of the new venture group is that the personnel assigned to manage it are totally committed to the one goal of successfully introducing the new item. Bates (1967) claimed the venture group approach was better, faster, and cheaper than other methods of new product development. It has its dangers, however, in that the enthusiasm and emotional commitment generated in the group's members by the entrepreneural concept and hope of reward may lead them to ignore or even to conceal substantial negative factors which may come to light during the course of the program. In addition, the very nature of the method leads to a relatively heavy outlay of high-quality manpower and funds on behalf of a product at a relatively early stage in its development—when its potential, if any, has not been fully evaluated.

The venture teams may consist of one or more persons. Three seems to be a common number. Bates summarized their functions as follows. (1) The manager liaises with top management, develops and matures the concept, makes marketing strategy decisions, presents the product to potential clients in and out of the firm, and administers the team. (2) The marketing/economic/financial representative assists with concept development, draws up product profitability statements, conducts market research, plans marketing action, performs sales tests, does product administration, and takes charge of venture cost control. (3) The technical/research/production representative assists with concept development, evaluates technical feasibility, supervises technical research programs, develops the product and its package, makes co-packer arrangements, and oversees production through the test marketing stage.

PACKAGING DEVELOPMENT

Development of the package should take place simultaneously with the development of the foodstuff, if at all possible. There is not only the obvious reason for this approach that timeliness of product introduction is generally improved, but the direction of the development of the food may be altered midway in the project to take advantage of some special interaction of the food and package. In many cases it is impossible to develop the product independently of the package. If the R&D team includes a packaging technologist, completion can often be greatly expedited and mistakes prevented. Small companies may have food technologists who double as packaging specialists—this is a sure way to obtain complete integration of the development effort. In any case, the packaging development scientist should be under technical direction. Companies who delegate the packaging development function to nontechnical departments such as purchasing or marketing exhibit a poor understanding of the true nature of the process and may experience costly errors as well as fail to achieve any creativity.

The course of a packaging development project will, in general, follow the same pattern as that of a new foodstuff. Schier (1974) recommends establishing a "Packaging Profile" at the start of a project. This outline of objectives (1) describes the new product, including its food chemistry, (2) includes organizational marketing objectives, (3) details protective requirements, (4) sets company merchandising goals, (5) estimates package costs, based on profit structure, (6) establishes graphic and label requirements, and (7) schedules each step of the development.

Schier recommends that the system be initiated and followed by a coordinator who is competent in establishing packaging criteria. The coordinator consults with all involved departments and outside suppliers before submitting the packaging profile to management for approval. The profile should be initiated at the time of the new product's conception, not after its completed development. This enables the packaging coordinator to develop prototypes in various materials and forms. As time passes, the profile is updated taking advantage of knowledge acquired in the interim. Changes should be specified in written form and circulated among personnel who will be affected by them.

COMPLIANCE

Assuming ever-increasing importance in the guidance of new product development is the constantly expanding activity of the national government in regulating the characteristics of foodstuffs offered for sale, as well as their packaging, labeling, advertising, and means of production and distribution. It is absolutely essential that the persons involved in developing new products have ready access to reliable sources of information on existing regulations as well as some source of counsel on specific questions which may arise during the course of the research. It is useless and even dangerous to rely on personal opinions of what seems to be morally right or technically effective, since laws and regulations in this field are often based on doctrinaire or political considerations which have little or nothing to do with common sense or logic, much less ethics.

The most obvious solution is to retain a lawyer who specializes in this type of work. Only the largest companies will be able to afford a staff lawyer specializing in these problems. Although numerous law firms are now assigning one or more of their members to food regulatory work, it is often difficult to locate legal advisors who are sufficiently technically oriented to be able to understand the scientific problems connected with new product development. Nonetheless, such assistance is highly desirable for helping to resolve stalemates resulting when technical limitations seem to be incompatible with legal requirements.

For day-to-day evaluation of the legal questions concerning proposed formulas, processing methods, package structures, graphics, distribution methods, advertising campaigns, etc., it will be necessary to assign one member of the product development team to follow the regulations as issued and proposed by agencies of the Federal government and localities where it is expected the product will be sold, processed, or stored. Sources of information will be the Federal Register (complete, but sometimes difficult to interpret), advisory services such as Food Chemical News (valuable for interpretation, predictions, analysis, etc.), the news media (generally late, inaccurate, and biased), technical and semi-technical publications such as Food Product Development (often very good in their analyses but delayed by their printing schedules and incomplete due to editorial exigencies), trade organizations such as Grocery Manufacturers of America (generally quite good—if you are a member), and advice from suppliers (quality of information from good to very bad and almost always biased).

BIBLIOGRAPHY

ALBRECHT, J. J. 1974. The importance of new product development. Mfg. Confectioner *54*, No. 2, 21–23, 25.

ALLEMAN, H. J. 1967. The practical development of new products. Proc. Am. Soc. Bakery Engrs. *1967*, 287–292.

AMMERMAN, G. R. 1963. New product development. Presented at the September 27, 1963 meeting of the Wisconsin Sect., Inst. Food Technologists.

ANON. 1963A. Reviews of data on research and development. Natl. Sci. Found. Bull. *41*.

ANON. 1963B. Will your notebook stand up in court? Chem. Eng. Progr. *59*, No. 7, 12–12.

ANON. 1968. How much should your company spend on R&D? Business Management *34*, No. 2, 57–60, 62.

AUBER, R. P. 1965. Outside ideas—dynamite! Res. Management *8*, 183–190.

BALDERSTON, J. 1969. Successful administration of a research laboratory. Res. Develop. *20*, No. 6, 24–25, 27–28, 30.

BASS, L. W. 1967. The planning of innovation. Chem. Ind. *1967*, 1671–1675.

BATES, B. 1967. Venture group approach to new products. Food Prod. Develop. *1*, No. 5, 34–42.

CAMPBELL, G. J. 1974. Organizing for new product development. Machine Design *46*, No. 12, 116–119.

CARLSON, A. W. 1961. Planning a laboratory. Cereal Sci. Today *6*, 236–238.

CHMIEL, D. F., WOGHIN, G., and SCHWARZ, R. K., Jr. 1972. PERT—A management control system for launching new products. Marketing Dynamics *1972* (Oct./Nov.) 18–21, 30–31.

COTTON, R. H. 1970. Management of product development in a large diversified corporation. Food Technol. *24,* No. 4, 36–38.

DESROSIER, N. W., and DESROSIER, J. N. 1971. Economics of New Food Product Development. Avi Publishing Co., Westport, Conn.

EINBINDER, D. 1974. Guesswork-free development plan pays off in packages that sell. Package Eng. *19,* No. 12, 56–59.

GUILL, J. H., JR. 1965. What the bakery production man should know about food and drug labeling and regulations. Proc. Am. Soc. Bakery Engrs. *1965,* 214–219.

HAYDEN, A. J. 1968. New product development as viewed by the production head—a major challenge in coordination. Mfg. Confectioner *48,* No. 8, 25–27, 33.

HEID, J. L. 1963. Research and development. *In* Food Processing Operations, Vol. 1, M. A. Joslyn, and J. L. Heid (Editors). Avi Publishing Co., Westport, Conn.

HEID, J. L. 1971. New products from smaller companies. Food Prod. Develop. *5,* No. 5, 47–49.

HILTON, P. 1971. Exploit new product technology. Food Prod. Develop. *5,* No. 5, 38, 40.

HOLBHAN, J. L. 1963. Bringing new products and processes into the plant. Food Technol. *17,* 395–396, 398.

HOLLANDER, M. B. 1970. Get the most out of R and D. Res. Develop. *21,* No. 12, 18, 20, 22.

KILBORN, R. H., and AITKEN, T. R. 1961. Baking laboratory layout and procedures. Cereal Sci. Today *6,* 253–254, 257–259.

KILL, J. F. 1965. New product development in cakes, sweet yeast raised products, and cookies. Proc. Am. Soc. Bakery Engrs. *1965,* 244-248.

KONING, J. W., JR. 1974. The Scientist Looks at Research Management. Amacom, New York.

KRAMER, A. 1965. The effective use of operations research and EVOP in quality control. Food Technol. *19,* 37–39.

LANG, D. W. 1970. Critical Path Analysis. Dover Publications, New York.

LAPORTE, L. 1969. R&D: Its Makeup, Management, and Measurement. National Industrial Conference Board, New York.

LOWE, W. C. 1961. Identifying and evaluating the barrier problems in technology. *In* Technological Planning on the Corporate Level, J. R. Bright (Editor). Harvard Business School, Boston.

MARTING, E. 1964. New products—New Profits. American Management Assoc., New York.

MATTSON, P., and LOTT, D. 1971. What do users and prospective users think of external product development services? Food Prod. Develop. *5,* No. 5, 50, 52.

NEBESKY, E. A., and PETERSON, M. S. 1966. Packaging research: A challenge to industry. Package Eng. *11,* No. 12, 84–88.

RIVES, R. C. 1964. What is your new product policy? *In* New Products—New Profits, E. Marting (Editor). American Management Assoc., New York.

ROBERTS, E. A. 1968. The myths of research management. Sci. Technol. No. *80,* 40–46.

SCHIER, J. 1974. Developing a package for the short market test. Food Prod. Develop. *8,* No. 8, 76.

SCHWARTZ, D. A. 1971. Conducting meaningful preference research for less than $1,000. Food Prod. Develop. *5,* No. 5, 26, 29, 42.

SEILER, R. E. 1965. Improving the Effectiveness of Research and Development. McGraw-Hill Book Co., New York.

SOUDER, W. E. 1970. Pitfalls in R&D. Business Horizons *1970* (June) 54–56.

STANLEY, A. O., and WHITE, K. K. 1965. Organizing the R&D Function. American Management Assoc., New York.

TABER, A. P. 1968. Evaluation of R&D. Res. Develop. *19,* No. 10, 22–27.

TAUBER, E. M. 1971. Systematic generation of ideas for new foods. Food Prod. Develop. *5,* No. 2, 58–59, 62.

TERRY, H., and NAGY, S. F. 1974. A closely managed activity. Industrial Res. *16,* No. 8, 39–41.

WADE, W. 1965. Industrial Espionage and Mis-use of Trade Secrets, 2nd Edition. Advance House Publishers, Ardmore, Penn.

316   SNACK FOOD TECHNOLOGY

WALTERS, J. E. 1965. Research Management: Principles and Practice. Spartan Books, New York.
WILLIAMSON, M. A. 1965. How to manage R&D innovation. Res. Develop. *16,* No. 8, 26–27.
WILLIAMSON, M. A. 1968. R&D management tomorrow. Res. Develop. *19,* No. 8, 69, 70, 72.
WOLF, A. 1959. From an idea to a finished product. Baking Ind. *111,* No. 1407, 23–28.
ZIEMBA, J. W. 1966. R&D guides Green Giant. Food Eng. *37,* No. 1, 98–100.

CHAPTER 25

Quality Control

INTRODUCTION

The intensity and extent of quality control in the snack industry is extremely variable, ranging from the entirely subjective testing procedures performed on rare occasions by one untrained person to series of elaborate instrumented analyses performed on a more or less continuous basis by a team of professionals. The former situation is becoming increasingly less common as even the smaller, less formally organized companies become aware that maintenance of uniform quality is essential to the very existence of their business. Complex and inflexible government regulations, bearing severe penalties and enforced by an army of civil servants, are more effective arguments than customer displeasure for rigid adherence to a set of rules designed to ensure maintenance of satisfactory quality, and no company of any size can afford to ignore this factor.

The purpose of this chapter is to review the opinions of some specialists in the field as to the proper role of quality control and then to add some general observations on the relationship of quality control to other departments in the snack food plant. Some of the administrative and procedural principles involved in the management of a quality control department are discussed, but details of sampling or testing procedures are not reviewed.

QUALITY CONTROL OR QUALITY ASSURANCE

"Quality Control" is slowly being displaced by "Quality Assurance" as the term expressing the objectives of the group of people who do the sampling, testing, and other activities related to the evaluation of product conformance to predetermined standards. Quality Assurance is intended to express an expanded responsibility of the group for correcting problems (or recommending corrective action) which could result in defective product, as well as the long recognized responsibility for screening out nonspecification products after they have been made. The most effective and conscientious managers in this field have always accepted the broader responsibility, but it is certainly worthwhile to acknowledge that it is a valid concern of the department by adopting the more appropriate name. In this chapter, the two terms are used more or less interchangeably.

MISSION OF QUALITY CONTROL

Kramer and Twigg (1970) defined the mission of quality control as the maintenance of quality at levels and tolerances acceptable to the buyer while minimizing costs to the vendor. Amerine *et al.* (1965) defined quality control as the "application of sensory, physical, and chemical tests in industrial production to prevent undue variation in quality attributes, such as color, viscosity, flavor, etc." These authors particularly emphasize customer satisfaction, certainly an important point. It is also important, however, to give much attention to the third parties in all transactions involving food—the Federal, State, and local regulatory bodies.

In addition to satisfying the consumer, the product must meet the requirements of governmental regulatory agencies. A food which does not meet the minimum acceptability requirements of all customers may result only in a few complaints, but a product which does not meet all of the applicable legal requirements may be confiscated and the manufacturers subjected to punitive fines and other penalties. This situation must be emphasized because it is becoming more critical every day; yet many food producers are not aware of the changing climate. A product may look good, taste good, have good texture, and perform well in its intended application, and still be unfit for distribution. To be suitable for sale, it must also be wholesome, conform to all applicable labeling and packaging requirements, and be prepared and stored under conditions tending to prevent contamination by noxious or esthetically undesirable materials (actual contamination need not be shown). Quality Control and all other departments of food handling companies must remain aware of these restrictions at all times.

POSITION OF QUALITY CONTROL IN THE
TABLE OF ORGANIZATION

It is agreed by all authorities that the quality control function must be completely separated, administratively, from the production and purchasing departments because of conflicting goals of these functions. Production properly aims for maximum output with minimum rejections, and purchasing attempts to procure raw materials at the lowest possible price, while quality control must be relied upon to maintain some fixed level of quality regardless of the effect on production rate or ingredient costs. Although the aims of marketing and quality control are somewhat alike, the general administrative principle that scientific personnel should not be managed by persons who are not scientists dictates that quality control must remain separate from marketing.

Most companies have found that the best administrative arrangement is to have the chief of quality control report directly to the president or executive vice president. Alternatively, there can be a vice president of technical services who administers quality control operations as well as research, development, and similar activities.

FUNCTIONS OF A QUALITY CONTROL DEPARTMENT

The basic definitions quoted previously can be supplemented by a statement of the functions encompassed by statistical quality control (slightly paraphrased from Allan 1959): (1) Acceptance component (the traditional quality control function and based on fixed standards)— approves good product and sets aside defective product for rework or scrap. (2) Prevention component—uses data generated by inspection to indicate areas where an out-of-control situation is starting to develop. (3) Assurance component—uses data from customer complaints and quality audits or reviews to assure that the extent of conformance is maintained at a satisfactory level.

The implementation process undertaken in order to see that the functions are properly carried out would include the following. (1) Preparing specifications for raw materials, including ingredients and packaging materials. (2) Preparing specifications for finished products, including the package. (3) Preparing outline procedures, in which the essential conditions for manufacture are listed. (4) Performing or supervising sampling of raw materials and finished products. (5) Performing tests necessary to determine compliance with specifications and applicable Federal, State, and local regulations, including sanitation inspections. (6) Determining compliance of finished products and production and storage areas with all applicable Federal, State, and local regulations. This includes determining the compliance of label statements with packaging laws. (7) Conducting a continuous program of devising new and improved tests. (8) Assisting production and engineering in trouble-shooting when an out-of-control situation exists. (9) Performing supervisory and administrative functions applicable to personnel engaging in the above activities.

It does not include (1) performing sanitation procedures (e.g., fumigations, etc.), (2) performing maintenance procedures, (3) performing technical service functions for marketing, production, or procurement.

Certainly the quality control department should never be regarded as being responsible for production and engineering functions such as checking the accuracy of meters, instruments, thermometers, timers, etc., except in conjunction with production and engineering during troubleshooting operations triggered by an out-of-control situation.

Wolf (1970) describes how quality control functions in a large organization, starting with the development of a new product. The quality control manager directs the chemical and the bacteriological testing of all new ingredients that are used in the product. Specifications are written for each ingredient that has been selected for the new product. These specifications are used by the purchasing department in selecting suppliers for the ingredients. Before a supplier is approved, the quality control sanitation inspector checks the facilities of the supplier to assure that the facility meets the sanitation and microbiological standards.

The quality control department actively assists the process engineering group in setting and evaluating process standards. The optimum processing temperatures are specified by the pathogenic laboratory manager to eliminate the danger of bacterial contamination and growth. All processes are checked to assure compliance with all Federal and State regulations.

If new equipment has to be built for the process, the quality control department is consulted to assure that the new equipment meets all U.S. FDA and Good Manufacturing Practice directives. After the equipment is built and tested, the quality control department takes samples of products at different processing stages to determine whether or not there is bacterial growth during processing.

All incoming ingredients for the products are checked for bacterial, chemical, and physical properties before use.

While the new product is placed into production, the plant quality control manager works with the new product and process engineering managers to become familiar with all phases of production. Checksheets are prepared listing all critical processing specifications that have to be monitored by the quality control process inspector during mass production. The quality control process inspector's job is to notify the production foreman if a standard is not followed.

Before a new product is authorized for sale, the executive committee approves the quality standards for the product. The standards for color, volume, texture, symmetry, flavor, and weight are established. The quality control line inspectors guide the production people to assure that all substandard products are rejected.

The shipping quality control inspector checks the temperature of the products and the condition of the shipping cases before shipment. The shipping inspector must also check the truck or rail car, to assure that the product is shipped at the correct temperature in a sanitized vehicle.

The sanitation inspector conducts daily plant inspections and notifies the production foreman if a sanitation deficiency exists. Once a month the plant is thoroughly inspected and graded. Twice a year an independent outside agency inspects the facilities and grades the plant sanitation level. Immaculate sanitary conditions in a food manufacturing plant are essential to produce quality products.

Specifying ingredients, processing, finished products, and sanitation standards are meaningless unless these standards are observed.

Some of the standards, such as texture or flavor of a product, are subject to interpretation. For example, if a quality control line inspector rejects a product, he places a "hold tag" on it. The "hold tag" cannot be removed from the product without quality control authorization. In addition, no product with a "hold tag" on it can be shipped. The production foreman, whose product has been rejected by a quality control line inspector, can appeal the decision if he disagrees with the decision of the quality control line inspector. The production foreman then notifies the production manager. If the production manager agrees with the production foreman, he will try to convince the quality control manager that the product in question is up to standard. If the quality control manager disagrees, the vice president of production and vice president of R&D decide the fate of the product. If no agreement is reached, the president of the company makes the final decision. This system of checks and balances minimizes the chance of accepting a substandard product or rejecting an acceptable product.

Allen (1968) lists the following "Ten Commandments" of quality control, particularly applicable to biscuit factories. (1) Make up complete formulas showing all the ingredients and any allowable tolerances in sugar, water, leavening, etc. (2) Clearly show on the formulas the complete mixing conditions, including temperature, times, etc. (3) Have mixer personnel record observed mixing conditions for each batch. Insist that it be filled out as the batch is mixed, not before nor long after. (4) Post the correct dough weight for each variety by the machine and have the machine man record weights periodically—every 15 min. (5) Post the heat pattern, time of bake, size, and count for each variety at the delivery end of the oven, and have the baker record these measurements periodically. Be sure the light is good at the oven inspection doors and the delivery end of the oven. Before each run, give the baker a sample to duplicate in color. (6) Have the laboratory run the moisture and pH on each batch and report the results to the foreman or baker immediately. (7) Establish and post the standard weight and tolerance for each package. (8) Package weights should be taken at least every 15 min and recorded. The records are invaluable, not only for quality control, but in any underweight controversy with a government inspector. (9) Be sure that size and count specifications are accurately made on new varieties before the packing material is ordered. (10) Have random samples of product picked off the line by a disinterested party for inspection. It is often amazing how different a product looks in the package than you thought it looked coming out of the oven or on the packing table.

Technical Services

From time to time, members of the quality control staff may be called upon to provide advice and assistance to other departments of the company. They may be asked to assist in solving urgent processing problems or to act as expert witnesses in legal proceedings, for example. In well-run companies, these extramural tasks should be relatively infrequent and it will not be necessary to have a separate group to handle them, but in companies which sell ingredients, many man-hours may be needed for handling customer complaints, advising on the best conditions for use of product, evaluating specifications, etc. In these companies, it may be desirable to have a separate technical service branch, to which all of the customer inquiries can be directed, thus eliminating the need for frequent interruptions of the experiments and analytical work constituting the principle duties of research and quality control personnel.

Persons selected for such a technical service group should have familiarity with the methods and equipment of the customer industry. It is not essential that they have an extensive background in their own manufacturing techniques. Field work is often required, and a willingness and ability to work with the customer's production personnel is essential.

The technical service group will usually not have extensive laboratory or pilot plant facilities of their own, and they should be able to call upon quality control or research personnel for specific short-term work to solve those few problems which cannot be solved by verbal consultations or field studies.

Records

It should be obvious that accurate, permanent records must be kept showing all significant actions taken by the quality control department, since legal or administrative questions may arise concerning products shipped many months before.

Detailed record keeping is assuming increasing importance in preventing liability losses due to presumably defective products. Recent emphasis in the courts of the legal doctrine of "strict liability," which holds that a plaintiff need not prove negligence in manufacture but simply that a defect existed at the time of use or consumption, has created added hazards for companies not maintaining adequate quality control records. The records should show that all defective products are destroyed or reworked, and any evidence that the recommendations of a quality control manager have been overridden is likely to have very unfortunate effects if the product in some way causes injury or disease. Any company that feels that destruction of product is too costly should consider the effect on their finances of a few losses in the courts, where awards of $100,000 and up are commonplace. Judgments assigning criminal liability for defective products are also becoming more frequent.

Courts no longer allow the manufacturer to hold confidential the company files and records. Disclosure and interrogatory procedures are now intended for the benefit of the plaintiff and permit almost total examination at any time.

All records relating to quality control should thus be in condition for review of the company's files. The claims representative of the company's insurance carrier and its corporate legal counsel can provide advice on the establishment of the records system.

Two important documents that belong in a company's quality control file are a "Certificate of Insurance" (issued by a responsible carrier) and a "Hold Harmless Agreement" provided by each vendor who supplies ingredients, packaging materials, equipment that contacts foods, etc. The validity and scope of these certificates should be approved by the company's insurance representative and its legal counsel.

Documentation through all steps in handling consumer complaints is highly important in protecting against broad liability. Admission of fault where there is some doubt, merely in order to calm down an angry customer, is very dangerous. Examination of the supposedly defective product, questioning the offended customer, and entering the results in the records are minimum steps.

There should be a well-developed plan for tracing the distribution or sale of a company's product in case a faulty or dangerous condition is discovered.

ESTABLISHING STANDARDS AND WRITING SPECIFICATIONS

It is the finished product quality which is of prime importance. All other requirements must be directed toward this goal.

When a new product is being developed, marketing or an executive committee must decide upon the minimum essential limits of product quality. Quality control and manufacturing then set the quantitative in-process and finished product specifications which will ensure that the product as offered for sale meets these requirements. The product development group acts in an advisory capacity both for raw materials and processing descriptions. A similar division of responsibilities is desirable when it is necessary to write specifications for existing products.

If the finished product quality which marketing regards as essential cannot be met within the existing technological limitations, either concessions must be made by them or the project must be abandoned as being nonfeasible. It is important to obtain the written concurrence of all interested parties to the finished specification. If there are demands for changes which cannot be met, the conflict must be resolved by top management.

The philosophies or motivations behind the establishment of specifications for raw materials (including packaging materials) and for finished products are quite different. The ultimate criterion of the quality of the finished product is its acceptability to the consumer, and this may include its performance in equipment of the user when that requirement is applicable. Determining these limitations is properly a function of market research and, once they are determined, the requirements are embodied in a finished product specification (as modified by considerations of cost and available facilities).

The essential limitations for raw materials (ingredients) must be set by R&D on the basis of their knowledge of what is necessary in order to allow the finished product to meet its specifications. Raw materials specifications should be made as broad as possible so that purchasing may procure the lowest-priced ingredients which will function satisfactorily. Redundant limitations should be scrupulously avoided. All raw materials should be covered by some sort of specification and testing procedure. Frequently, a supplier will submit a specification which covers an ingredient known to be satisfactory. Such specifications should be carefully screened to remove any limitations which would restrict the source of supply unnecessarily.

In addition to the specifications for raw materials and finished product, limitations may be applied at various stages in the manufacturing process. The establishment and enforcement of in-process standards is justified by the need for preventing excess scrap and must be based on technological considerations alone.

Finished Product Standards

Ideally, measurement of the acceptability of a product should be based on evaluations of it, in the way it is usually consumed, and by panels of testers representative of potential buyers in the marketing area. When appropriate, home preparation trials should be used. For institutional and manufacturing materials, the tests might include an evaluation of the performance of the product in the types of equipment used by potential customers. As a practical matter, such consumer and user tests are clearly impossible for routine quality control purposes, and the buyers' requirements must be expressed by a workable specification arrived at through the development or selection of tests which translate the targeted consumers' desires into chemical, physical, or organoleptic terms.

The relationship of the results of these tests to the fundamental factor of consumer acceptability depends upon the wise choice of the conditions of the test as well as on some uncontrollable circumstances. In no case, however, can a physical or chemical test be expected to give a perfect correlation with consumer ratings of a food.

Tests made after the sample has been stored for some time may be more meaningful than tests conducted on fresh product, although the two sets of results should be capable of being related by some constant factor. Nonetheless, retaining a product in stock until long-term storage tests have been completed is impractical. To prevent unnecessary increases in inventory, the finished product should be tested as expeditiously as possible. Tests which require several days for completion should be avoided.

Specifications should be subject to change if improved evidence becomes available. Case histories of consumer complaints on existing items are useful guides in revising the standards. A system of tabulating these complaints together with a periodic review is very worthwhile. The availability of new equipment (plant or laboratory) or improved ingredients may make it possible to institute more rigorous standards, leading to a better product at equal or lower cost.

Ingredient Specifications

Ingredient requirements are initially established by R&D, and they are based on information accumulated during new product development or process improvement studies. These requirements may be modified as experience is gained in the manufacture of the product and should be reviewed frequently during the early history of its commercial production.

There are many functional qualities of ingredients which are not subject to evaluation except by performance tests conducted under conditions resembling as closely as possible those existing in the plant. It is generally agreed that it is preferable to substitute tests of specific characteristics for performance tests when the former become available with the advance of technology.

Quality control must assure itself that the tests proposed by R&D can be performed in a timely and reproducible manner with the available facilities. The manner and frequency of sampling should be left to the decision of quality control unless there are clear-cut reasons why they must be prescribed by R&D.

TEST METHODS

The limits given for a quality factor listed in a specification must list the reference source of the test procedure which can be conducted under standard conditions. These tests may be developed in the laboratory to meet a particular need, but it is more common to adapt methods described in reference works devoted to analytical procedures or to refer directly to these published techniques. Many thoroughly verified test methods for foods are given in the latest edition of AOAC Methods (Anon. 1975). Other societies concerned with special types of food have published collections of analytical procedures appropriate to their special field of interest. See, for example, Cereal Laboratory Methods (Anon. 1970). Individual tests are described in scientific journals, but these are not likely to be as well-seasoned as those reviewed and investigated by the committees of the various societies.

The ways in which new analytical methods are developed, verified, and standardized are generally known to scientists and technologists and will not be discussed here. Special tests should be described in complete detail in the specifications themselves, or in a handbook of laboratory procedures maintained by the quality control administration.

The tendency in modern quality control testing is to rely more and more on instrumental methods, as on gas chromatography for testing for off-odors, spectrophotometry for color standardization, etc. Organoleptic tests cannot be entirely replaced by instrumental methods at the present state of our knowledge, however, because of insufficient understanding of the fundamental factors affecting acceptability and uniformity of foods. Consumer panels for quality control are expensive, slow, and unwieldy; while expert panels are not entirely satisfactory because of the difficulty in relating their scores to consumer evaluation and in assembling the panel on short notice. Expert panels can be of great value in detecting deviations from a standard. As a practical matter, individual judgments of flavor, texture, and appearance made by the quality control technologist must often be relied upon both for finished products and many raw materials.

SAMPLING

It is an old axiom that a test can only be as good as the sample. The development of statistically valid sampling procedures is sometimes more difficult and complex than developing a new analytical technique, and it is quite frequently much harder to administer and police. Statistical quality control textbooks should be relied upon for guidance. Two of the publications relating to the application of statistically valid sampling methods are MIL-STD-105D and ASTME-122-S6T (Anon. 1964). Steiner (1971) offers some sampling procedures for controlling the average quality of lots or the proportion lying outside specifications. Peach (1964) gives a detailed review of the statistical justification for sampling methods.

SANITATION INSPECTIONS

Sanitation operations are not usually under the jurisdiction of the quality control department. They are usually part of the functions managed by the Production or Manufacturing Department, although, as Bakka (1973) states, top management, quality control, research and development, engineering, production, sales, and product distribution must all be concerned with sanitation. Sanitation procedures are never a proper function of quality control personnel except in those emergency situations when all hands must cooperate in an effort to correct an out-of-control condition. The sanitarian and the cleanup crew, not quality control, must be held responsible for sanitation. In plants large enough to maintain a full-time sanitarian and assistants, these personnel should be placed under the jurisdiction of production. Quality control should have an opportunity to pass on the acceptable nature of sprays, powders, poisons, etc., used in disinfestation procedures.

Inspections of the plant designed to detect unsanitary processing conditions which could lead to product contamination or a citation by one of the government regulatory bodies are operations justifiably assignable to quality control. When an official inspection is made by a representative of one of the government regulatory bodies, he should be accompanied in his rounds by the quality control chief or his assistant. A member of the production staff should also participate in this inspection. Comments of the official inspector should be recorded and forwarded by the quality control representative to the appropriate department for action. Performance of the corrections directed by the inspector should be policed by the quality control department, and further contacts (if they are necessary), should be handled by the chief.

HACCP ANALYSES

Hazard Analysis and Critical Control Point studies are undertaken to identify all the points in a food processing operation which could present potential dangers to consumers' health and welfare. Every step in the process from receiving of raw materials to distributing the product must be covered by the study. Three types of hazards are considered in the study: (1) microorganisms—bacteria and molds including *Salmonella, Shigella,* coagulase positive staphylococci, *Clostridium botulinum, Escherichia coli,* and *Aspergillus flavus;* (2) chemicals including toxic materials such as mercury, lead, bacterial toxins, pesticide residues, and aflatoxin; and (3) foreign matter such as stones, glass, and tramp metal.

In January 1973, FDA instituted a pilot inspection program based on a Hazard Analysis and Critical Control Point system. HACCP inspectors identify and analyze critical control points in the food processing plant and evaluate the adequacy of process control operations, attempting to bring potential dangers to the attention of management for corrective action before defective foods are made. The FDA says this approach has proved to be highly successful and its use will be expanded. Among the early achievements was the identification of *botulinum* toxin in packs of mushrooms, with subsequent intensive clean-up and revision of operations in the industry.

As of February 1975, over 600 HACCP inspections had been made, mostly in the fields of low-acid canned foods and frozen "heat and serve" items. As a result of the inspections, a number of firms had to close down until major equipment improvements could be made and satisfactory plant quality control procedures were begun.

Findings from the HACCP inspections are evaluated by technical experts at FDA headquarters. Plants are categorized as (1) needing only routine follow up, (2) having major deviation from good manufacturing practice and needing prompt correction, and (3) having a critical deviation from good manufacturing practice needing immediate correction. Firms in the second category are notified and given a time limit for correcting the deviation. Firms having a critical deviation from good manufacturing practice may face a seizure or recall of products already marketed.

The FDA regards the HACCP system as giving them a basis for predicting a plant's performance in the future. Unlike earlier regulations and inspections, it surveys the entire production process and is designed to prevent blunders which would result in unnecessary expenses to industry and health threats to the consumer.

COMPLIANCE

The compliance function can be regarded as the interface between the company and regulatory agencies. Its place in the organization charts is not agreed upon by all managements, but there seems to be a tendency to combine it with the technical operations and to create a Vice President of Technology and Compliance. There is some justification for assigning this function to the legal department, if the company has one. Other departments are less suitable as recipients of the compliance responsibility.

Most of the contacts of the compliance manager will be with the Food and Drug Administration, but the U.S. Department of Agriculture, and the Federal Trade Commission as well as State and municipal regulatory agencies can be involved. If sales are made to military agencies or PX's, inspections by Armed Forces personnel will occur.

The Compliance Manager should always be consulted about specifications, new ingredients, new processing or packaging methods, label design, advertising claims, and anything else that may concern the wholesomeness of the food as it reaches the consumer and that may affect the consumers' expectations (nutrition, weight, etc.). There is some question as to whether or not Compliance should also encompass OSHA, labor relations, and nonfood regulations. These responsibilities are generally delegated to other departments—OSHA to Manufacturing, and labor law compliance to an Industrial Relations Manager, for example. In any case, Compliance personnel should have ready access to a capable legal counsel who specializes in these problems.

RECALL

A recall is the removal of product from the marketplace in an attempt to prevent deteriorated, unwholesome, misrepresented, or otherwise unsatisfactory food from reaching the consumer. Recalls may be voluntary, that is initiated by the firm, or involuntary, in which case a court order or FDA directive is being followed. Recalls can be disastrous to the manufacturer, not only because of the economic impact of searching out, collecting, shipping, and destroying (or reprocessing) defective material, but also as a result of the adverse publicity that usually accompanies the action. The effects can be minimized by setting up a plan to deal with such an emergency before it occurs.

A recall plan should include the designation of both a person to act as a coordinator and an emergency staff group to assist him during the program, the preparation of outlines of actions to be followed in each type of recall, the assigning of a public spokesman to present the company's viewpoint to media, and predetermination of cost allocations.

An essential basis for making decisions on details of the plan is an understanding of the proclaimed policy of the FDA on recall procedure. This can be found in the recall chapter of the FDA Regulatory Procedures Manual.

ROLE IN PROMOTING A QUALITY-CONSCIOUS ATTITUDE IN OTHER DEPARTMENTS

The quality control department can be an effective agency in promoting a positive attitude toward quality improvement. The attitude of employees engaged in producing the goods offered for sale is a factor affecting quality, although the relative contribution of employee attitude versus machine limitations is probably heavily weighted in favor of the latter. Nonetheless, sloppy or haphazard work can result when employees are not aware of the importance of quality to the maintenance of the company's financial strength, and, therefore, to their jobs. If it is known that quality control will consistently detect departures from the norm, an alert attitude is fostered in production people. Customer complaints which are received by quality control can be brought to the attention of the responsible operator and his foreman.

One approach to securing emotional involvement of personnel in quality improvement is the use of such flamboyant and ephemeral programs as "Zero Defects" with variants POP, POW, PRIDE, etc. The net value of these programs in general, and the Zero Defects program in particular is likely to be very small. The usual experience is that the average production worker tends to be very cynical about these programs and lets superficial acquiescence cover a complete lack of interest. Even those workers who can be motivated initially suffer a letdown after the high-pressure campaign terminates and may lose what real enthusiasm they originally had.

Zero Defects is especially bad, in the opinion of some, because it sets an impossible goal. No one likes to shoot at a target he knows he can't hit. Furthermore, the strong pressure to reduce defects may very well lead to concealment of errors with the results that quality control cannot perform its function properly.

EXTENT OF THE CONCERN OF QUALITY CONTROL

As a technological matter, the concern of quality control must start with raw materials. Quality of finished product is inseparable from quality of components. Seldom are there processing variations which will compensate for raw material inadequacies. Conversely, use of top quality ingredients is of no avail if the processing conditions are not adequately controlled.

The question of where the concern of quality control ends is partly a matter of policy. In most cases there is no effective control over some stages in the distribution cycle, and none over the conditions in the store or in the consumer's pantry. The easy answer is to say that product quality is a quality control concern until the material is consumed, but, realistically, if there is no control over conditions encountered by the product, an attempted evaluation of these conditions or of the product after it encounters them, is of only academic interest, or at best, the concern of R&D who might find in the data guidance for designing products to be stable under some presumed average set of conditions. It is an elementary scientific principle that testing when one or more important conditions are unknown is a waste of time. That is why the practice of securing samples from many different outlets and examining them for guidance in quality control is fundamentally unsound.

Where a limited number of controlled storage outlets are in question, the problem is slightly more clear-cut. Then, each storage area must be sampled and it must be sampled on a routine basis. The results of the evaluation are applicable only to that storage area and only to the set of conditions existing during the history of the sample. If enough person- nel are available to do the necessary sampling, the testing, and the eval- uation of results, some information subject to analysis and/or extrapola- tion can often be obtained.

If a program of sampling product in retail outlets is to be undertaken, it is very important to set down at the beginning exactly what is to be determined. The goals must be defined in precise terms. When this is attempted, it is often found that the desired answer cannot be obtained using the contemplated procedure or, in fact, any conceivable proce- dure. If whoever designs the project is allowed to use generalities and nonscientific nomenclature in setting the goals, then a program which has a built-in failure potential may be put into effect.

In the author's opinion, it is far better to have a policy of testing product after storage in the laboratory under conditions of controlled temperature and humidity than it is to waste funds in accumulating data which cannot be extrapolated to new situations. The conducting of such tests on existing products is a proper function of quality control.

CONSTRUCTION OF CONTROL CHARTS

The following discussion is adapted from a paper by Stewart (1964).

Statistical quality control can be applied to the two types of quality factors, variables and attributes. It is important to understand the dif- ference between these two types of data. When a record is made of actu- al measured quality characteristics, such as the weight, diameter, length, width, or thickness of a cookie or cracker, the quality is said to be expressed by variables. When a record shows the number of articles conforming and the number of articles failing to conform to any speci- fied requirement, it is said to be a record by attributes. Examples of at- tributes are: packages with good seal or no seal; broken or unbroken cookies; and packages with code date or no code date.

The Shewhart Control Charts that are widely used in statistical quali- ty control are either measurement charts or attribute charts. The ad- vantages of Shewhart Charts over some other methods of graphing data are that they show the measurements in chronological sequence and fa- cilitate visualizing the relationship of data to quality limits.

The most widely-used measurement charts are the average and range charts, commonly called X-Bar and R charts. The average and range charts are used when the quality characteristic of interest is actually measured each time a sample is inspected. For each characteristic stud- ied, separate X-Bar and R charts must be made. Only one kind of mea- surement at a time can be recorded on these charts.

A useful quality control chart cannot be made without reliable data. Before any charts are made, study the line from which the samples will be taken and make some preliminary decisions based on equipment limitations.

Control charts can be applied to finished products or to any phase of the baking operation. Control charts can be applied to scaling operations, dough weights, temperatures, oven times, oven temperatures, etc. They are wonderful tools for spot-checking the weaknesses of the three M's—Materials, Manpower, and Machinery.

BIBLIOGRAPHY

ALLAN, D. H. W. 1959. Statistical Quality Control. Van Nostrand Reinhold Co., New York.

ALLEN, M. C. 1968. The most important thing we have to sell. Snack Food 57, No. 12, 33–35.

ALLGAUER, A. J. 1970. Quality assurance: methods. Proc. Am. Soc. Bakery Engrs. 1970, 191–200.

AMERINE, M. A., PANGBORN, R. M., and ROESSLER, E. B. 1965. Principles of Sensory Evaluation of Food. Academic Press, New York.

ANON. 1964. ASTM Standards. American Society for Testing and Materials, Philadelphia.

ANON. 1970. Cereal Laboratory Methods, 7th Edition. American Association of Cereal Chemists, St. Paul.

ASSOC. OF OFFICIAL AGRICULTURAL CHEMISTS. 1975. Methods of Analysis, 12th Edition. Assoc. of Official Agricultural Chemists, Washington, D. C.

BAKKA, R. 1973. Sanitation—a four step program. Snack Food 62, No. 3, 42–43, 51.

CURTIS, J. E., and HUSKEY, G. E. 1974. HAACP analysis in quality assurance. Food Prod. Develop. 8, No. 3, 19, 24–25.

DALBY, G., and HILL, G. 1960. Quality testing of bakery products. In Bakery Technology and Engineering, S. A. Matz (Editor). Avi Publishing Co., Westport, Conn.

DUNTLEY, J. M. 1961. Q.C. laboratory operations. Cereal Sci. Today 6, 286–287.

FEIGENBAUM, A. V. 1961. Total Quality Control—Engineering and Management. McGraw-Hill Book Co., New York.

HERSCHDOERFER, S. M. 1967. Quality Control in the Food Industry. Academic Press, N. Y.

HOWER, R. K. 1964. Some reflections on flour control. Cereal Sci. Today 9, 314.

KILBORN, R. H., and AITKEN, T. R. 1961. Baking laboratory layout and procedures. Cereal Sci. Today 6, 253–254, 257–259.

KRAMER, A., and TWIGG, B. A. 1970. Quality Control for the Food Industry, 3rd Edition, Vol. 1. Avi Publishing Co., Westport, Conn.

KRAMER, A., and TWIGG, B. A. 1973. Quality Control for the Food Industry, 3rd Edition, Vol. 2. Avi Publishing Co., Westport, Conn.

KURTZ, O. L., and HARRIS, K. L. 1974. Run a tight shop. Snack Food 63, No. 2, 40–41, 47.

LAWLER, F. K. 1973. FDA launches new quality program. Food Eng. 45, No. 4, 64.

MARKEL, M. F. 1969. The impact of food laws on the function of bakery engineers. Proc. Am. Soc. Bakery Engrs. 1969, 53–59.

MEEK, D. C. 1969. A total quality system. Cereal Sci. Today 14, 189–192.

PEACH, P. 1964. Quality Control for Management. Prentice-Hall, Englewood Cliffs, N. J.

PRATT, D. B. JR., 1960. Management looks at quality control—a communications problem. Cereal Sci. Today 5, 90–92, 99.

RICHARDS, E. 1972. Quality control of food products. J. Soc. Dairy Technol. 25, No. 4, 195–199.

SEELEY, R. E. 1974. Establishing an incoming quality control program for packaging. Package Develop. 1974 (Nov./Dec.) 8, 16.

STEINER, E. H. 1971. Sampling schemes for controlling the average quality of consignments or the proportion lying outside specification. Brit. Food Mfg. Industries Res. Assoc. Sci. Tech. Surveys 65.

STEWART, T. J. 1964. Practical Quality Control. Schulze-Burch Biscuit Co., Chicago.

STRAUSS, A. 1964. Specifications for ingredients. Proc. Am. Soc. Bakery Engrs. *1964*, 96–102.

TENGQUIST, H. N. 1968. A relationship between laboratory and production. Proc. Am. Soc. Bakery Engrs. *1968*, 181–185.

WOLF, A. 1970. Quality assurance: principles and organization. Proc. Am. Soc. Bakery Engrs. *1970*, 183–188.

CHAPTER 26

Nutritional Supplementation

INTRODUCTION

Snack foods have become a prime target of some consumer advocates and nutritionists who have castigated them as sources of "empty calories." In an attempt to counter this criticism, a number of manufacturers of snack foods have undertaken the development of products which would be supplemented by vitamins, minerals, or proteins. Anyone considering the initiation of such development work should clearly understand that the composition, labeling, and advertising claims are very rigidly controlled by Federal laws and regulations. The legal problems are likely to be far more formidable than the technical ones, even though the latter are difficult enough.

RECOMMENDED DAILY ALLOWANCES

The amount of nutrient required over a given period of time by an individual may be affected by his previous nutritional status, age, weight, health, and genetic constitution, among a much larger group of factors. Average figures for daily nutrient requirements have been published by various scientific authorities, quasi-official bodies, and regulatory agencies. Typical of these lists is the Recommended Daily Dietary Allowance published by the Food and Nutrition Board, National Academy of Sciences–National Research Council. The 1974 version of this authoritative compilation, which includes recommendations for infants, children, various age groups of men and women, pregnant women, and lactating women is given in Table 26.1. Changes in this table may be made as new data are obtained regarding human nutritional needs.

Several years ago the Food and Drug Administration set up a list of Minimum Daily Requirements on which food manufacturers could base claims of the nutritional contribution made by their products. The Minimum Daily Requirements have been superseded by the U.S. Recommended Daily Allowances, published in 1973. These include vitamins and minerals, as shown in Table 26.2, as well as protein. The U.S. RDA of protein in a food product is 45 gm if the Protein Efficiency Ratio (PER) of the total protein in the product is equal to or greater than that of casein, and 65 gm if the PER of the total protein in the product is less than that of casein. Total protein with a PER less than 20% of casein may not be stated on the label in terms of a percentage of the U.S. RDA.

The U.S. RDA's described above are for the ordinary, nondietetic general purpose foods. Products for special dietary use and some nutritional supplements have different recommended daily allowances. The Federal Register, Vol. 38, No. 13, and subsequent modifications should be consulted for additional information.

VITAMINS

Vitamin supplementation is somewhat simpler than mineral or protein supplementation because of the lower amounts of vitamins required. The cost is reasonable, a small fraction of a cent per portion if supplementation is restricted to the five vitamins in the list of seven "commonly analyzed nutrients." Cost becomes more of a factor if vitamin E and biotin are included at high percentages of the U.S. RDA.

Vitamins can definitely affect flavor. Thiamine is the primary offender, although other vitamins can contribute noticeable off-flavors. Riboflavin is highly colored, but its yellow hue is compatible with many snacks. Storage deterioration of these labile substances must be compensated for by addition of extra amounts so that the consumer receives the full claimed quantity.

Some of the salt producers offer vitamin-fortified snack toppings. Morton Salt Company's "standard formula" uses dendritic salt as a carrier for five vitamins. When applied to potato chips at a minimum level of 1.5% sodium chloride, it will provide at least one U.S. RDA of vitamins A, C, B-1, B-2, and niacin per pound of product. Sufficient overage is present to offset normal processing and storage loss. An adhesive substance (e.g., polysorbate 80) is included to bind the vitamins to the salt crystals and prevent segregation and dusting in application. The ingredient exhibits a light yellow color which is not readily distinguishable after application to the snack. Custom formulation to give different vitamin (or mineral) levels will also be provided by Morton and some other producers.

When supplementing with any nutritional factor, the manufacturer must establish a quality control program which will insure that every lot contains the claimed amount. The necessary analyses can add substantially to the overall cost of the program.

MINERALS

The large amounts of calcium and phosphorus needed to supply appreciable fractions of the U.S. RDA can lead to seemingly insuperable problems in snack supplementation. These nutrients are usually added as some form of calcium phosphate (e.g., tricalcium phosphate) which does not contain equimolar proportions of the two nutrients, leading to an unnecessary overage of one of them. Even $\frac{1}{10}$ of the U.S. RDA of

TABLE 26.1

RECOMMENDED DAILY DIETARY ALLOWANCES[1]

Designed for the maintenance of good nutrition of practically all healthy people in the U.S.A.

									Fat-Soluble Vitamins		
	Age	Weight		Height		Energy	Protein	Vitamin A Activity		Vitamin D	Vitamin E Activity[5]
	(yr)	(kg)	(lb)	(cm)	(in)	(kcal)[2]	(gm)	(RE)[3]	(IU)	(IU)	(IU)
Infants	0.0–0.5	6	14	60	24	kg × 117	kg × 2.2	420[4]	1,400	400	4
	0.5–1.0	9	20	71	28	kg × 108	kg × 2.0	400	2,000	400	5
Children	1–3	13	28	86	34	1,300	23	400	2,000	400	7
	4–6	20	44	110	44	1,800	30	500	2,500	400	9
	7–10	30	66	135	54	2,400	36	700	3,300	400	10
Males	11–14	44	97	158	63	2,800	44	1,000	5,000	400	12
	15–18	61	134	172	69	3,000	54	1,000	5,000	400	15
	19–22	67	147	172	69	3,000	54	1,000	5,000	400	15
	23–50	70	154	172	69	2,700	56	1,000	5,000		15
	51+	70	154	172	69	2,400	56	1,000	5,000		15
Females	11–14	44	97	155	62	2,400	44	800	4,000	400	12
	15–18	54	119	162	65	2,100	48	800	4,000	400	12
	19–22	58	128	162	65	2,100	46	800	4,000	400	12
	23–50	58	128	162	65	2,000	46	800	4,000		12
	51+	58	128	162	65	1,800	46	800	4,000		12
Pregnant						+300	+30	1,000	5,000	400	15
Lactating						+500	+20	1,200	6,000	400	15

Source: Natl. Acad Sci.–Natl. Res. Council (1974).
[1] The allowances are intended to provide for individual variations among most normal persons as they live in the United States under usual environmental stresses. Diets should be based on a variety of common foods in order to provide other nutrients for which human requirements have been less well defined. See text for more detailed discussion of allowances and of nutrients not tabulated.
[2] Kilojoules (kJ) = 4.2 × kcal.
[3] Retinol equivalents.
[4] Assumed to be all as retinol in milk during the first six months of life. All subsequent intakes are assumed to be half as retinol and half as β-carotene when calculated from international units. As retinol equivalents, three fourths are as retinol and one fourth as β-carotene.

these minerals (0.1 gm calcium and 0.1 gm phosphorus) can thus require the addition of about ½ gm of the phosphate. Textural and visual defects can result from the application of such amounts to portion quantities of most snacks.

Iron, which is a "commonly analyzed nutrient" along with calcium and five vitamins, can also pose problems connected with adverse organoleptic properties. Sodium iron pyrophosphate is bland, white, and relatively insoluble but there is doubt as to its physiological availability. Ferrous sulfate and reduced iron are apparently utilized more completely by the body, but may affect appearance and flavor. In some cases, they also increase the rate of rancidity development.

Water-Soluble Vitamins							Minerals					
Ascorbic Acid (mg)	Folacin[6] (μg)	Niacin[6] (mg)	Riboflavin (mg)	Thiamin (mg)	Vitamin B-6 (mg)	Vitamin B-12 (μ)	Calcium (mg)	Phosphorus (mg)	Iodine (μg)	Iron (mg)	Magnesium (mg)	Zinc (mg)
35	50	5	0.4	0.3	0.3	0.3	360	240	35	10	60	3
35	50	8	0.6	0.5	0.4	0.3	540	400	45	15	70	5
40	100	9	0.8	0.7	0.6	1.0	800	800	60	15	150	10
40	200	12	1.1	0.9	0.9	1.5	800	800	80	10	200	10
40	300	16	1.2	1.2	1.2	2.0	800	800	110	10	250	10
45	400	18	1.5	1.4	1.6	3.0	1,200	1,200	130	18	350	15
45	400	20	1.8	1.5	2.0	3.0	1,200	1,200	150	18	400	15
45	400	20	1.8	1.5	2.0	3.0	800	800	140	10	350	15
45	400	18	1.6	1.4	2.0	3.0	800	800	130	10	350	15
45	400	16	1.5	1.2	2.0	3.0	800	800	110	10	350	15
45	400	16	1.3	1.2	1.6	3.0	1,200	1,200	115	18	300	15
45	400	14	1.4	1.1	2.0	3.0	1,200	1,200	115	18	300	15
45	400	14	1.4	1.1	2.0	3.0	800	800	100	18	300	15
45	400	13	1.2	1.0	2.0	3.0	800	800	100	18	300	15
45	400	12	1.1	1.0	2.0	3.0	800	800	80	10	300	15
60	800	+2	+0.3	+0.3	2.5	4.0	1,200	1,200	125	18+[8]	450	20
80	600	÷4	+0.5	+0.3	2.5	4.0	1,200	1,200	150	18	450	25

[5] Total vitamin E activity, estimated to be 80% as α-tocopherol and 20% other tocopherols.

[6] The folacin allowances refer to dietary sources as determined by *Lactobacillus casei* assay. Pure forms of folacin may be effective in doses less than ¼ of the recommended dietary allowance.

[7] Although allowances are expressed as niacin, it is recognized that on the average 1 mg of niacin is derived from each 60 mg of dietary tryptophan.

[8] This increased requirement cannot be met by ordinary diets; therefore, the use of supplemental iron is recommended.

Storage loss of mineral supplements is generally not observed, except perhaps for iodine.

PROTEIN

Proteins—or the amino acids of which proteins are composed—are required in the diet for building and repairing body tissues. Most snacks are poor sources of protein and the protein which is present is often of poor nutritional quality.

Protein supplementation takes two forms: (1) increasing the PER (a measure of nutritional quality) of the protein already present in the food raw materials by adding small amounts of the limiting amino acids, and (2) increasing the total amount of protein present by adding some

TABLE 26.2

U.S. RDA OF VITAMINS AND MINERALS[1]

Vitamins	
Vitamin A	5000 International Units
Vitamin C	60 mg
Thiamin	1.5 mg
Riboflavin	1.7 mg
Niacin	20.0 mg
Vitamin D	400.0 International Units
Vitamin E	30.0 International Units
Vitamin B-6	2.0 mg
Folic acid	0.4 mg
Vitamin B-12	6.0 μg
Biotin	0.3 mg
Pantothenic acid	10.0 mg
Minerals	
Iron	18.0
Calcium	1.0 gm
Phosphorus	1.0 gm
Iodine	150.0 μg
magnesium	400.0 mg
Zinc	15.0 mg
Copper	2.0 mg

[1] The reader should be warned that these figures may be changed at any time by action of the Food and Drug Administration.

purified nitrogenous material such as casein, isolated soy protein, egg white, etc.

Some discussion of protein efficiency is necessary at this point, though it is a very complex subject incompletely understood in its more arcane aspects. An enormous number of nitrogenous compounds meeting certain structural requirements are called proteins. They are of large and variable molecular weight and are composed mostly of alpha-amino acids joined by peptide linkages.

The proteins in the human body contain at least 20 different amino acids. These must be obtained entirely from food proteins (or smaller molecules containing amino acids) since man cannot convert inorganic nitrogen to amino compounds. Some kinds of amino acids can be synthesized if a sufficient quantity of suitable precursors are available; these are called "nonessential" amino acids. For the human adult, eight amino acids are "essential," i.e., they must be present in the protein consumed as food. These are isoleucine, leucine, lysine, methionine, phenylalanine, threonine, tryptophan, and valine. Part of the phenylalanine requirement can be met by tyrosine, and part of the methionine requirement can be met by cystine. The infant also needs histidine.

The value of any given protein as a nutrient for man is related to the congruency of its distribution of amino acids with human requirements. Experimental measurement of this fit is difficult. One test that has been fairly well accepted is the evaluation of the PER, usually determined by measuring the amount of weight gained per weight of protein consumed by weanling rats when the protein being tested is the sole source of dietary nitrogen. Various authors have pointed out defects in the concept, but the test is still widely used in the absence of a simpler test having wide acceptance (McLaughlan 1972).

Casein is often used as a standard or control in PER tests, with a figure of 2.50 being commonly observed. Proteins from whey and egg rate even higher, soy protein is good, while gelatin, bone stock, and other collagen-derived proteins (e.g., the protein in pork skins) are very poor.

Supplementation of snack foods with significant amounts of protein is difficult because most available additives (casein, soy protein, egg white, etc.) alter the appearance, flavor, or texture of the product. Protein supplementation is expensive; unlike vitamins which, though costly on a per pound basis, are added in micro-amounts, several grams of protein must be added for each portion. When all factors are considered, defatted soy flour is probably the best candidate as a protein additive. If the material is carefully chosen, between 15 and 20% total protein content can be attained in a snack based on puffed corn meal without causing grossly objectionable off-flavors and colors. Possibly somewhat higher levels could be reached using purified soy protein, but the added cost as well as the textural effects (hardness, tackiness, effect on expansion) are negative considerations.

It has been known for a long time that severe heat treatment, such as applied in the manufacture of puffed cereals, has a detrimental effect on the nutritive quality of proteins. Addition of lysine seems to be an effective means of restoring the protein quality of these heat-damaged cereals even though destruction of lysine cannot always be proven by chemical analyses. This seeming paradox was explained by showing that severe heat treatment caused a resistance to release of lysine from the protein by enzyme action. It appears that gun-puffing may cause an actual decrease in lysine, however. Because lysine is the limiting amino acid in native cereal proteins, its further decrease in availability due to high heat treatment can bring the PER of the finished food to a very low level. The sulfur-containing amino acids, cystine and methionine, are also present at relatively low levels in most cereals and are heat sensitive.

Certain amino acids are available as synthetic, substantially pure compounds. Lysine and methionine can be purchased in this form and may be useful for the supplementation of snack foods. Because of the effects discussed above, it would be advisable to add them in powdered form to the cooked food, after all heat treatment has been completed. For supplementation purposes there are some significant drawbacks.

Among these are high cost and undesirable flavor. The cost increase may be acceptable in certain specialty items or if the consumer can be convinced of the added value, but the flavor defects may be more difficult to overcome.

A typical approach is given in the patent by Waitman *et al.* (1973), who described the preparation of high protein potato snacks by adhering a coating of proteinaceous material to raw potato slices before they are deep-fat fried. From the patent text, it is evident that the chips have a dull brown surface. It is not known how this atypical appearance will affect consumer acceptability.

Blessin *et al.* (1971) showed that the protein quality of popcorn could be increased by infusing the grain with aqueous solutions of L-lysine monohydrochloride. About 61% of the added amount could be recovered after popping. Although flavor was affected slightly or not at all, expansion was reduced as much as 45%.

OTHER NUTRIENTS

Lipids

Experiments have shown that diets rich in polyunsaturated fatty acids lower plasma cholesterol. High levels of cholesterol maintained over a long period of time are thought to cause, or at least to be associated with, degenerative changes in blood vessels leading to circulatory disease marked by such untoward consequences as death. It would appear, then, that replacement of saturated fats by polyunsaturated fats would be desirable from a health standpoint. This reasoning has led to the development of dietetic foods containing mostly polyunsaturated fats. Stringent Federal regulations govern the label claims which can be made for such foods, and any manufacturer intending to market them should seek legal guidance.

Many of the fats used by snack food manufacturers are highly saturated. Coconut oil is a prime example. Corn oil is moderately high and safflower oil is very high in polyunsaturated fatty acids. Susceptibility to oxidative rancidity is closely associated with the degree of unsaturation.

Arachidonic acid or its precursor, linoleic acid, is essential for growth and dermal integrity in human infants. Symptoms of the deficiency state in experimental animals include dermatitis, impairment of growth, decreased efficiency of energy utilization, etc. Spontaneous deficiency is very rare in infants and has never been documented in adult human subjects. To prevent a subclinical physiologic deficiency and to provide reserves against stress, it has been recommended that infants should receive 3% of calories as linoleic acid. Supplementation of snacks or other foods for adults does not seem to be indicated.

Fiber

Publications by some medical researchers and nutritionists seem to indicate that gastrointestinal and cardiovascular diseases are correlated with a decrease in dietary fiber intake. Transit time of food residues in the colon, stool weight/concentration, and the binding of bile salts are directly affected by the amount of dietary fiber, and these factors singly or in combination are thought to result in the more generalized physiological symptoms. The rationale presented by the proponents of the beneficial nature of dietary fiber is rather complex, in many cases, but there is experimental evidence that increased use of whole cereal grains (which are relatively high in fiber) can decrease cholesterol levels in rats. Statistical and demographic evidence has been assembled for some of the other claimed benefits (Spiller and Amen 1974; Anon. 1975).

Snack foods prepared from whole cereal grains would be good sources of dietary fiber. Addition of a few percentages of bran to purified cereal fractions such as cornmeal or starch would considerably improve the fiber content of the finished product. Dairy and meat products have negligible fiber content, and potatoes are rather low in this component.

Although there is little published information about the effect of processing on the fiber content of foods, substances comprising this fraction are known to be relatively inert, and it is to be expected that the manufacturing techniques used in the snack industry would not destroy fiber. Nearly all milling processes are designed to separate fibrous parts of the grain, however.

BIBLIOGRAPHY

ANDERSON, R. H. 1973. Nutritional value of proteins of cereal grains. Cereal Sci. Today *18*, 330–333.

ANON. 1973. Vitamin fortified toppings for fat-fried snacks. Morton Salt Company Tech. Serv. Bull. *72-5B*.

ANON. 1975. Fiber a "new" nutritional ingredient. Food Process. *36*, No. 1, 24–26.

BEDENK, W. T., and PURVES, E. R. 1972. Production of high protein ready-to-eat breakfast cereals containing soy and malt. U.S. Pat. 3,682,647. Aug. 8.

BLESSIN, C. W., CAVINS, J. F., and INGLETT, G. E. 1971. Lysine-infused popcorn. Cereal Chem. *48*, 373–377.

CONWAY, H. F., and ANDERSON, R. A. 1973. Protein-fortified extruded food products. Cereal Sci. Today *18*, No. 4, 94–97.

GAGE, J. 1971. A marketing guide to snack food fortification. Snack Food *60*, No. 8, 60, 63.

GAGE, J. 1972. Food fortification: some visible and invisible cost considerations. Food Prod. Develop. *5*, No. 8, 20–21.

HARRIS, R. S., and KARMAS, E. 1975. Nutritional Evaluation of Food Processing, 2nd Edition. Avi Publishing Co., Westport, Conn.

McLAUGHLAN, J. M. 1972. Nutritional evaluation of proteins by biological methods. Cereal Sci. Today *17*, 162–165.

342   SNACK FOOD TECHNOLOGY

OHREN, J. A. 1973. Protein fortification with cottonseed flour. Snack Food 62, No. 10, 38–39.

ROSENFIELD, D. 1973. Utilizable protein: quality and quantity concepts in assessing foods. Food Prod. Develop. 7, No. 3, 20–22.

SCHROEDER, H. A. 1971. Losses of vitamins and trace minerals resulting from processing and preservation of foods. Am. J. Clin. Nutr. 24, 562–573.

SENTI, F. R. 1972. Guidelines for the nutrient composition of processed foods. Cereal Sci. Today 17, 157–161.

SPILLER, G. A., and AMEN, R. J. 1974. Role of dietary fiber in nutrition. Food Prod. Develop. 8, No. 7, 30, 32, 61.

WAITMAN, R. H., KELLY, N. H., and HOLLIS, F. H., JR. 1973. High protein potato snacks. U.S. Pat. 3,754,931. Aug. 28.

Index